# Social Information Transmission and Human Biology

# Society for the Study of Human Biology Series

**10 Biological Aspects of Demography**
*Edited by W. Brass*

**11 Human Evolution**
*Edited by M. H. Day*

**12 Genetic Variation in Britain**
*Edited by D. F. Roberts and E. Sunderland*

**13 Human Variation and Natural Selection**
*Edited by D. Roberts (Penrose Memorial Volume reprint)*

**14 Chromosome Variation in Human Evolution**
*Edited by A. J. Boyce*

**15 Biology of Human Foetal Growth**
*Edited by D. F. Roberts*

**16 Human Ecology in the Tropics**
*Edited by J. P. Garlick and R. W. J. Keay*

**17 Physiological Variation and its Genetic Base**
*Edited by J. S. Weiner*

**18 Human Behaviour and Adaption**
*Edited by N. J. Blurton Jones and V. Reynolds*

**19 Demographic Patterns in Developed Societies**
*Edited by R. W. Horns*

**20 Disease and Urbanisation**
*Edited by E. J. Clegg and J. P. Garlick*

**21 Aspects of Human Evolution**
*Edited by C. B. Stringer*

**22 Energy and Effort**
*Edited by G. A. Harrison*

**23 Migration and Mobility**
*Edited by A. J. Boyce*

**24 Sexual Dimorphism**
*Edited by F. Newcome et al.*

**25 The Biology of Human Ageing**

**26 Capacity for Work in the Tropics**
*Edited by K. J. Collins and D. F. Roberts*

**27 Genetic Variation and its Maintenance**
*Edited by D. F. Roberts and G. F. De Stefano*

**28 Human Mating Patterns**
*Edited by C. G. N. Mascie-Taylor and A. J. Boyce*

**29 The Physiology of Human Growth**
*Edited by J. M. Tanner and M. A. Preece*

**30 Diet and Disease**
*Edited by G. A. Harrison and J. Waterlow*

**31 Fertility and Resources**
*Edited by J. Landers and V. Reynolds*

**32 Urban Ecology and Health in the Third World**
*Edited by L. M. Schell, M. T. Smith and A. Billsborough*

**33 Isolation, Migration and Health**
*Edited by D. F. Roberts, N. Fujiki and K. Torizuka*

**34 Physical Activity and Health**
*Edited by N. G. Norgan*

**35 Seasonality and Human Ecology**
*Edited by S. J. Ulijaszek and S. S. Strickland*

**36 Body Composition Techniques in Health and Disease**
*Edited by P. S. W. Davies and T. J. Cole*

**37 Long-term Consequences of Early Environment**
*Edited by C. J. K. Henry and S. J. Ulijaszek*

**38 Molecular Biology and Human Diversity**
*Edited by A. J. Boyce and C. G. N. Mascie-Taylor*

**39 Human Biology and Social Inequality**
*Edited by S. S. Strickland and P. S. Shetty*

**40 Urbanism, Health and Human Biology in Industrialized Countries**
*Edited by L. M. Schell and S. J. Ulijaszek*

**41 Health and Ethnicity**
*Edited by H. Macbeth and P. Shetty*

**42 Human Biology and History**
*Edited by M. T. Smith*

**43 The Changing Face of Disease: Implications for Society**
*Edited by C. G. N. Mascie-Taylor, J. Peters and S. T. McGarvey*

**44 Childhood Obesity: Contemporary Issues**
*Edited by Noel Cameron, Nicholas G. Norgan, and George T. H. Ellison*

**45 The Nature of Difference: Science, Society and Human Biology**
*Edited by George T. H. Ellison and Alan H. Goodman*

**46 Social Informatin Transmission and Human Biology**
*Edited by Jonathan CK Wells, Simon Strickland, and Kevin Laland*

Numbers 1–9 were published by Pergamon Press, Headington Hill Hall, Headington, Oxford OX3 0BY. Numbers 10–24 were published by Taylor & Francis Ltd, 10–14 Macklin Street, London WC2B 5NF. Numbers 25–40 were published by Cambridge University Press, The Pitt Building, Trumpington Street Cambridge CB2 1RP. Further details and prices of back-list numbers are available from the Secretary of the Society for the Study of Human Biology.

# Social Information Transmission and Human Biology

Edited by
**Jonathan CK Wells, Simon Strickland, and Kevin Laland**

CRC Taylor & Francis
Taylor & Francis Group
Boca Raton London New York

CRC is an imprint of the Taylor & Francis Group, an informa business

The cover illustration shows a limestone stela from Masjid-i Soleiman in the kingdom of Elymais, of the ancient Parthian empire. The figure seems to have wings on his head and carries a bag in one hand and a staff, and probably represents Hermes the messenger of the Greek gods, the bringer of dreams, and the god of music, also well known for his cunning.

Published in 2006 by
CRC Press
Taylor & Francis Group
6000 Broken Sound Parkway NW, Suite 300
Boca Raton, FL 33487-2742


Printed in the United States of America on acid-free paper
10 9 8 7 6 5 4 3 2 1

International Standard Book Number-10: 0-8493-4047-0 (Hardcover)
International Standard Book Number-13: 978-0978-0-8493-4047-5 (Hardcover)



informa
Taylor & Francis Group
is the Academic Division of Informa plc.

**Visit the Taylor & Francis Web site at**
**http://www.taylorandfrancis.com**

**and the CRC Press Web site at**
**http://www.crcpress.com**

# Acknowledgments

We are very grateful to the Society for the Study of Human Biology, and in particular to its President Nick Norgan, for supporting the concept of this symposium from the outset and providing both financial and organizational help throughout. It was with great sadness that we learned of Nick's death from cancer while this book was in production. We dedicate this volume in memory of him.

# Editors

**Jonathan Wells** works at the Institute of Child Health, London. His research interests include many aspects of pediatric nutrition, such as energetics and body composition, as well as evolutionary perspectives on maternal–offspring dynamics and their significance for human health. This work has stimulated his interest in alternative pathways for the transmission of biological information between generations.

**Simon Strickland** has worked as an anthropologist in Nepal, Malaysian Borneo, the Philippines, and South India. His research interests span a variety of issues including human biology and social inequality, energetics and energy flow studies, the nutritional anthropometry of adults, the biological and social significance of the areca nut, and the nature of oral tradition and ritual languages.

**Kevin Laland** is professor in biology at the University of St. Andrews. His research interests include experimental studies of social learning, innovation, diffusion, and tradition using fish, birds, and primates. This work is complemented by theoretical explorations of the adaptive value of social learning, models of gene-culture coevolution, cultural evolution, and evolutionary models of niche construction.

# Contributors

**Nadine Allal**
Department of Anthropology
University College London
London, UK

**Robert Aunger**
Hygiene Centre
London School of Hygiene
and Tropical Medicine
London, UK

**Virginia Berridge**
Public & Environmental Health
Research Unit
London School of Hygiene
and Tropical Medicine
London, UK

**Gillian Brown**
School of Psychology
University of St. Andrews
St. Andrews, Fife, UK

**John Cleland**
Centre for Population Studies
London School of Hygiene
and Tropical Medicine
London, UK

**Therese Hesketh**
Centre for International Child Health
Institute of Child Health
London, UK

**Alasdair I. Houston**
Centre for Behavioural Biology
Department of Mathematics
University of Bristol
Bristol, UK

**Kevin Laland**
Centre for Social Learning
and Cognitive Evolution, School
of Biology
University of St. Andrews
St. Andrews, Fife, UK

**Roland Littlewood**
Department of Anthropology
University College London
London, UK

**Kelly Loughlin**
Public & Environmental Health
Research Unit
London School of Hygiene
and Tropical Medicine
London, UK

**Ruth Mace**
Department of Anthropology
University College London
London, UK

**John M. McNamara**
Centre for Behavioural Biology
Department of Mathematics
University of Bristol
Bristol, UK

**Steven Mithen**
School of Human & Environmental
Sciences
The University of Reading
Reading, UK

**Daniel Nettle**
Psychology, Brain and Behaviour
University of Newcastle
Newcastle, UK

**John Odling-Smee**
Institute of Biological Anthropology
University of Oxford
Oxford, UK

**Andrew Prentice**
Nutrition and Public Health
Intervention Research Unit
London School of Hygiene
and Tropical Medicine
London, UK

**Rebecca Sear**
Department of Social Policy
London School of Economics
London, UK

**Stephen Shennan**
Institute of Archaeology
University College London
London, UK

**Simon Strickland**
Society for the Study of Human Biology
Department of Human Sciences
Loughborough University
Leics, UK

**Susan Watkins**
Department of Sociology
University of Pennsylvania
Philadelphia, PA

**Andrew Whiten**
Centre for Social Learning and
Cognitive Evolution, and Scottish
Primate Research Group
School of Psychology
University of St. Andrews
St. Andrews, Fife, UK

**Jonathan Wells**
MRC Childhood Nutrition
Research Centre
Institute of Child Health
London, UK

# Abstract

The social transmission of information has rapidly become seen as a major component of biology in many organisms, especially in humans. In recent years, social behavior has increasingly been investigated using evolutionary approaches. At the same time, it is becoming apparent that the limitations of our current understanding of human behavior often hinder our ability to improve health and well-being. This volume brings together a diverse range of perspectives on human social information transmission linked by a number of themes. These themes include: (1) how behavioral strategies are formulated by individuals interacting with others; (2) how behavior constructs the population's ecological niche, thereby influencing selective pressures; (3) how the social transmission of information influences the health and reproductive fitness of individuals and populations; and (4) the evolutionary consequences of social information transmission. To illustrate these themes, the chapters draw on models and data ranging from observations on chimpanzee populations in the wild and on the human archaeological record, to studies of contemporary humans in both developing and industrialized countries. The broad approach is intended to highlight the extent to which social information transmission pervades human biology. Therefore many of the chapters address areas of behavior that are familiar to social scientists, such as social anthropologists and demographers, but do so using a variety of cross-disciplinary perspectives. This multidisciplinary approach should prove stimulating for researchers in a range of academic subject areas and facilitate closer collaboration between biological and social scientists.

# Contents

# 1 Introduction

*Jonathan Wells, Simon Strickland, and Kevin Laland*

**CONTENTS**



## 1.1 HISTORICAL OVERVIEW

Anthropology as a discipline is the study of humans, in the widest sense. However, most contemporary anthropologists align themselves with one or the other of two subdisciplines, social/cultural anthropology or biological/physical anthropology, which use very different theoretical bases and modes of investigation. For several decades, social anthropologists have rarely focused on evolutionary parameters such as fitness, even when addressing the cultural components (i.e., marriage systems) of fundamentally biological processes such as reproduction. Equally, biological anthropologists have a relatively brief history of addressing the social transmission of information, a phenomenon that is so central to the behavior patterns studied by social anthropologists. A comprehensive evolutionary approach to the study of human social behavior is therefore still maturing.

For the majority of the 20th century, evolutionary biological models remained largely focused on the gene, both as the principal unit of selection and as the primary mode of inheritance. Environments were considered to present problems, which organisms must ultimately solve through genetic adaptation (Lewontin, 1983). How well individual organisms solved such problems could be assessed in terms of reproductive success, taking into account both quality and quantity of the offspring. From this perspective, successful problem solving leads directly to the replication of genetic information. The advent of Hamilton's concept of inclusive fitness (Hamilton, 1964), Trivers' conflict theories (Trivers, 1972, 1974), and the popularity of the gene's-eye view advocated by Williams (1966) and Dawkins (1976, 1982) placed further emphasis on the transmission of genetic information over time.

Historically, this gene-based evolutionary approach to human behavior has had comparatively little appeal to social scientists. While this might reflect a prevalent view that the vast proportion of information transmitted by humans of all ages is nongenetic, there are additional reasons. Social anthropology was actively steered away from biological issues as the field underwent conceptual development in the mid-20th century, but this divergence was secondary to a prior discomfort with approaches to human culture that were deemed racist, or to engender racism (Kuper, 1999). As discussed in detail by Wolf (1994), the racial categorization of human populations that can be dated back to many ancient civilizations became incorporated into scientific thought in the 18th century through the work of specialists such as Linné and led subsequently to propositions of relationships between physical type and temperamental or moral disposition. The discipline of phrenology, initiated by the Austrian physician Franz Gall in the early 19th century, focused specifically on the notion that skull morphology predicted variability in both brain shape and personality. Evolution-minded anthropologists of the 19th century, including Lubbock, Tylor, and Morgan, viewed human societies as progressing through a series of stages, with some races more advanced than others, and with the former possessing larger and more efficient brains (Oldroyd, 1983). Such work, persisting into the 20th century, was firmly opposed by Boas, perhaps the most influential figure in American anthropology, who argued against attempting to link cultures with biological populations (Boas, 1940), and the vast majority of biologists would now agree with this view (Plotkin, 2002). Nevertheless, during the early and middle parts of the 20th century, eugenicists actively sought in genetics a biological mechanism for such supposed population-specific dispositions. Given that several repressive political regimes in the 20th century have justified themselves through criteria of "biological superiority" and given the multitude of other negative ramifications of erroneous progressive views of human evolution, it is perhaps not surprising that social anthropologists have continued strongly to reject gene-centered models of human behavior. Distinguishing progressive from branching views of evolution, eugenics from genetics, and gene-centered from broader biological approaches are therefore key issues in encouraging interdisciplinary collaboration, as discussed further below.

In the first half of the 20th century, the discipline of physical anthropology was closely associated with phrenology. However, in the first of the present series of SSHB symposium volumes, a marked shift in emphasis was described (Roberts and Weiner, 1957). The revised discipline, termed *biological anthropology* in the UK, could be said to date from this period, focusing on a core of related issues including biological variability within and between populations, the concept of adaptation, and the evolutionary history of the human species. Within this remit, biological anthropologists until recently tended to remain focused on those areas that were amenable to a gene-based approach. Much attention has been directed to the "suite" of traits that emerged relatively recently in human evolution, and which are therefore considered to be the defining features of our genus, in most cases preserved in our species. These traits include bipedality, major expansion of the brain, meat eating, language, and the progressive development of technology. For example, much effort has been directed to understanding the large brains of humans — the factors that may have favored encephalization, the timescale of the expansion in volume, the energetic consequences, the universal behavioral capacities (language, tool use, etc.) that derived from it, and so on — from the perspective of natural selection.

Although human behavior has been subject to evolutionary analyses dating back to Darwin's *The Descent of Man*, notably through the work of human ethologists, the application of modern methods began in earnest with the advent of human sociobiology. Systematic attempts to consider human behavior in relation to Darwinian fitness began in the late 1970s and early 1980s (Wilson, 1975; Blurton et al., 1978; Chagnon and Irons, 1979) using the theoretical developments of Hamilton (1964) and Trivers (1972, 1974) and have continued subsequently (Betzig, 1997; Cronk et al., 2000). This work has in general followed the conventional evolutionary approach, of assessing fitness in terms of differential genetic replication. Many areas of social behavior have remained relatively unexplored from an evolutionary perspective, in particular those that resist a gene-orientated approach. Thus, evolutionary biologists are still only beginning to understand what humans achieve using their large brains, including the capacity to transmit information between individuals (Richerson and Boyd, 2004).

The dominant emphasis on the gene in biology is now being challenged from several directions simultaneously, as evolutionary biologists increasingly appreciate the more complex ways in which organisms receive, contain, and use information. For example, extragenetic sources of information include cytoplasmic inheritance and other maternal effects, social information transmission, and plasticity in response to the environment during ontogenesis (Schlichting and Pigliucci, 1998; West-Eberhard, 2003). The importance of extragenetic sources of information in evolution is now widely acknowledged, to the extent that some have suggested that a broadening of the neo-Darwinian synthesis is required (Pigliucci, 2005). Such complications are reinforced by the findings of molecular genetics, which reveals that the relationship between genes and proteins are infinitely more complicated and bidirectional than hitherto conceived (Gilbert, 2003). A further challenge comes from the widespread recognition and acceptance of multilevel selection within evolutionary biology (Rose and Lauder, 1996).

Behavior is also studied from many different perspectives. Early work by ethologists focused on the concept of instinct — "behavior hard-wired into the genes" (Tinbergen, 1951; Lorenz, 1966) — although the categorization of any behavior as instinctive (or innate) is now considered extremely problematic (Lehrman, 1953). At the same time, comparative psychologists downplayed evolved predispositions and placed emphasis on associative learning processes. Modern behavioral research, which draws from ethology, comparative psychology, physiology, and behavioral ecology, is multifaceted and recognizes a multitude of integrated influences on behavior (Bateson and Martin, 1999). Over the last 20 years, within the field of behavioral research, interest has also been increasingly directed to processes of social learning and transmission, such as imitation, emulation, and other simple forms of copying (Heyes and Galef, 1996; Meltzoff and Prinz, 2002; Fragaszy and Perry, 2003). Building on such findings, evolutionary biologists and anthropologists began formally to model the role played by "culture" in the dynamics of human behavioral change over time (Lumsden and Wilson, 1981; Cavalli-Sforza and Feldman, 1981; Boyd and Richerson, 1985; Durham, 1991; Feldman and Laland, 1996; Richerson and Boyd, 2004).

## 1.2 CULTURE AND MEMETICS

The increase in, and diversity of interest in, cultural information that developed among biologists from the 1960s to the 1980s is illustrated in Table 1.1, which lists a variety of terms proposed by different authors to address the circulation of socially transmitted cultural information units outside the genome. Following publication of Dawkins' influential book *The Selfish Gene*, many researchers have converged on adopting the term *meme* for modeling transmission of units of nongenetic information. However, the field of "memetics" remains controversial. There is considerable debate over what qualifies as memes, how memes are stored and transmitted, and more generally over the validity of models that attempt to express nongenetic information in terms of discrete units. We note that evolutionary models of cultural transmission are not restricted to those of memetics, and there are a variety of other approaches, many of which are considerably more sophisticated than memetic analyses. For instance, formal "cultural evolution" and "gene-culture co-evolutionary" models avoid use of the term *meme* and are often developed without assuming culture is packaged into units (Cavalli-Sforza and Feldman, 1981; Boyd and Richerson, 1985).

Perhaps the major contribution achieved by the emergence of memetics has been to focus attention specifically on nongenetic information itself. At a more specific level, the close analogy that Dawkins suggested between the meme and the gene is regarded by many to be problematic, although often such putative problems emerge from an oversimplistic conception of the gene (Laland and Brown, 2002). The correspondence between biological and cultural evolution is more impressive than the uninitiated typically assume (Mesoudi et al., 2004). Furthermore, evolutionary investigation of the social transmission of information does not require the inheritance of informational units. Richerson and Boyd (1985, 2004) have shown that much social transmission fails to satisfy the concept of information units, because beliefs and values are not necessarily discrete information and may not replicate faithfully in the manner of genes; yet this does not prevent cultural evolution from occurring or being modeled. Richerson and Boyd (2004) recommend focusing on phenomena

**TABLE 1.1**
**Units of Cultural Information, Reviewed in Lumsden and Wilson (1981)**

| Name | Author |
|---|---|
| Idea | Huxley 1962, Cavalli-Sforza 1971 |
| Mnemotype | Blum 1963 |
| Idene | Hoagland 1964 |
| Sociogene | Swanson 1973 |
| Instruction | Cloak 1975 |
| Cultural Type | Boyd and Richerson 1976 |
| Meme | Dawkins 1976 |
| Concept | Hill 1978 |
| Culturgen | Lumsden and Wilson 1981 |

such as ideas, values, skills, beliefs, and attitudes, while waiting for other disciplines to improve our understanding of how information is actually stored in brains.

Conceptualizing the transmission of learned information more broadly as "culture," as Richerson and Boyd suggest, is not without its own problems. The definition of culture is a notoriously slippery issue (Durham, 1991) — for instance, is it ideational information in the brain, material objects outside the body, behavioral practices that link the two, or some combination of these? Focusing on social information transmission is one way to approach this dilemma, allowing us to consider both how ideas, conventions, and traditions are created and maintained and also how information transmission underlies the production and utilization of artefacts, which may then act as extracorporeal information stores (Durham, 1991). At least within those circles, applying evolutionary principles to the study of cultural transmission, there is increasing agreement that what matters in culture is the information held in and processed by brains (Plotkin, 2002; Richerson and Boyd, 2004), although there is still debate as to the biology of how that information is stored and acquired.

This discussion suggests that biologists are now getting to grips with the role of socially transmitted information in evolution. Indeed it is a rapidly growing area of research, and this volume cannot hope to do justice to the many contemporary debates. We elected to focus on aspects of information transmission that elucidate the pursuit of behavioral strategies, which are recognized to be a primary focus of selection (Houston and McNamara, 1999). This is a challenging area, for strategies typically lack direct external manifestation or material evidence and must be inferred from indirect evidence. However, such a focus is critical for improving our understanding of communication from a biological evolutionary perspective.

## 1.3 THE AIMS OF THIS SYMPOSIUM VOLUME

In this volume, we therefore aimed to incorporate a variety of theoretical perspectives and a wide-ranging selection of biological issues in which behavior and socially transmitted learned information is relevant to the "mechanics" of differential survival and reproduction, through which selection acts. Such an approach will have two main benefits.

First, we hope those addressing contemporary social and biological issues in humans will appreciate the value of adopting an evolutionary perspective. As we have entered the 21st century, it is increasingly apparent that our ability to reduce mortality from diseases is hampered not just by our knowledge of physiology, but also by our poor understanding of human behavior. Although many human populations remain most influenced by infectious diseases, in industrialized societies the big killers are "lifestyle" diseases such as heart disease, hypertension, type 2 diabetes, lung cancer, HIV/AIDS, and vehicle accidents. These diseases are furthermore becoming common in rapidly modernizing countries such as India, which is currently experiencing an epidemic of type 2 diabetes. Progress in reducing the prevalence of these diseases is heavily dependent on improving our understanding of how to induce beneficial behavioral change and ensure the ecological sustainability of practices. Public health practitioners and biologists need to understand the dynamics and mechanisms underlying behavioral variability and its development. Just as economists need to take into account history, so those concerned with biological outcomes would benefit by taking into account evolutionary processes and perspectives. Critically, such an understanding requires recognition of how our evolved characters respond behaviorally to acquired information.

Second, we hope that those developing theoretical models of human behavior will themselves gain from an improved understanding of how the social transmission of information affects biological outcomes. Mathematical models are intentionally constructed to provide a simplified approach in which some factors are held constant in order better to understand what happens when other factors vary. Such models have played a key role in eliciting widespread interest in the role of culture in evolution and often identify important factors that defy intuition or empirical observation (Boone, 2002). However, the confidence with which their conclusions can be accepted depends on their being sufficiently realistic and on whether empirical data support them. An evolutionary approach requires particular attention to outcomes such as mate choice, fertility, and mortality that underlie differential survival and reproductive success. Such issues are central to many chapters in this volume, which explore the relationship between behavior and fitness. This approach comes of course with the caveat that contemporary humans are themselves variable: they are not the same as past humans and may be even less representative of hominin biology in earlier time periods.

There can be little doubt that human behavior has exerted a powerful influence on the evolutionary dynamic of our species (Richerson and Boyd, 2004). For instance, Boone (2002) has recently calculated that small changes in human mortality rates can translate into near-exponential population growth, which may in turn allow rapid colonization of new environments and the accompanying spread of cultural information. Thus, behavior that influences mortality schedules can represent a potent factor in human biology and demographics. Cultural transmission is also key to the extraordinary variability that characterizes human behavior. The magnitude of difference in ecological, technical, material, and symbolic traits demonstrated by diverse human populations illustrates the capacity of the mind to both accommodate and generate different contexts (see Odling-Smee, this volume). More generally, variability and change are of course fundamental issues for all three disciplines of biological anthropology, social anthropology, and archaeology.

## 1.4 THE STRUCTURE OF THIS BOOK

The aim of this book is therefore to address the role of social information transmission in human biology. To begin with, it is necessary to summarize the different evolutionary approaches to human behavior — to provide a firm basis for investigating our characteristic behavioral strategies — in the context of which social information transmission occurs. In Chapter 2, Laland and Brown review the five main recent approaches and highlight their relative strengths, their weaknesses, and their potential misapplication. They go on to discuss how the various approaches differ in the emphasis placed on the role of social transmission in explaining human behavioral variation, while also suggesting that the importance of historical factors (as opposed to current ecological factors) in accounting for human behavior may have been underestimated by evolutionary biologists in general.

One of the problems addressed by Laland and Brown is the relationship between cultural and genetic variability. Whereas some evolutionary approaches such as memetics pay minimal attention to genetics, others focus on the interaction between genetic and cultural information. In Chapter 3, Odling-Smee focuses specifically on the concept of niche construction, an approach highlighting how the tendency of organisms to manufacture and alter their environment impacts on selection pressures, which in turn act back on the population as well as on other populations. The vast majority of human behavior is related to the physical and social niches constructed by earlier and current human populations, and the positive feedback cycle relating current behavior and niche construction is a major characteristic of social information transmission. This theme is further addressed in several subsequent chapters, in particular that of Hesketh who discusses the effect of a socially engineered niche (state regulation of fertility) on human biology in China.

A critical issue in the transmission of nongenetic information is the rapid timescale in comparison with changes in gene frequencies. Game theory is already a well-established component of the evolutionary biologist's toolkit, able to address the evolution of behavioral strategies according to the criteria of maximizing fitness. However, a limitation of this approach is that it has restricted flexibility with which to take into account how organisms either respond to the strategy of those with whom they interact or even preempt or second guess others' strategies. Social organisms, such as humans, have multiple dynamic relationships with other individuals — indeed, in humans, our conspecifics are perhaps the dominant feature of our ecological environment. Behavioral interactions are rarely one-off occurrences but are, rather, characterized by continued modifications to strategy, as in a chess game. The dynamic transmission and processing of information that occurs in the generation of individuals' strategies must therefore be incorporated into our understanding of human behavior. In Chapter 4, McNamara and Houston describe how dynamic mathematical modeling can be used to evaluate how individuals formulate strategies and decisions. Such models are already proving extremely informative in many areas of zoology and are destined to play a major role in the study of human behavior. One evolutionary approach to human behavior, human behavioral ecology (see, for example, Chapter 11 below), uses variability in the number of offspring as its main outcome. McNamara and Houston highlight a challenge to this approach by demonstrating that strategies optimal in the long term are not necessarily those with the highest payoff within one or even two generations.

As discussed above, focusing on the transmission of ideational units, similar to that of genetic units, has played an important role in opening up the phenomenon of human culture to evolutionary biologists. The difficulty of this approach lies in developing a model that is sufficiently robust to underpin a hypothesis-testing program of empirical research. In Chapter 5, Aunger approaches the major challenge facing the discipline of memetics, the nature of the meme itself. Aunger argues that neural information in brains should be distinguished from signals that instigate meme reconstruction when they are received by other brains. Such an argument is particularly relevant to the difficult issue of whether cultural information does indeed spread through replication. Aunger remains cautious about memetics and notes that even if the meme does prove a valid concept, meme transmission would be only one of several processes underlying human culture. However, the benefit of his approach is that it can be aligned with broader biological models of signaling and is thus likely to make hypothesis-testing research more achievable.

As several of the chapters in this volume demonstrate, the social transmission of information can have major effects on reproductive success, through its central roles in mate choice, and fertility and mortality patterns. However, one reason social anthropologists have not favored biological approaches to human behavior is their rejection of the hypothesis that our behavior systematically maximizes reproductive success. Few biologists have explicitly argued in defense of this hypothesis (though see Alexander, 1979), but biologists have traditionally regarded human behavior that reduces reproductive success as maladaptive and have tended to focus disproportionately on those traits considered adaptive. More recently, dual-inheritance theorists have considered how behavior may spread within populations at the expense of reproductive fitness. What has not been addressed is what human behavior is teleologically for, if not for reproduction. In Chapter 6, Wells and Strickland argue that humans can maximize both genetic and nongenetic ends. In both cases, these ends comprise the transmission of information, but whereas reproduction transmits genes, humans can also transmit other information about themselves in the form of their reputation. This approach emphasizes the importance of personal identity for understanding human behavior. Personal identity is a major focus of social anthropological study but has been relatively ignored by biologists with their primary interest in genetically or game-theory derived strategies (although see biological market theory, Bshary and Bronstein, 2004). An enhanced understanding of identity and the self could also contribute to many aspects of medicine, including improving the success of public health education campaigns. More generally, the notion of different fitness currencies broadens the range of human behavior that can be incorporated within an evolutionary biological perspective.

Biological anthropologists have traditionally learned much about human biology by studying our closest living relatives, the great apes. Chimpanzees, gorillas, and orangutans diverged from humans over the last 14 million years or so since sharing a common ancestor. It is often assumed, perhaps erroneously, that contemporary differences between apes and humans can be attributed disproportionately to the evolutionary trajectory of our own species, on account of the succession of novel niches that hominids and humans have constructed and occupied. To what extent, then, is the biological role of cultural information peculiar to humans as opposed to apes in general? In Chapter 7, Whiten explores whether traditions can be identified in ape populations and, if so, whether they can be considered adaptive. Certain behavior patterns, such as use of specific tools or grooming behaviors, appear to be present in some chimpanzee populations but not in others, implying their local discovery and transmission within populations. Whiten also shows how some behavior patterns have spread between populations until natural boundaries have hindered their continued transmission, thus illustrating the flow of learned information between primate populations that is clearly so fundamental to human technological uptake. This analysis suggests that all apes are likely to have exploited social information transmission for adaptation to local environments throughout their recent evolutionary history and that contemporary humans and other hominins have simply done so on an unprecedented scale. Of particular interest is the finding that social learning is implicated in chimpanzee "healthcare" (self-medication in response to nematode infection), highlighting the diversity of pathways through which social information transmission can influence fitness in this group of animals.

The majority of human social information transmission is achieved through verbal language, although in recent decades technology has increased the possible pathways of transmission and modes of storage. Yet human language is an enigmatic feature of human biology. It has often been remarked that other social animal species function satisfactorily without language, so why do humans need it? In Chapter 8, Nettle considers the costs and benefits of communicating through a symbolic language system that allows not only honest communication but also deceit. All communication may be considered manipulative, for any that evokes no receiver response (i.e., has zero manipulatory effect) could be said to contain no significant information (Haig, 1996). However, information may also benefit receivers, who must therefore assess the quality of incoming information and guard against being manipulated against their interests. This is one reason the development of language is often thought to have coevolved with theory of mind, the ability to conceptualize and evaluate mental states (Malle, 2002). Nettle argues that verbal communication is closely related to reciprocity and may be considered unreliable in scenarios where reciprocity is also vulnerable to exploitation and manipulation. To reduce the penalties of deceit or miscomprehension, individuals must complement social information transmission with learning through direct experience. This argument is consistent with other models of information sharing, since, if all members learn by social learning from others rather than direct learning, learned information may become detached from environmental variation and its use may cease to be adaptive (Richerson and Boyd, 2004). The role of information transmission as manipulation is an issue that is explored in several other chapters, in particular those addressing public health campaigns that may seek either to encourage, or to discourage, particular behaviors and also the state regulation of individuals' behavior for the manipulation of population demographics.

Verbal language is however one of several means, characterized by a high degree of intentionality, by which humans transmit information between individuals. Two other means are body language and material culture, each of which may predate the evolution of complex vocal communication. In Chapter 9, Mithen highlights some of the key developments in communication that have occurred during the period of human evolution. He argues that body language, such as posture, movement, and gesture, is closely associated with the evolution of bipedalism, which began with the Australopithecines and was largely completed with the emergence of *Homo ergaster*, 2 million years BP. Mithen further argues that bipedalism, through its demand for improved muscular control and its effects on cranial anatomy, may have favored both the increased brain size and the changes in larynx structure that enabled the emergence of early verbal language. Modeling the change from "protolanguage" to "language" is more challenging. Here, Mithen argues that though fossil evidence suggests an early development of language capability, illustrated by evidence of the appropriate oral anatomy in Neanderthals and hence perhaps in *Homo heidelbergensis*, there is minimal evidence of symbolic representation until well after the emergence of modern *Homo sapiens*. The fossil and archaeological record suggests that the emergence of the modern human mind dates not from 160,000 years BP, when *Homo sapiens* evolved, but from around 70,000 years BP. Whether this late date can be attributed simply to the poor fossil record for earlier periods, or whether subsequent selective pressures generated either cultural or genetic change, remains an important area of investigation.

The archaeological record represents a rich source of evidence for many further issues, in particular providing evidence over many generations with which to test theoretical models of social information transmission. In Chapter 10, Shennan describes how evolutionary theory can help archaeologists address the histories of human cultural traditions. Theoreticians (e.g., Boyd and Richerson, 1985) have shown that many factors and processes may influence the transmission of knowledge, beliefs, values, and practices within and between generations. Examination of the archaeological record allows these competing hypotheses to be evaluated, elucidating which process was important in a given case. For example, Shennan argues that whereas the initial spread of farming from the Near East into Europe 11,000–7,000 years BP involved the westwards dispersal of populations practicing a more efficient mode of subsistence, subsequent demographic dynamics appear to have comprised disproportionately the movement between groups of women, indicating the emergence of territorial patrilineal corporate groups. The issue of whether populations or their practices dispersed through time has many implications for evolutionary biology is a current focus of molecular geneticists attempting to reconstruct the pattern of migrations throughout the period of *Homo sapiens* (Jobling, Hurles, and Tyler-Smith, 2003; Oppenheimer, 2003).

Behavior clearly has an effect at many levels on human reproduction, and in pursuing this end many different types of information require processing. One type of information that has become increasingly important in recent decades is knowledge of new methods by which control of fertility can be enhanced. Humans naturally regulate reproduction through a variety of physiological and behavioral mechanisms. Some of these mechanisms, such as maternal nutritional status (fat stores, leptin levels) operate involuntarily. Others, such as the frequency of intercourse, duration of lactation, behavioral taboos, and infanticide, may be more intentional. Contemporary humans can now have access to further methods of contraception, through recent technological developments, with major effects on two areas of human biology — pregnancy and the risk of contracting a sexually transmitted disease. Knowledge and use of contraception is therefore a critical issue for both fertility rates and rates of morbidity and mortality. In Chapter 11, Mace and colleagues consider how the use of new contraception methods spread in a rural Gambian population after they were first made available several decades ago. Their analysis considers both the factors that affected its uptake and how it influenced fertility in users compared with nonusers. Consistent with Cleland and Watkins (see below), Mace and colleagues highlight the importance of informal discussion in determining individuals' choices. Far from simply copying other individuals, women appear to have exploited the process of social transmission to incorporate dynamic evaluation of the information.

A nonintuitive finding of this study is that those using contraception had more surviving children than those who eschewed it.

The uptake of behavioral practices intended to improve population health is likewise addressed by Cleland and Watkins in Chapter 12. Here, the focus is the broader issue of whether national governmental and nongovernmental programs designed to promote safer sexual behavior are indeed achieving their objective, with specific reference to HIV/AIDS. Comparing current "safe-sex" programs with previous programs intended to reduce fertility, Cleland and Watkins find much common ground and suggest that the apparent lack of success of HIV/AIDS prevention programs may in part be an artefact of the time that is required for them to become effective. Population health programs are characterized by two distinct parties, and an important issue is whether the recipients of apparently beneficial information trust the senders. Like Mace and colleagues, Cleland and Watkins emphasize the importance of informal discussion as a way of evaluating health advice. Such discussion enables individuals to interpret their own experience in the context of others who they do trust and appear to play a key role in the acceptance of public health messages. Recent work in South Africa suggests that earning the trust of the target population is critical for the acceptance of health advice and that an important aspect of this process is the adaptation of the advice to local circumstances. Given that changes in health-related behavior may, in the context of the current HIV/AIDS pandemic, translate into major decreases in morbidity and mortality, understanding this issue is clearly of major importance.

The transmission of HIV/AIDS by sexual activity is one of a broader range of activities that may be pleasurable in the short term yet harmful to subsequent health. Where costs to health take time to emerge, individuals may either initially be unaware of them, uncertain of their likelihood, or discount them against the immediate pleasure. Public health campaigns are not only potentially beneficial but also manipulative, and an appropriate approach needs to be developed to maximize acceptance of the advice and achieve constructive changes in behavior. In Chapter 13, Berridge and Loughlin review the development of public health campaigns in the UK in the mid-twentieth century with a focus on smoking. Berridge and Loughlin observe that although the potential contribution of mass education to public health was recognized in the 1930s, it took several decades for large-scale public health programs to emerge. On the one hand, it was necessary to overcome the population's tendency to reject centrally directed "propaganda," while on the other, campaigns became effective only when they exploited the mechanics of mass consumption, in particular, commercial advertising, that boomed from the 1960s. Successive campaigns adopted increasingly targeted and aggressive strategies, switching from the neutral presentation of information to shock tactics and advocacy, and found visual images to provoke the strongest response. The use of the same media to promote and attack a single activity such as smoking has analogies elsewhere with contemporary attempts to combat the burden of malaria by transgenically modifying, as opposed to eradicating, anopheles mosquitoes (Moreira et al., 2004). Whether mass advertising is effective in inducing behavioral change in relation to health remains debated, but there is no doubt that the issue of how to increase individuals' responsibility for their health through their behavior is critical for contemporary public health management.

The role of information transmission by the state in constraining human behavior is further considered by Hesketh in Chapter 14. The application of the One Child Family policy in China represents the rapid construction of a new social niche in which many aspects of human biology have been affected. The short timescale over which this niche has been constructed renders it questionable as to whether significant time has elapsed for a genetic response to occur (although see Kumm et al., 1994; Laland et al., 1995). For example, Fisher (1930) argued that in the long term, distortions of the sex ratio tend to resolve; however, this analysis ignores the potency of socially transmitted information (Kumm et al., 1994). In contemporary China there is a significant excess of young males, resulting from socially transmitted norms and practices that value sons over daughters. At the same time, the niche can be considered an important source of selection on the gene pool. First-born children are now known to make different metabolic adaptation in utero compared to later-born offspring, with such adaptations associated with increased risk of developing the metabolic syndrome in later life (Gluckman and Hanson, 2005), which may in turn influence metabolic traits in the next generation. According to existing data, Hesketh finds that effects on the age and gender structure of the population are far greater than those on child health, but this situation requires monitoring and the actual consequences of the policy for later adult health in this and subsequent generations remain unknown. That the policy will have long-term effects is demonstrated by its effects on family structure. In general, young adult humans contribute resources disproportionately to both the young and the old; however, the One Child Policy is generating the challenging situation where each nuclear family may have two grandparents to care for but only one child to help them do so.

Whereas most chapters in this book have addressed the capacity of social information transmission to exert largely beneficial effects on human biology, some issues are more controversial. Personal identity is a particularly intriguing area, where predispositions influence the response to social factors. For example, conventionally, humans are considered either male or female, with the distinction initially based on physiology and then reinforced through socially derived gender roles. A very small minority of children show ambiguous sexual anatomy at birth, requiring clinicians to contribute to the development of gender identity and sexual function. Gender identity cannot be forced on such children, rather it needs to be elicited gradually, with physical treatment delayed for a period until behavioral tendencies mature. To paraphrase Ingold (2001), human identity is never made but is always in the making. Throughout their lives, individuals may also conceal information, or deliberately impart incorrect information. By definition, Machiavellian strategies depend on the *manipulation* of information, and the role of social information transmission in human behavior is therefore extremely complex.

In Chapter 15, Littlewood addresses a related theme by considering the contemporary "epidemic" of multiple personality disorders within the context of westernization and globalization. The critical role of personal identity and the self in human behavior has already been alluded to in Chapter 6, however the self is itself a plastic entity and reflects social influences. Littlewood notes that European history has been characterized by a steady trend toward individualization of the self, reflected for example in a shift toward attributing mental illness to internal anxieties rather than external demonic possession. This trend now appears to be going into reverse, with increasing fragmentation of the self on account of exposure to the diverse influences emerging in an increasingly globalized world. This approach emphasizes once more the importance of social processes in the generation of apparent individuality and agency, but it also highlights an intriguing issue for evolutionary biologists. Some evolutionary approaches (though not all) incorporate the assumption that individual humans, whether consciously or not, pursue rational strategies; and such models are likely to consider the self as the expression of more fundamental tendencies to behave in a manner that maximizes fitness. How can more fragmented selves be incorporated into such models when, as Littlewood notes, "to have a hundred personalities is perhaps to have none?" The nature of the self during human evolution is an extremely complex issue (see, for example, Humphrey, 1986) but not one that, on the basis of our understanding of its contemporary diversity, can be entirely ignored by biologists.

Finally, it is important to ask what the benefits of increased biological inquiry into the issue of social information transmission might be for the other disciplines that study human behavior. Archaeology represents a discipline that has rapidly incorporated a Darwinian perspective over the last decade (Maschner, 1996; Shennan, 2002), with many benefits (though see also Smith, 2000). Biological anthropology has received attention through application of models such as optimal foraging and other behavioral ecological theory (Smith and Winterhalder, 1992; Cronk et al., 2000), and there is increasing recognition of the critical role of social information transmission (Richerson and Boyd, 2004). Memetic models have also been applied, most explicitly in the form of "cultural virus theory" (Cullen, 1996), as have approaches based on evolutionary psychology (Mithen, 1996; Donald, 1991) and gene-culture coevolution (Shennan 2002). Archaeologists have a privileged position in being able to evaluate multigenerational time periods during which behavior may nevertheless have changed rapidly. Their data may illustrate the differential survival of both human populations and cultural practices — for example, just as Neanderthal populations went extinct around 30,000 years ago, so Tasmanian populations that colonized the island around 34,000 years ago appear subsequently to have lost various types of technology that are evident in the earlier archaeological record (Henrich, 2004). Each of these examples illustrates the exploitation by archaeologists of approaches developed within the framework of evolutionary biology.

Whereas many, perhaps most, archaeologists recognize the benefits to archaeology of adopting an evolutionary perspective, the equivalent scenario for social anthropology is more ambivalent. The relationship between the natural and social sciences is often portrayed as antagonistic and counterproductive. To some extent this representation is misleading simply because the tendency of different disciplines to address different questions is not necessarily evidence of inconsistency. On the other hand, it is not difficult to find instances of social or biological anthropologists denigrating specific approaches or studies conducted by the "opposing" school. Nevertheless, we believe that here too there are many encouraging signs.

One sign is shown simply by academic and institutional practices. Social anthropologists are increasingly involved in medical research and public health campaigns, whether in industrialized, modernizing, or relatively traditional societies.

Like archaeologists, social anthropologists are also interested in change and do not see cultures as static; they, therefore, share with other disciplines an interest in how behavior changes over time. A recent anthropological review of health in Haiti demonstrated the importance of adopting a historical perspective to highlight the continuing impact of "structural violence" on the distribution of diseases and the provision of healthcare (Farmer, 2004). This contextual approach is highly relevant to the differential survival and reproduction that are so central to evolutionary approaches. In clinical practice, too, there is increasing awareness of the contribution of ethnicity to both physiology and behavior, as well as the establishment of a subdiscipline, medical anthropology, that considers different cultural models of diseases and their treatment. The role of cultural traits is likewise increasingly investigated in biological studies of behavior, particularly those testing evolutionary models. Examples include examinations of the extent to which cultural phylogeny can explain regional distribution of behavior patterns (Mace and Holden, 1999) and consideration of the value of wealth in relation to fertility (Mace, 1997). At many levels, therefore, those studying human behavior are broadening their approaches and benefiting from cross-disciplinary perspectives.

Another sign is the development of less divisive theoretical perspectives. Recently, Ingold (2001) presented a radical approach to the issue of interdisciplinary relations, arguing that a developmental perspective could help dissolve the boundaries between social and biological anthropology, archaeology, and psychology. Each of these disciplines is traditionally defined in relation to specific goals. For example, the two branches of anthropology have traditionally focused on individuals as either biological organisms or cultural persons, while psychology has focused on the mind of the individual.

Ingold argues that such approaches have inevitably represented as fixed or transmitted structures and phenomena that are, in fact, relations and processes continuing through the lifetime of the organism. For example, individuals do not acquire cultural information as complete organisms but rather incorporate it continually through their lifespan, whether, for example, the style of walking or the behaviors of kinship. The evolved structures of the mind that are central to psychology (especially the "evolved architecture" beloved of evolutionary psychologists) are likewise difficult to separate from the developmental processes undergone by each mind within a lifecycle as it constructs diverse capacities. Above all, the "biological" as relevant to behavior is "developmental" rather than "genetic."

We agree that emphasizing contrasts between natural and social, body and mind, or biological and cultural will ultimately hinder our understanding of human behavior. We interpret Ingold's criticism that the "notion of culture … transmissible from one generation to the next as a corpus of knowledge, independent of its application in the world, is untenable for the simple reason that its rests on the impossible precondition of a ready-made cognitive architecture" not as a refutation of the concept of the social transmission of information per se but as a warning that without understanding how each organism continually receives, appraises, and incorporates information during its development and lifetime, our models of social transmission will remain simplistic. Children should be a special concern of those studying socially transmitted information flow but are conspicuous by their absence in many biological and social anthropological investigations of human behavior.

A developmental approach reminds us that how learned information is transmitted, obtained, processed, stored, recalled, and expressed by humans is extremely complex. In the final chapter of this volume, Strickland provides an epilogue in which he considers several of the themes prominent in the preceding chapters that have particular relevance to social anthropologists. To a biologist such as Alexander (1989), memory and intention might seem little more than the cognitive expression of evolved predispositions for maximizing reproductive success. Strickland notes the importance of understanding the evolution of human capacities for, and uses of, memory, including innovative methods of representation and their biological consequences. Important issues here are the extent to which memory is a reliable means of transmitting beneficial information over time under varying conditions and the extent to which, through the vehicles of tradition and socialization, memory acts as a means steadily to draw individuals more tightly into constraining social relationships. Finally, both biological and social sciences could helpfully reappraise their uses of teleological argument in explaining human purposive behavior. Accommodating a wider range of "ends" than reproductive success could bring these hitherto somewhat disparate fields closer.

These are all issues that would prove rewarding for biological anthropologists to pursue and that incorporate evolutionary perspectives. We hope that one reaction to this volume will be for some to do so.

## 1.5 CONCLUSION

If this volume has successfully demonstrated the multiple roles and effects of social information transmission in human biology, it is clear that the next area to address is how such information is stored and processed. The transmission of learned information may be implicated in numerous biological processes as these chapters have shown, but biologists will only benefit from this knowledge if they can improve their understanding of how it shapes human behavior. Information enters not just brains but also minds that are themselves constructed through the medium of social interaction. To understand the role of the mind in such social processes, the skills and experience of social anthropologists will be critical. A recent study of the development of adolescent gender identity in the context of teenage pregnancy demonstrated this by emphasizing the need to approach health-related behavior within the context of local social relations, cultural constructs, and regional history (Mkhwanazi, 2004). Such a perspective is crucial for, but lacking in, public health programs intended to address the behavioral basis of HIV transmission. More generally, we believe that biological models of human behavior are destined to remain simplistic if they fail to take into account the symbolic component of information storage and transmission. In pursuing a more sophisticated understanding of human behavior, we believe many of the apparent inconsistencies between different disciplines will prove redundant.

This volume was above all intended to stimulate those studying human behavior by juxtaposing a variety of approaches and areas in which the issue of social information transmission is important. If readers find themselves making connections between ideas addressed by different contributors, then it will have served its purpose.

## REFERENCES


Alexander, R.D., *Darwinism and human affairs,* Pitman, London, 1979.

Alexander, R.D., *The biology of moral systems,* Aldine de Gruyter, New York, 1989.

Bateson, P., and Martin, P., *Design for a life—how behaviour develops,* Jonathan Cape, London, 1999.

Betzig, L.L., (Ed.) *Human nature: a critical reader,* Oxford University Press, Oxford, 1997.

Blum, H.F., On the origin and evolution of human culture, *Am. Sci.*, 51, 32–47, 1963.

Blurton Jones, N., and Reynolds, V., *Human behaviour and adaptation*, Taylor and Francis, London, 1978.

Boas, F., *Race, language, and culture*, Free Press, New York, 1940.

Boone, J.L., Subsistence strategies and early human population history: an evolutionary ecological perspective, *World Archaeol.*, 34, 6–25, 2002.

Boyd, R., and Richerson, P.J., A simple dual inheritance model of the conflict between social and biological evolution, *Zygon*, 11, 254–262, 1976.

Boyd, R., and Richerson, P.J., *Culture and the evolutionary process*, University of Chicago Press, Chicago, 1985.

Bshary, R., and Bronstein, J.L., Game structures in mutualistic interactions: What can the evidence tell us about the kind of models we need? *Adv. Study Behav.*, 34, 59–101, 2004.

Cavalli-Sforza, L.L., Similarities and dissimilarities of sociocultural and biological evolution, In: *Mathematics in the archaeological sciences*, Hodson, F.R., Kendall, D.G., and Tatu, P., Eds., Edinburgh University Press, Edinburgh, 535–541, 1971.

Cavalli-Sforza, L.L, and Feldman, M.W., *Cultural transmission and evolution: a quantitative approach*, Princeton University Press, Princeton, NJ, 1981.

Chagnon, N.A., and Irons, W. (Eds.), *Evolutionary biology and human social behaviour*, Duxbury Press, North Scituate, 1979.

Cloak, F.T., Is a cultural ethology possible? *Hum. Ecol.*, 3, 161–182, 1975.

Cronk, L., Changon, N.A., and Irons, W., *Adaptation and human behaviour: an anthropological perspective*, Aldine de Gruyter, New York, 2000.

Cullen, B.R.S., Cultural virus theory and the Eusocial pottery assemblage, In: *Darwinian Archaeologies*, Maschner, H.D.G., (Ed.), Plenum Press, New York, 43–59, 1996.

Dawkins, R., *The selfish gene*, Oxford University Press, Oxford, 1976.

Dawkins, R., *The Extended Phenotype,* Freeman, Oxford, 1982.

Donald, M., *Origins of the modern mind*, Harvard University Press, Cambridge, MA, 1991.

Durham, W.H., *Coevolution: Genes, culture and human diversity*, Stanford University Press, Stanford, CA, 1991.

Farmer, P., An anthropology of structural violence, *Curr. Anthropol.*, 45, 305–325, 2004.

Feldman, M.W., and Laland, K.N., Gene-culture coevolutionary theory, *Trend. Ecol. Evol.* 11, 453–457, 1996.

Fisher, R.A., The genetical theory of natural selection, Clarendon, Oxford, 1930.

Fragaszy, D., and Perry, S. (Eds.), *The biology of traditions: models and evidence*, Cambridge University Press, Cambridge, 2003.

Gilbert, S., *Developmental Biology,* 7th ed., Sinauer, Sunderland, MA, 2003.

Gluckman, P., and Hanson, M., *The fetal matrix: evolution, development and disease*, Cambridge University Press, Cambridge, 2005.

Haig, D., Placental hormones, genomic imprinting, and maternal-fetal communication, *J. Evol. Biol.*, 9, 357–380, 1996.

Hamilton, W.D., The genetical evolution of social behaviour (I and II). *J. Theor. Biol.*, 7, 1–16; 17–52, 1964.

Henrich, J., Demography and cultural evolution: why adaptive cultural processes produced maladaptive losses in Tasmania, *Am. Antiquity*, 69, 197–21, 2004.

Heyes, C.M, and Galef, B.G., (Eds.), *Social learning in animals: the roots of culture*, Academic Press, New York, 1996.

Hill, J., The origin of sociocultural evolution, *J. Soc. Biol. Structures*, 1, 377–386, 1978.

Hoagland, H., Science and the new humanism, *Sci.*, 143, 111–114, 1964.

Houston, A.I., and McNamara, J.M., *Models of Adaptive Behaviour*, Cambridge University Press, Cambridge, 1999.

Humphrey, N., *The inner eye*, Faber and Faber, London, 1986.

Huxley, J., Evolution: biological and human, *Nature*, 196, 203–204, 1962.

Ingold, T., From complementarity to obviation: on dissolving the boundaries between social and biological anthropology, archaeology, and psychology, In: *Cycles of contingency: developmental systems and evolution*, Oyama, S., Griffiths, P.E., and Gray, R.D., (Eds.), MIT Press, Cambridge MA, 255–79, 2001.

Jobling, M.A., Hurles, M.E., and Tyler-Smith, C., *Human evolutionary genetics: origins, peoples and disease,* Garland, New York, 2003.

Kumm J., Laland, K.N., and Feldman, M.W., Gene-culture coevolution and sex ratios: the effects of infanticide, sex-selective abortion, sex selection, and sex-biased parental investment on the evolution of sex ratios, *Theor. Pop. Biol.,* 46, 249–278, 1994.

Kuper, A., *Culture: the anthropologists' account*, Harvard University Press, Cambridge MA, 1999.

Laland, K.N., and Brown, G.R., Sense and nonsense: evolutionary perspectives on human behaviour, Oxford University Press, Oxford, 2002.

Laland, K.N., Kumm, J., and Feldman, M.W., Gene-culture coevolutionary theory: a test case, *Curr. Anthropol.,* 36, 131–156, 1995.

Lehrman, D.S., A critique of Konrad Lorenz's theory of instinctive behaviour, *Q. Rev. Biol.*, 337–363, 28, 1953.

Lewontin, R.C., Gene, organism and environment. In: *Evolution from molecules to man*, D.S. Bendall, (Ed.), Cambridge University Press, Cambridge, 273–285, 1983.

Lorenz, K., *On aggression* (Transl. Latzke, M.), Methuen, London, 1966.

Lumsden, C.J, and Wilson, E.O., *Genes, mind and culture: the co-evolutionary process*, Harvard University Press, Cambridge, MA, 1981.

Mace, R., The co-evolution of human fertility and wealth inheritance strategies, *Philos. Trans. Roy. Soc. Lond. Series B*, 353, 389–397, 1997.

Mace, R., and Holden, C., Evolutionary ecology and cross-cultural comparison: the case of matrilinearity in sub-Saharan Africa, In: *Comparative primate socioecology*, Lee, P.C., (Ed.), Cambridge University Press, Cambridge, 387–405, 1999.

Malle, B.F., The relation between language and the theory of mind in development and evolution, In: The evolution of language out of pre-language, Givón, T., and Malle, B.F., (Eds.), John Benjamins, Amsterdam, 265–284, 2002.

Maschner, H.D.G., (Ed.), *Darwinian archaeologies,* Plenum, New York, 1996.

Meltzoff, A., and Prinz, W., *The imitative mind: development, evolution, and brain bases*, Cambridge University Press, Cambridge, 2002.

Mesoudi, A., Whiten, A., and Laland, K.N., Perspective: is human cultural evolution Darwinian? Evidence reviewed from the perspective of *The Origin of Species*, *Evol.*, 58, 1–11, 2004.

Mithen, S., *The prehistory of the mind*, Thames and Hudson, London, 1996.

Mkhwanazi, N.N., *Teenage pregnancy and gender identities in the making in a post-apartheid South African township*, unpublished PhD thesis, University of Cambridge, 2004.

Moreira, L.A., et al., Fitness of anopheline mosquitoes expressing transgenes that inhibit Plasmodium development, *Genet.* 166, 1337–41, 2004.

Oldroyd, D.R., *Darwinian Impacts*, The Open University Press, Milton Keyes, 1983.

Oppenheimer, S., *Out of Eden: the peopling of the world,* Constable, London, 2003.

Pigliucci, M., Expanding evolution: a broader view of inheritance puts pressure on the neo-Darwinian synthesis, *Nature*, 435, 565–566, 2005.

Plotkin, H.C., *The imagined world made real*, Allen Lane, London, 2002.

Richerson, P.J., and Boyd, R., *Not by genes alone: how culture transformed human evolution*, Chicago University Press, Chicago, 2004.

Roberts, D.F., and Weiner, J.S., (Eds.), The scope of physical anthropology and its place in academic studies, Church Army Press, Oxford, 1957.

Rose, M.R., and Lauder, G.V., (Ed.), *Adaptation*, Academic Press, New York, 1996.

Schlichting, C.D., and Pigliucci, M., Phenotypic evolution: a reaction norms perspective, Sinauer Associates Inc, Sunderland, MA, 1998.

Shennan, S.J. *Genes, memes and human history: Darwinian archaeology and cultural evolution,* Thames and Hudson, London, 2002.

Smith, E.A., Three styles in the evolutionary analysis of human behaviour, In: *Adaptation and Human Behaviour,* Cronk, L., Chagnon, N., and Irons, W., Aldine de Gruyter, New York, 27–46, 2000.

Smith, E.A., and Winterhalder, B., *Evolutionary Ecology and Human Behavior*, Aldine de Gruyter, New York, 1992.

Swanson, C.P., *The natural history of man*, Prentice-Hall, Englewood Cliffs, NJ, 1973.

Tinbergen, N., *The study of instinct*, Clarendon Press, Oxford, 1951.

Trivers, R.L., Parental investment and sexual selection, In: *Sexual Selection and the Descent of Man, 1871–1971*, Campbell, B., (Ed.), Aldine, Chicago, 136–179, 1972.

Trivers, R.L., Parent-offspring conflict, *Am. Zool.* 14, 249–264, 1974.

West-Eberhard, M.J., *Developmental plasticity and evolution*, Oxford University Press, Oxford, 2003.

Williams, G.C., *Adaptation and natural selection: a critique of some current evolutionary thought*, Princeton University Press, Princeton, NJ, 1966 (reprinted 1996).

Wilson, E.O., *Sociobiology: the new synthesis*. Harvard University Press, Cambridge, MA, 1975.

Wolf, E.R., Perilous ideas: race, culture, people, *Curr. Anthropol.*, 35, 1–12, 1994.

# 2 An Introduction to Evolutionary Models of Human Social Behavior

*Kevin Laland and Gillian Brown*

## CONTENTS



## ABSTRACT

We provide brief overviews and critical analyses of five evolutionary approaches to the study of human behavior: human sociobiology, human behavioral ecology, evolutionary psychology, cultural evolution (memetics), and gene-culture coevolution. We go on to discuss how the five approaches differ in the emphasis they place on the role of social transmission in explaining human behavioral variation.

## 2.1 FIVE EVOLUTIONARY ACCOUNTS OF HUMAN BEHAVIOR

### 2.1.1 HUMAN SOCIOBIOLOGY

Evolutionary accounts of human behavior date back to Darwin, particularly to *The Descent of Man, and Selection in Relation to Sex* (1871) and *The Expression of the Emotions in Man and Animals* (1872). However, modern evolutionary approaches to the study of human behavior took off in the mid-1970s, following a series of theoretical and conceptual advances within behavioral biology. Among these new methods and ideas were William Hamilton's *inclusive fitness* (later termed *kin selection*), Robert Trivers's *reciprocal altruism* and *conflict theories*, George Williams's *gene's-eye-view perspective*, and the use of optimality theory and related mathematical approaches, including John Maynard-Smith's *evolutionary game theory.* In this section, we will provide a short background to these important advances within evolutionary biology and highlight the controversy surrounding their application to our own species. A more extensive account can be found in Laland and Brown (2002).

In 1975, Harvard biologist Edward O. Wilson brought together the recent, ground-breaking advances in evolutionary biology in his book *Sociobiology: The New Synthesis*. Wilson's important contributions were to define the field of *sociobiology* as a coherent new approach to the study of behavior and to demonstrate its feasibility and importance (Segerstråle, 2000). The lasting significance of Wilson's book is well illustrated by the fact that it was voted in 1989, by members of the USA's *Animal Behavior Society*, the most important book on animal behavior of all time. A year after Wilson's *Sociobiology* was launched, Oxford zoologist Richard Dawkins published *The Selfish Gene*, one of the most popular scientific books of the 20th century. Both of these books are a celebration of the "gene's-eye view," the notion that if we wish to understand evolution by natural selection it is a convenient and useful heuristic to look at the problem from the perspective of the gene and ask which traits would be most likely to increase in frequency in the next generation.

Few contemporary biologists would contest the assertion that the sociobiological perspective was a major advance over the ethological, individual- and group-selection based methodologies that it superseded. Indeed, there is broad overlap between the tools and approach of sociobiology and those of the contemporary schools that are described below. There is no doubt that sociobiological methods are insightful and effective and that they have reaped great reward in the study of other animals. Virtually all of the debate over sociobiology relates to the legitimacy of its application to humans. Why then, when Wilson had seen the behavior of ants, gulls, and monkeys fall into place, should he not apply these new methods to our own species? In the final chapter of his book, Wilson turned his thoughts to human nature, offering bold and speculative evolutionary hypotheses for controversial topics such as gender roles, aggression, and religion. He stated quite openly that one of the goals of sociobiology was to "reformulate the social sciences in a way that draws these subjects into the Modern [evolutionary] Synthesis" (1975, p. 4). Wilson's book was to catalyze the appearance of a stream of works utilizing and extending the theme of human sociobiology. However, those biologists who rushed into the territory of the human sciences received an extremely hostile reception from resident sociologists and anthropologists. The unprecedented tumult over sociobiology was to prove the biggest scientific controversy of the decade.

Immediately prior to the emergence of sociobiology, the only evolutionary approach to the study of animal behavior was ethology. Ethologists, such as Konrad Lorenz and Niko Tinbergen, stressed species-specific behavior while paying little attention to the behavioral variability observed within populations. This occasionally led them to lapse into group-selectionist explanations. For example, many animal vocalizations, displays, and aggregations were thought to be means by which individuals could assess population density so as to influence their decision of whether or not to reproduce. In *On Aggression* (1966), Lorenz described highly restrained and ritualized disputes between animals and depicted them as fought in a manner that allowed the population's best representatives to reproduce. The most overt group selectionism came from Scottish ecologist V.C. Wynne-Edwards, whose *Animal Dispersion in Relation to Social Behavior* (1962) argued that limitation of population growth could be achieved by some individuals altruistically restraining their reproduction.

Associated with sociobiology was a rejection of such group selectionist arguments. George C. Williams, in his classic 1966 book *Adaptation and Natural Selection*, pointed out that behavior that benefited the group was unlikely to occur where individuals would be able to cheat the system for their own benefit, as such cheaters would outcompete other members of the population and increase in numbers. He also pointed out that the movement of individuals between groups would erode group differences and weaken group selection further. Williams maintained that a "gene is selected on one basis only, its average effectiveness in producing individuals able to maximize the gene's representation in future generations" (1966, p. 251). This is the evolutionary perspective that was brought to prominence by Richard Dawkins (1976) in *The Selfish Gene*.

The main difficulty facing evolutionary biologists opposed to group selection was to explain instances of what appeared to be true altruism and cooperation in nature. For example, in many colonies of ants, bees, and wasps, the workers do not reproduce and instead devote their efforts to raising the offspring of queens. In *The Origin of Species*, Charles Darwin described the presence of these workers as "the one special difficulty, which at first appeared to me insuperable, and actually fatal to my whole theory" (1859, p. 257). Why should an individual behave in a way that decreases its own chances of surviving and reproducing while increasing another individual's reproductive success? Over a hundred years later, in 1964, William Hamilton devised a satisfactory solution: the answer was kinship. Close relatives share copies of many of the same genes, and hence individuals may increase the frequency of these common genes in the next generation by helping closely related kin to reproduce.

Hamilton's solution was based on an idea proposed by British geneticist J.B.S. Haldane, who had joked that he would lay down his life for two brothers or eight cousins. This quip reflects Haldane's reasoning that for altruistic behavior to evolve it must result in a net increase in the frequency of any underlying genes in the next generation. Hamilton (1964a, 1964b) devised a mathematical model that predicted that altruistic behavior may be selected if the fitness cost to the altruist ($c$) is less than the benefit to the relative ($b$) multiplied by the probability that the relative possesses the same gene (i.e., the degree of relatedness, $r$), or $c < br$. The theory, which became known as *kin selection*, was to revolutionize our understanding of animal social behavior.

As an expert on ants, Wilson was quick to appreciate the significance of Hamilton's work. In *Sociobiology*, he showed how kin selection could be used to explain altruism in primates, social carnivores, and birds, as well as in insects. Not surprisingly, it was to kin selection that Wilson turned to explain why human homosexuals should apparently be prepared to forego direct reproductive opportunities. Wilson speculated that "the homosexual members of primitive societies may have functioned as helpers.... Freed from the special obligations of parental duties, they could have operated with special efficiency in assisting close relatives. Genes favoring homosexuality could then be sustained at a high equilibrium level by kin selection alone" (1975, p. 555). This hypothesis is typical of the final chapter of *Sociobiology*: bold, speculative, backed by little empirical evidence, and naïvely insensitive to political connotations.

The influential work of Hamilton was followed by equally important papers by Robert Trivers. In a few years at Harvard University in the early 1970s, Trivers single-handedly devised a wealth of sociobiological theory, building on his mastery of the gene's-eye view perspective. For instance, in a paper published in 1971, Trivers introduced the idea of *reciprocal altruism*. He suggested that, even among unrelated individuals, altruistic behavior could be selected if there was a high probability that the behavior would be reciprocated and if the chances of cheating were reduced by individuals remembering previous interactions. As well as explaining various altruistic acts among animals, Trivers argued that a history of reciprocal altruism among our ancestors could have led to the evolution of human characteristics such as friendship, moralistic aggression, gratitude, and a sense of justice.

Trivers also contended that differences in degrees of relatedness would result in conflicts of interest between individuals. For instance, he reasoned that parents may invest equally in all offspring as parents are equally related to each of them, while each offspring should favor increased investment in itself rather than in current or future siblings, leading to *parent-offspring conflict* (Trivers, 1972, 1974). He proposed that natural selection would favor manipulative traits in offspring, such as begging and tantrums, that helped them to get more resources from their parents than their parents may otherwise provide and suggested that the temper tantrums of young children were an attempt to manipulate their parents into giving further investment. While the exact dynamics of parent-offspring conflict have proved to be more complex than Trivers envisaged (Bateson, 1994), his ideas provided huge impetus for further work by biologists and led to a fresh interpretation of parent-offspring relations in humans. Trivers's work on conflicts between the sexes and on differential investment in sons and daughters has also been seminal (Trivers, 1972; Trivers and Willard, 1973) and launched a vibrant industry of studies that applied such reasoning to humans.

Another important sociobiological development was *evolutionary game theory*, although this was never emphasized in Wilson's *Sociobiology*. Building on earlier work by economists, John Maynard-Smith pioneered a novel mathematical approach for predicting the evolution of social behavior in situations where the optimal behavioral strategy depends upon what others are doing. The method allowed researchers to work out whether a behavior is an *evolutionarily stable strategy*, on the assumption that over millions of years of evolution this is what would have evolved. These methods, originally designed to explore animal conflicts, were soon being used to interpret human behavior (e.g., Dawkins, 1976).

Many anthropologists, psychologists, sociologists, and some prominent biologists (notably Steven Gould and Richard Lewontin) responded negatively to human sociobiology. Critics rejected the sociobiologists' findings, accused sociobiology of "reductionism" and "biological determinism," denied that society reflected biological imperatives, and linked sociobiology to former disturbing applications of evolutionary theories, including the eugenics movement. Wilson, and other biologists countered, describing the critics as political extremists and cultural determinists, and the result was the massive furore now well known as the "sociobiology debate."

Given the widely acknowledged utility of the aforementioned tools within biology, why was there such controversy regarding their application to human social behavior? This is a complex issue for which there is no simple answer and to which many books have been dedicated (see, for instance, Segerstråle, 2000). In our view (Laland and Brown, 2002), the controversy was not justified by any "reductionism" or "determinism" on the part of the sociobiologists, although it was almost certainly fuelled by their works being portrayed in this manner. As far as we are concerned, all of sociobiologists' major protagonists have bent over backwards to be clear that they do not believe in genetic determinism. Their use of the "gene for" language does not imply any belief that single genes are responsible for behavior, that genes are the sole or even the major causal influence on behavior, or that such behavior patterns are fixed and inevitable. To the extent that the cause of the debate can be attributed to a single major factor, we believe it is that human sociobiology too often was not scholarly. Too much human sociobiology was weak, speculative, superficial evolutionary storytelling, ignorant of relevant social science literatures and unsupported by empirical findings.

Rightly or wrongly, and notwithstanding the recent success of daughter disciplines, human sociobiology was resoundingly rejected by the social sciences. Even today, it remains the case that most social scientists see little use for evolutionary theory in their work, many are deeply suspicious of evolutionary arguments and *sociobiology* is an unfashionable word. There are respects in which sociobiology can legitimately be portrayed as "triumphant" (Alcock, 2001), but these do not yet extend to any winning over of the social science community. Nonetheless, the sociobiological revolution, and the ensuing debate over the application of these methods to our own species, has provoked a number of contemporary evolutionary approaches to human behavior, including evolutionary psychology, human behavioral ecology, cultural evolution (memetics), and gene-culture coevolution (Figure 2.1).

### 2.1.2 Human Behavioral Ecology

In the late 1970s, while the human sociobiology debate was raging, a small number of anthropologists, notable Napoleon Chagnon, William Irons, and Kristen Hawkes, began to test adaptivistic sociobiological ideas with real data from human populations. From these early beginnings, a small but highly productive field of *human behavioral ecology* has emerged, which can be characterized as anthropologists applying the methods of animal behavioral ecology to remote human populations as if they were any other species of animal. Human behavioral ecologists explore, for instance, the extent to which the behavioral differences observed between human groups are responses to particular environments. Their aim is to determine how ecological and social factors affect behavioral variability within and between populations and to predict patterns of behavior using optimality and fitness-maximization models.

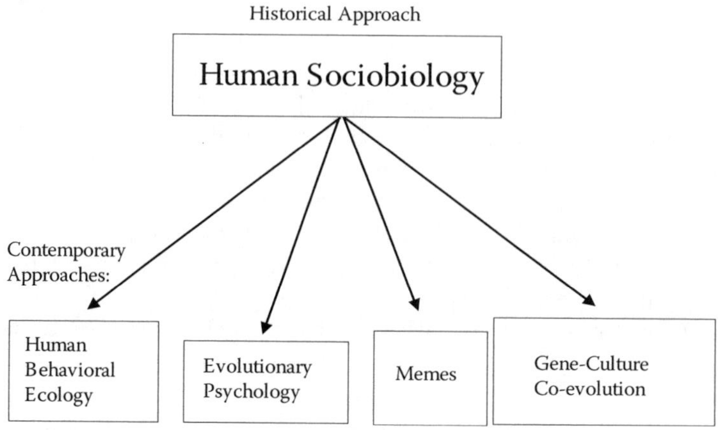


**FIGURE 2.1** Five evolutionary approaches to the study of human behavior. Human sociobiology gave rise to, or triggered the emergence of, the contemporary schools of human behavioral ecology: evolutionary psychology, cultural evolution and memetics, and gene-culture coevolution.

A key assumption of human behavioral ecology is that human beings are able to alter their behavior flexibly in response to environmental conditions in a manner that optimizes their lifetime reproductive success. This does not necessarily mean that these researchers believe that humans are deliberately or consciously aiming to leave as many descendants as possible. Rather, the assumption is that a history of selection has endowed our species with a tendency to respond to the environments in which we find ourselves by weighing up the costs and benefits of adopting particular strategies and carrying out the optimal solution. How humans, or indeed any other animals, do this is not a concern to the behavioral ecologists (Grafan, 1984). For these researchers, it simply doesn't matter whether humans behave in an adaptive manner as a consequence of their genes, their psychological mechanisms, their learning, or their culture — as long as their behavior is adaptive it can be predicted with formal models.

Another defining characteristic of human behavioral ecology is the testing of hypotheses derived from formal or mathematical evolutionary theory. Usually the predictions of the models are tested with data gathered on small communities in remote regions of the world, such as the Ache of Paraguay or the Kipsigis of Kenya. An example is Monique Borgerhoff Mulder's (1990) study of marriage practices in the Kipsigis. Where resources are critical to the reproductive success of females but can be monopolized by males, behavioral ecologists predict that polygynous matings will occur — and in this respect humans may be no different. The theory suggests that, in some circumstances, women may raise more children by marrying an already-married man with high quality resources than by marrying a man with lower quality resources. In fact, there is predicted to be a threshold over which females are expected to mate polygynously as predicted by a "polygyny threshold model" (Verner, 1964). Borgerhoff Mulder collected data on the Kipsigis, who cultivate maize and herd livestock, and found that the number of surviving offspring born to women increased monotonically with the size of their husband's plot of land but decreased with the number of wives he had. Borgerhoff Mulder also found that there were indeed cases of women that married an already-married man with a big plot in preference to an unmarried man with a small plot, as the model predicts. The uninitiated could be forgiven for being surprised that a mathematical model, assuming optimal behavior and originally designed for birds, should be able to predict human mating behavior so effectively.

In the late 1980s, human behavioral ecology came under attack from the founders of the newly forming discipline of *evolutionary psychology.* The most hostile remarks came from Donald Symons (1987), who argued that the research program of the human behavioral ecologists was seriously misguided because it did not formulate or test hypotheses concerning human *adaptations* or shed light on the human mind where such adaptations would be found, but merely established which behavior patterns appeared *adaptive* by correlating human behavioral traits with reproductive success. Symons (1987) claimed "Darwinism can be 'applied' to traditional social science phenomena only insofar as it illuminates the psychological adaptations that underpin those phenomena." Other evolutionary psychologists, notably Leda Cosmides and John Tooby, joined in Symons's attack (Tooby, and Cosmides 1987, 1990). Not surprisingly, the human behavioral ecologists defended their position, and a vigorous and sometimes bitter debate continues to the present day (e.g., Smith et al., 2001).

The alternative perspectives of human behavioral ecology and evolutionary psychology revolve around the distinction between "adaptation" and "adaptive behavior" (Figure 2.2). An adaptation is a character favored by natural selection for its effectiveness in a particular role. It is something that has evolved to fulfill that function, and it may or may not be adaptive in the current environment. Adaptive behavior is functional behavior that increments reproductive success, and it may or may not be an adaptation. By and large, human behavioral ecologists have been concerned with whether behavior is adaptive while evolutionary psychologists are

The Difference between Adaptive Behavior and Adaptations

| | | Is the behavior adaptive?<br>*Adaptive behavior* is functional behavior that increments reproductive success. | |
|---|---|---|---|
| | | Yes | No |
| Is the character an adaptation?<br>*An adaptation* is a character favored by natural selection for its effectiveness in a particular role. | Yes | Current Adaptation<br>A *current adaptation* is an adaptation that has remained adaptive due to a continuity in the selective environment. | Past Adaptation<br>A *past adaptation* is an adaptation that is no longer adaptive due to a change in the selective environment. |
| | No | Exaptation<br>An *exaptation* is a character that now enhances fitness but was not built by natural selection for its current role. | Dysfunctional By-Product<br>A *dysfunctional by-product* is a character that neither enhances fitness nor was built by natural selection. |

**FIGURE 2.2** The difference between adaptive behavior and adaptations.

concerned with whether particular psychological characters are adaptations. The two schools are asking different questions.

One reason many evolutionary psychologists have reservations about the human behavioral ecology approach is that they regard modern environments, with their cities, computers, hospitals, and contraceptives, as very different from the ancestral environment to which they believe humans are adapted. As a consequence of this "adaptive lag," evolutionary psychologists expect that much of human behavior is not adaptive at the current time and see no reason to measure reproductive success in modern populations. The observation that most human behavioral ecology is carried out on small-scale, preindustrial societies is interpreted by such critics as tacit admittance that this argument carries some weight. Perhaps the strongest counterpoint, highlighted above, is the observation that behavioral ecological models are remarkably effective in predicting human behavior (Hill and Hurtado, 1996; Cronk et al., 2000; Smith et al., 2000).

### 2.1.3 Evolutionary Psychology

The term *evolutionary psychology* can be used broadly to describe the work of many researchers with quite different views, ranging from behavioral ecologists (Barrett et al., 2001) to evolutionary epistemologists (Plotkin, 1994, 1997) to nativistic psychologists (Tooby and Cosmides, 1989, 1990). Nonetheless, globally, and particularly in North America, evolutionary psychology is dominated by a modular, adaptationist school, led by researchers at the University of California at Santa Barbara (Tooby, Cosmides, Symons), who seek to identify human mental adaptations, the "evolved psychological mechanisms" that they regard as universal characteristics of human nature. For these researchers, adaptations are not found at the behavioral level but in the regulatory procedures that control behavior. A second emphasis is on modularity of mind, or domain specificity; most evolutionary psychologists believe that minds are composed of a large number of psychological mechanisms dedicated to finding quick and efficient solutions to particular problems that were of significance to our ancestors. A third emphasis is on the *environment of evolutionary adaptatedness* (or EEA) — the past selective environment, commonly associated with a Pleistocene African savannah, in which the principal psychological characteristics of human nature are deemed to have evolved. These researchers reason that if they can isolate the problems our stone-age ancestors faced, they might be able to predict the psychological mechanisms necessary to solve these problems and gain clues as to how the mind is set up to function.

A representative example of contemporary evolutionary psychology is a study of sex differences in human mate choice by David Buss of the University of Texas (Buss et al., 1990). Buss reasoned that, from the perspective of our ancestors in the EEA, women faced the burdens of internal fertilization, a nine-month gestation, and lactation and consequently would have benefited by selecting mates who possessed and were willing to provide food, protection, and other resources. Conversely, Buss argued that selection should have favored the evolution of psychological mechanisms in males that render the prospect of many sexual partners desirable and females of high fertility attractive. In order to test hypotheses concerning sex differences in human mating behavior and investigate whether such characteristics are universal,

Buss and his collaborators obtained questionnaire data from almost 10,000 people in 33 countries. They repeatedly found sex differences consistent with their evolutionary arguments. For instance, on average women placed more value than men on good financial prospects in a partner, while men placed more value than women on the physical attractiveness of their partner. While there are both weaknesses to the study and other ways of interpreting the data (Eagly and Wood, 1999; Laland and Brown, 2002), Buss et al. can take credit for one of the most far-reaching attempts to evaluate the evolutionary psychologists' hypothesis of a universal, evolved human psychology.

Evolutionary psychology has come in for criticism on a number of counts. Many critics have pointed to problems with how the EEA concept is currently being used (Foley, 1996; Lloyd and Feldman, 2001). The argument is that too little is known about how our ancestors behaved during the Pleistocene for researchers to be able to reconstruct selection pressures with any great reliability. As a result the EEA concept can easily lead to undisciplined speculation and storytelling in which virtually any attribute can be regarded as an adaptation to a bygone stone-age world.

There is also a feeling among some evolutionary biologists that evolutionary psychology has become detached from the primary evolution literature and that it is perpetuating an oversimplified and outdated conception of the evolutionary process (Coyne and Berry, 2000; Lloyd and Feldman, 2001). Much evolutionary psychology neglects recent developments within evolutionary biology and fails to reference seminal evolutionary biology texts and papers. For instance, a common argument from evolutionary psychology is that there has been insufficient time in the 10,000 years since the end of the Pleistocene for significant evolutionary change to occur in our cognitive architecture (see, for instance, Barkow et al., 1992 and articles therein). However, observations of rates of natural selection in the wild have, over the last 20 years, led evolutionary biologists to the conclusion that biological evolution may typically be faster than previously thought (Endler, 1986). Kingsolver et al. (2001) published a review in which they collated over 2,500 estimates of the response to selection of natural populations of organisms. They concluded that the average rate of selection would cause a quantitative trait to change by one standard deviation in just 25 generations, which for humans would be around 500 years. Such rates leave ample time for substantive biological evolution since the Pleistocene. Evolutionary psychologists who simply use a stereotype of a Pleistocene, hunter-gatherer ancestor to apply an evolutionary angle to their data deserve the criticisms of storytelling that are levelled against them. However, it would be unfair to dismiss a vibrant and creative discipline on the basis of these weaker studies. Evolutionary psychology is an embryonic field that has brought the study of the mind well into the domain of evolutionary theory, bringing with it a welcome focus on proximate mechanisms.

### 2.1.4 Memes and Cultural Evolution

In *The Descent of Man*, Darwin (1871, p. 61) describes the evolution of language, stating boldly that "the survival or preservation of certain favored words in the struggle for existence is natural selection." Darwin's intuition that natural selection may be a general law for how a multitude of processes change has received both empirical and theoretical support. There is a respectable scientific and philosophical tradition known as evolutionary epistemology that stresses the universal nature of natural selection, backed by luminary philosophers such as Karl Popper, David Hull, and Daniel Dennett and Nobel Prize winning scientists such as Konrad Lorenz and Gerry Edelman (Plotkin, 1982, 1994). Moreover, many researchers have noted parallels between the processes of biological evolution and cultural change, which are, in fact, stronger than commonly portrayed (Mesoudi et al., 2004).

The idea that aspects of cultural change may exhibit parallels to biological evolution by natural selection was brought to prominence by Richard Dawkins (1976) in *The Selfish Gene*. Dawkins pointed out that cultural traits possess variation, heredity, and differential fitness, the three characteristics that are necessary for evolution. He suggested that fashions, diets, customs, language, art, and technology evolve over historical time and devised the notion of the "meme," a cultural analogue of the gene. Memes are virus-like cultural replicators that "propagate themselves in the meme pool by leaping from brain to brain via a process which, in the broad sense, can be called imitation" (Dawkins, 1976, p. 206).

The principal approach of memetics is to reason from a "meme's eye view." In her book *The Meme Machine*, Susan Blackmore challenges us to "[i]magine a world full of hosts for memes (e.g., brains) and far more memes than can possibly find homes. Now ask, which memes are more likely to find a safe home and get passed on again?" (1999, p. 37). The answer given is the eye-catching or high-profile memes, rather than the memes we might objectively judge to be in our interests, or worthy of our attention. From an evolutionary perspective, the only virtue is replication. Meme enthusiasts believe that cultural traits evolve, not because the traits are of utility to individuals (although they may be) but because their characteristics aid meme propagation. Memes are there for the good of the memes.

The principal problem with memetics is that it hasn't yet developed a serious empirical program of research (Aunger, 2000). Thus far, meme enthusiasts have largely been content with using a meme's-eye view to devise new evolutionary hypotheses rather than with conducting empirical research. In this respect, memetics is also vulnerable to the criticism of storytelling. However, a study by Nan Li and colleagues (2000) illustrates the kind of investigation that more quantitatively minded memeticists could undertake. Li et al. (2000) used cultural evolution theory to develop models that predict how the sex ratio at birth in China is likely to change in the face of fluctuations in fertility and a strong, culturally transmitted preference for sons. This son preference is often manifest in excess female mortality through infanticide of daughters, underinvestment in daughters, or female-biased sex-selective abortion, which skews the sex ratio toward males. The researchers conducted a survey to measure rates of cultural transmission across generations, equivalent to measures of the infectiousness of the meme for preferring sons, in the two rural areas of China, chosen as low and high projections for the country at large. Entering these estimates into their models, they were able to predict that across the whole of China in 2020 the sex ratio at birth would range between 1.1 and 1.34, that is, between 110 and 134 males for every 100 females, depending upon the strength of the transmission of son preference. Currently, the birth sex ratio in China is around

1.14 (Clarke, 2000). To the extent that phenotypic traits and preferences are similar to memes, this study serves as an illustration of how the fidelity of meme transmission can be estimated with empirical studies and how meme frequency changes can be modeled mathematically and put to use to make valuable demographic projections. The analysis suggests that reducing the distortion in the sex ratio will depend critically upon weakening the transmission of the preference for sons.

Recent years have witnessed some extremely promising developments that suggest that an empirical science of cultural evolution is emerging, albeit not necessarily under the auspices of "memetics" (see, for example, Shennan, this volume). A strong case can now be made for culture exhibiting key Darwinian properties (Mesoudi et al., 2004). For instance, evidence that cultural variants are competing like biological variants comes from testing the prediction that over time one variant will increase in frequency while another shows a corresponding decrease. This has been demonstrated by archaeologists using the method of "frequency seriation," in which the frequencies of excavated cultural artifacts are recorded at different time periods, allowing the reconstruction of lineages of competing artifacts. For example, O'Brien and Lyman (2000) have detailed how lineages of prehistoric projectile points from the southwestern United States show an increase in functional efficiency over time, demonstrating competition of successive forms in one or more lineages, with each new, more efficient variant outcompeting and replacing the older form.

A further illustration of the analysis of cultural variation using Darwinian principles is provided by studies employing comparative statistical models. Mace and Pagel (1994) presented a phylogeny for nine Kenyan pastoralist cultures, suggesting that the distribution of camel keeping can be explained most parsimoniously by just two independent adoptions of camel keeping, with all other similarities being due to diffusion or descent. Gray and Jordan (2000) similarly found that the distribution of 77 Austronesian languages could most parsimoniously be accounted for with a phylogenetic tree branching from a single common ancestor in Taiwan. Barbrook et al. (1998) and Tehrani and Collard (2002), also using the phylogenetic method, found evidence for cultural descent in manuscripts of *The Canterbury Tales* and in the decorative patterns of Turkmen textiles, respectively. While these researchers do not necessarily identify themselves with memetics, they nonetheless provide an illustration of the potential of the field of cultural evolution, broadly conceived. Further empirical methods for the study of cultural evolution are described in Laland and Brown (2002).

An interesting aspect of the memes literature is that many of the commonly cited "problems" of memetics are reservations that apply equally to biological evolution. For instance, various detractors draw attention to the difficulties associated with defining the boundaries of a meme, in comparison with which a layperson's concept of "gene" or "species" appears well specified. In reality, biologists and paleontologists continue to struggle to define and isolate species (Futuyma, 1998), and recent advances in molecular biology reveal bounteous complications that leave a molecular gene concept in virtual disarray (Gilbert, 2003). If these complications have not prevented evolutionary biology from being a progressive science, why should they necessarily hinder memetics?

### 2.1.5 Gene-Culture Coevolution

*Gene-culture coevolution* is a branch of theoretical population genetics that, in addition to modeling the differential transmission of genes from one generation to the next, incorporates cultural traits into the analysis (Feldman and Cavalli-Sforza, 1976; Cavalli-Sforza and Feldman, 1981; Boyd and Richerson, 1985; Feldman and Laland, 1996; Richerson and Boyd, 2005). Researchers in this field argue that the two inheritance systems cannot be treated independently, because i) what an individual learns may depend on its genotype expressed throughout development and ii) selection acting on the genetic system may be generated or modified by the spread of a cultural trait. This approach was pioneered by geneticists Marc Feldman and Luca Cavalli-Sforza, and was more recently brought to prominence by anthropologists Robert Boyd and Peter Richerson.

A key concept of gene-culture coevolution is an emphasis on a transmitted or inherited culture. Gene-culture coevolution enthusiasts treat culture as an evolving pool of ideas, beliefs, values, and knowledge that is learned and socially transmitted between individuals. They devise mathematical models, which are derived from the tools of theoretical population genetics but adapted to the peculiarities of modeling culture. The models are then used to explore how genes and culture might have interacted in the recent past as well as how they are currently interacting and to interpret patterns of variation in behavioral traits within and between human populations.

A good example is provided by mathematical analyses of the evolution of the ability of adult humans to consume dairy products without sickness. Adult humans vary considerably in their ability to digest milk as a result of differences in their physiology (Durham, 1991). In fact, if the entire world's population is considered, the majority of adult humans are made ill by consuming dairy products. This is because the activity level of the enzyme lactase in their bodies is insufficient to break down the lactose in dairy products, and whether they can do so depends on whether individuals possess appropriate genes. A strong correlation exists between the incidence of the genes for lactose absorption and a history of dairy farming in populations (Ulijaszek and Strickland, 1993). This has led to the hypothesis that dairy farming created the selection pressures that led to genes for lactose absorption becoming common in pastoralist communities.

Following work by Aoki (1986), Feldman and Cavalli-Sforza (1989) used gene-culture coevolutionary models to investigate this possibility. Their analysis showed that whether or not the allele allowing adult milk digestion achieved a high frequency depended critically on the fidelity of cultural transmission, that is, on the probability that the children of dairy-product users themselves became milk consumers. If this probability was very high then a significant fitness advantage to the genetic capacity for lactose absorption resulted in the selection of the absorption allele to high frequency within the 6,000 years, or 300 generations, available. However, if a significant proportion of the offspring of milk users did not exploit dairy products then the genes for lactose absorption would not spread even with a significant fitness advantage. In other words, differences in the strength of cultural transmission between cultures may account for genetic variability in lactose absorption. Thus, the analysis is able to account for both the spread of lactose absorption and the culturally related variability in its incidence. Recently, Feldman and Cavalli-Sforza's conclusions have received support from a comparative analysis by Holden and Mace (1997), which concluded that dairy farming spread prior to the genes for lactose absorption, not the other way around.

Some critics of gene-culture coevolution have questioned whether aspects of culture can legitimately be divided into units and analyzed with mathematical equations (Bloch, 2000). In fact, the goal of gene-culture coevolutionists is usually more modest than such detractors envisage. Gene-culture coevolutionists are not trying to model how entire cultures change over time, but rather to explore some of its general properties and to predict patterns of change in certain specific traits (Richerson and Boyd, 2004). Critics who argue that culture is too complex to be analyzed in such reductionist terms may be unaware that biological evolution involves similar levels of complexity. As long ago as 1963, Ernst Mayr criticized the "beanbag genetics" of theoretical population genetic models, to which J.B.S. Haldane (1964) responded with a vigorous defense. As with culture, gene interaction during development is enormously complex and interwoven, yet population genetics models have proven highly valuable, in spite of their simplifying assumptions (Crow, 2001). Gene-culture models have an equally impressive track record (Richerson and Boyd, 2004).

Gene-culture coevolution has associated with it a formal discipline and a progressive theoretical research program, which goes a long way toward justifying this approach. However, compared with, say, evolutionary psychology, which employs methods that are accessible to the vast majority of would-be researchers, gene-culture models are technical and mathematical and require a degree of training. Moreover, like human sociobiology and memetics, it currently suffers from a weak empirical foundation, although, once again, there are ways of doing empirical work that are currently underexploited (Laland and Brown, 2002; Baum et al., 2004).

## 2.2 SOCIAL TRANSMISSION AND THE ROLE OF CULTURE

We have seen above that, in the aftermath of the sociobiology debate, a number of distinct evolutionary schools of thought have emerged, with different methods and theoretical stances. One important difference between these evolutionary approaches that we have so far discussed indirectly is their differing views on the role of social transmission of information and their treatment of culture. In this section, we briefly overview how the approaches differ in the emphasis that they place on the role of transmitted culture in their explanations of human behavior.

Historically, the notion of culture has proven very difficult for academics to pin down. In a famous article published in 1952, two prominent anthropologists identified 164 different definitions of culture proposed by social scientists (Kroeber and Kluckholm, 1952), and that number has undoubtedly grown. However, while social scientists continue to differ in their views of culture, most would not dispute two points: first, culture is composed of symbolically encoded acquired information, expressed in behavior and artifacts and, second, it is socially transmitted within and between populations, largely free of biological constraints. Evolutionists regard culture very differently from social scientists.

For human sociobiologists, human culture was usually regarded as little different from any other aspect of the human phenotype. Yet for some, culture was behavior elicited by ecological conditions (Alexander, 1979), while for others (e.g., Wilson, 1975), it was composed largely of cultural universals closely tied to our biological nature. An additional perspective was that genetic diversity underlies the differences between cultures (Wilson, 1975; Lumsden and Wilson, 1981). This was one of the most controversial aspects of human sociobiology and a primary source of hostility from the social science community (Segerstråle, 2000). This view does not find favor among any of the contemporary evolutionary schools. All modern approaches accept that cultural change can occur without accompanying genetic change and that individuals from genetically distinct populations acquire the cultural traits of each other without difficulty (Flinn, 1997).

The principal sociobiological perspectives are, to some extent, represented by the contemporary evolutionary approaches of human behavioral ecology and evolutionary psychology. The human behavioral ecology tradition tends to regard the variability in human society and culture as, to a large extent, evoked by the ecological environment. That is not to say that culture is not learned nor socially transmitted but rather that humans are predisposed to learn that which maximizes their fitness by satisfying various proximal goals, such as obtaining food and mates and avoiding danger and disease. Culture is part of the unusually broad and flexible evolved mechanisms of behavioral adaptation that characterize humans, such that we are adapted to a wide range of conditions rather than to a particular environmental state.

In contrast, evolutionary psychologists are most interested in those aspects of human culture that are found universally among all populations, our human nature. According to this view, human minds are organized by complex, evolved information-processing structures that channel learning aptitudes toward that which is beneficial in our environment of evolutionary adaptedness. Compared to human behavioral ecologists, evolutionary psychologists have a view of the mind that is much more structured and prespecified by our genetic heritage and perhaps less flexible in the face of environmental variability.

The field of memetics holds that culture is a dynamic evolutionary system in its own right. Culture is phenotypic plasticity that acquired its own intrinsic capacity to change and is now, at least partly, out of genetic control. We don't expect a flu virus to operate to our advantage, so why should we expect a "mind virus" to always be in our interests? For some meme advocates, not only is cultural evolution largely unconstrained by genetic predispositions but genetic evolution may itself be driven by cultural imperatives (Blackmore, 1999). Others give greater credence to the notion of evolved structures in the mind, which filter incoming memes and affect which are adopted (Dennett, 1991).

Gene-culture coevolution asserts that cultural information is socially transmitted between individuals but that its acquisition is biased by evolved learning rules and motivational priorities. Advocates of memetics and gene-culture coevolution differ from most evolutionary psychologists and human behavioral ecologists in believing that cultural phenomena cannot be fully understood without recourse to the intrinsic processes of cultural change, which are at least partly independent of the processes of biological evolution. Such cultural processes do not necessarily act to maximize fitness and hence may result in arbitrary or even maladaptive outcomes. These researchers agree that cultural processes have a historical dependency, such that culture cannot be well predicted without knowledge of the population's traditions. In other words, there are processes besides evolutionary design of cognitive architecture that affect cultural content.

Evolutionary psychology would seem to have greater conflict with memetics, which denies any substantive filtering role for evolved psychological mechanisms, than with gene-culture coevolution, which sees evolved predispositions frequently being instrumental in decision making. However, evolutionary psychology's focus on genetically evolved, universal human cognitive mechanisms tends to make its practitioners relatively hostile to the view, shared by memetics and gene-culture coevolution, that humans acquire a substantial proportion of their beliefs and preferences through cultural inheritance and that these can change through cultural evolution (Smith, 2000).

It would seem that the aforementioned evolutionary schools can be distinguished by their differing views of social transmission. Are we predisposed to learn what is currently adaptive, guided by proximate motivational cues such as hunger or fear as the human behavioral ecologists maintain? Or is our brain set up to prioritize learning that which was important in the past, as the evolutionary psychologists suspect? Do we acquire whatever behavior or information just happens to be easiest or most compelling to learn, as memeticists would have it? Or is our learning partly dependent on evolved predispositions and partly on cultural processes, as the gene-culture coevolution theorists have it? In fact, each of these views could be true for different learned behavior patterns or on different occasions (Smith, 2000).

Ultimately, these issues will have to be worked out through experimentation and other forms of research, rather than through the polemical pronouncements that have thus far dominated most attempts to confront them (Smith, 2000). What kind of studies would tell us which of these views of culture is closest to the mark? There are two obvious places to start. First, researchers could carry out quantitative analyses across a multitude of behavioral traits to measure to what extent, or on what percentage of occasions, human behavior is currently adaptive. A good example of this type of study was carried out by anthropologist Robert Aunger on the food preferences of horticulturists and Pygmy foragers living in the Ituri forest of the Democratic Republic of Congo (Aunger, 1992, 1994). Aunger observed that different populations varied as to which foodstuffs they exploited and which they avoided, and asked whether their food avoidances were maladaptive. He found that individuals in one of the four ethnic groups suffered a selective disadvantage that resulted from their cultural beliefs about food. These maladaptive food avoidances generally reduced fitness by a few percent, mainly through their compromising female fertility. Aunger interprets these cases as likely to have been the outcome of cultural processes, rather than a consequence of adaptive lag. If Aunger's data can be regarded as representative, then a significant minority of cases of human behavior are maladaptive. It is, of course, conceivable that the amount of maladaptive behavior will differ in other societies or in relation to other domains.

A second, equally informative kind of analysis would measure, among diverse traits and across a broad range of populations, what percentage of the variance in behavior is explained by local ecology and what percentage is better predicted by cultural history. Just such an analysis was carried out by Guglielmino et al. (1995) on 277 African societies. They found that most traits correlated with cultural history rather than with ecology. If this study is representative, then socially transmitted cultural traditions are much more important than most evolution-minded researchers envisage.

We began this section with the observation that most social scientists regard culture as symbolically encoded acquired information that is socially transmitted between individuals unbound from biological constraints. Studies such as Guglielmino et al.'s (1995) suggest that evolutionary-minded researchers may have seriously underestimated the amount and significance of transmitted culture. A rejection of cultural determinism and of a *tabala rasa* model of the mind does not have to encompass a neglect of cultural processes. The role of social transmission in explaining behavioral variability may ultimately benefit from a close integration of evolutionary and social science approaches to the study of human behavior.

## ACKNOWLEDGMENTS

We are grateful to Jonathan Wells for helpful comments on an earlier draft of this article.

## REFERENCES

Alcock, J., *The Triumph of Sociobiology,* Oxford University Press, Oxford, 2001.

Alexander, R. D., *Darwinism and Human Affairs,* Pitman, London, 1979.

Aoki, K., A stochastic model of gene-culture coevolution suggested by the "culture historical hypothesis" for the evolution of adult lactose absorption in humans, *Proc. Natl. Acad. Sci. USA,* 83, 2929–2933, 1986.

Aunger, R., The nutritional consequences of rejecting food in the Ituri Forest of Zaire. *Hum. Ecol.,* 30, 1–29, 1992.

Aunger, R., Are food avoidances maladaptive in the Ituri Forest of Zaire? *J. Anthropol. Res.,* 50, 277–310, 1994.

Aunger, R., *Darwinizing Culture: the Status of Memetics as a Science*, Oxford University Press, Oxford, 2000.

Barbrook, A. C., et al., The phylogeny of the Canterbury Tales, *Nature,* 394, 839, 1998.

Barkow, J.H., Cosmides, L., and Tooby, J., *The Adapted Mind: Evolutionary Psychology and the Generation of Culture*, Oxford University Press, Oxford, 1992.

Barrett, L., Dunbar, R., and Lycett, J., *Human Evolutionary Psychology*, Macmillan, London, 2001.

Bateson, P.P.G., The dynamics of parent-offspring relationships in mammals, *Trend. Ecol. Evol.,* 9, 399–402, 1994.

Baum, W.M., et al., Cultural evolution in laboratory microsocieties including traditions of rule giving and rule following, *Evol. Hum. Behav.*, 25, 305–326, 2004.

Blackmore, S., *The Meme Machine,* Oxford University Press, Oxford, 1999.

Bloch, M., A well-disposed social anthropologist's problems with memes, In *Darwinizing Culture: The Status of Memetics as a Science,* Aunger, R., Ed., Oxford University Press, Oxford, 189–203, 2000.

Borgerhoff Mulder, M., Kipsigis women's preference for wealthy men: evidence for female choice in mammals? *Behav. Ecol. Sociobiol.,* 27, 255–264, 1990.

Boyd, R., and Richerson, P. J., *Culture and the Evolutionary Process*, Chicago University Press, Chicago, 1985.

Buss, D. M. et al., International preferences in selecting mates: a study of 37 cultures. *J. Cross-Cult. Psychol.,* 21, 5–47, 1990.

Cavalli-Sforza, L. L., and Feldman, M. W., *Cultural Transmission and Evolution: A Quantitative Approach,* Princeton University Press, Princeton, NJ, 1981.

Clarke, J. I., *The Human Dichotomy: The Changing Numbers of Males and Females*, Elsevier Science, Oxford, 2000.

Cosmides, L., and Tooby, J., From evolution to behavior: evolutionary psychology as the missing link, In *The Latest on the Best: Essays on Evolution and Optimality*, Dupré, J., Ed., MIT Press, Cambridge, MA, 277–306, 1987.

Coyne, J. A., and Berry, A., Rape as an adaptation: is this contentious hypothesis advocacy, not science? *Nature,* 404, 121–122, 2000.

Cronk, L., Chagnon, N., and Irons, W., *Adaptation and Human Behavior: an Anthropological Perspective*, Aldine de Gruyter, New York, 2000.

Crow, J. F., The beanbag lives on, *Nature,* 409, 771, 2001.

Darwin, C., *The Origin of Species by Means of Natural Selection, or the Preservation of Favoured Races in the Struggle for Life*, John Murray, London, 1859 (first edition reprinted by Penguin Books, London, 1968).

Darwin, C., *The Descent of Man and Selection in Relation to Sex*, John Murray, London, 1871 (first edition reprinted by Princeton University Press, Princeton, NJ, 1981).

Darwin, C., *The Expression of the Emotions in Man and Animals*, John Murray, London, 1872 (third edition reprinted by Harper Collins, London, 1998).

Dawkins, R., *The Selfish Gene,* Oxford University Press, Oxford, 1976.

Dennett, D.C., *Consciousness Explained,* Penguin, London, 1991.

Durham, W.H., *Coevolution: Genes, Culture and Human Diversity,* Stanford University Press, Stanford, CA, 1991.

Eagly, A.H., and Wood, W., The origins of sex differences in human behavior: evolved dispositions versus social roles, *Am. Psychol.,* 54, 408–423, 1999v.

Endler, J.A., *Natural Selection in the Wild,* Princeton University Press, Princeton, NJ, 1986.

Feldman, M.W., and Cavalli-Sforza, L.L., Cultural and biological evolutionary processes, selection for a trait under complex transmission, *Theor. Popul. Biol.* 9, 238–259, 1976.

Feldman, M.W., and Cavalli-Sforza, L.L., On the theory of evolution under genetic and cultural transmission with application to the lactose absorption problem, In *Mathematical Evolutionary Theory,* Feldman, M.W., Ed., Princeton University Press, Princeton, NJ, 145–173, 1989.

Feldman, M.W., and Laland, K.N., Gene-culture coevolutionary theory, *Trend. Ecol. Evol.* 11, 453–457, 1996.

Flinn, M.V., Culture and the evolution of social learning, *Evol. Hum. Behav.,* 18, 23–67, 1997.

Foley, R., The adaptive legacy of human evolution: a search for the environment of evolutionary adaptedness, *Evol. Anthropol.,* 4, 194–203, 1996.

Futuyma, D.J., *Evolutionary Biology,* 3rd edition, Sinauer, Sunderland, MA, 1998.

Gilbert S., *Developmental Biology,* 7th edition, Sinauer, Sunderland, MA, 2003.

Grafen, A., Natural selection, kin selection and group selection, In *Behavioural Ecology: An Evolutionary Approach*, 2nd edition, Krebs, J., and Davies, N.B., Eds., Blackwell Scientific Publications, Oxford, 62–84, 1984.

Gray, R.D., and Jordan F.M., Language trees support the express-train sequence of Austronesian expansion, *Nature,* 405, 1052–1055, 2000.

Guglielmino, C.R., et al., Cultural variation in Africa: role of mechanism of transmission and adaptation. *Proc. Nat. Acad. Sci. USA,* 92, 7585–7589, 1995.

Haldane, J.B.S., A defense of beanbag genetics, *Persp. Biol. Med.,* 7, 343–359, 1964.

Hamilton, W., The genetical evolution of social behaviour: I, *J. Theor. Biol.,* 7, 1–16, 1964a.

Hamilton, W., The genetical evolution of social behaviour: II, *J. Theor. Biol.,* 7, 17–32, 1964b.

Hill, K., and Hurtado, A.M., *Ache Life History: The Ecology and Demography of a Foraging People*, Aldine de Gruyter, New York, 1996.

Holden, C., and Mace, R., Phylogenetic analysis of the evolution of lactose digestion in adults, *Hum. Biol.* 69, 605–628, 1997.

Kingsolver, J.G., Hoekstra, H.E., Hoekstra, J.M., Berrigan, D., Vignieri, S.N., Hill, C.E., Hoang, A. Gilbert, P., and Beerli, P., The strength of phenotypic selection in natural populations. *Am. Nat.,* 157, 245–261, 2001.

Kroeber, A.L., and Kluckholm, C., Culture: a critical review of concepts and definitions, *Papers Peabody Museum Am. Archaeol. Ethnol.,* 47, 1–223, 1952.

Laland, K.N., and Brown, G.R., *Sense and Nonsense: Evolutionary Perspectives on Human Behaviour,* Oxford University Press, Oxford, 2002.

Li, N., Feldman, M.W., and Li, S., Cultural transmission in a demographic study of sex ratio at birth in China's future, *Theor. Popul. Biol.,* 58, 161–172, 2000.

Lloyd, E.A., and Feldman, M.W., Evolutionary psychology: a view from evolutionary biology, *Psychol. Enquiry,* 13, 150–156, 2001.

Lorenz, K., *On Aggression*, Routledge, London, 1963 (reprinted in 1996).

Lumsden, C.J., and Wilson, E.O., *Genes, Mind and Culture: The Coevolutionary Process*, Harvard University Press, Cambridge, MA, 1981.

Mace, R., and Pagel, M.D., The comparative method in anthropology. *Curr. Anthropol.,* 35, 549–564, 1994.

Mayr, E., *Animal species and evolution*, Harvard University Press, Cambridge, MA, 1963.

Mesoudi, A., Whiten, A., and Laland, K.N., Is human cultural evolution Darwinian? Evidence reviewed from the perspective of *The Origin of Species*, *Evolution,* 58, 1–11, 2004.

O'Brien, M. J., and Lyman, R.L., *Applying Evolutionary Archaeology: A Systematic Approach*, Kluwer Academic, New York, 2000.

Plotkin, H.C. (ed.), *Learning, Development, and Culture: Essays in Evolutionary Epistemology,* Wiley, Chicester, 1982.

Plotkin, H.C., *Darwin Machines and the Nature of Knowledge,* Penguin Books, London, 1994.

Plotkin, H.C., *Evolution in Mind: an Introduction to Evolutionary Psychology,* Penguin Books, London, 1997.

Richerson, P.J., and Boyd, R., *Not by Genes alone: How Culture Transformed Human Evolution,* Chicago University Press, Chicago, 2005.

Segerstråle, U., *Defenders of the Truth: The Sociobiology Debate*, Oxford University Press, Oxford, 2000.

Smith, E.A., Three styles in the evolutionary analysis of human behavior, In *Adaptation and Human Behavior: An Anthropological Perspective*, Cronk, L., Chagnon, N., and Irons, W., Eds., Aldine de Gruyter, New York, 27–46, 2000.

Smith, E.A., Borgerhoff Mulder, M., and Hill, K., Evolutionary analyses of human behaviour: a commentary on Daly and Wilson, Anim. Behav., 60, F21–F26, 2000.

Smith, E.A., Borgerhoff Mulder, M., and Hill, K., Controversies in the evolutionary social sciences: a guide for the perplexed, *Trend. Ecol. Evol.,* 16, 128–135, 2001.

Symons, D., If we're all Darwinians, what's the fuss about? In *Sociobiology and Psychology: Ideas, Issues and Applications*, Crawford, E., Smith, M., and Krebs, D., Eds., Erlbaum, Hillsdale NJ, 1987.

Tehrani, J.J., and Collard, M., Investigating cultural evolution through biological phylogenetic analyses of Turkmen textiles, *J. Anthropol. Archaeol.,* 21, 443–463, 2002.

Tooby, J., and Cosmides, L., Evolutionary psychology and the generation of culture, part I. Theoretical considerations, *Ethol. Sociobiol.,* 10, 29–49, 1989.

Tooby, J., and Cosmides, L., The past explains the present: emotional adaptations and the structure of ancestral environments, *Ethol. Sociobiol.,* 11, 375–424, 1990.

Trivers, R.L., The evolution of reciprocal altruism, *Q. Rev. Biol.,* 46, 35–57, 1971.

Trivers, R.L., Parental investment and sexual selection, In *Sexual Selection and the Descent of Man, 1871–1971*, Campbell, B., Ed., Aldine, Chicago, 136–179, 1972.

Trivers, R.L., Parent-offspring conflict, *Am. Zool.,* 14, 249–264, 1974.

Trivers, R.L., and Willard, D.E., Natural selection of parental ability to vary the sex ratio of offspring, *Science,* 179, 90–92. 1973.

Ulijaszek, S.J., and Strickland, S.S., *Nutritional anthropology. Prospects and perspectives*, Smith-Gordon, London, 1993.

Verner, J., Evolution of polygamy in the long-billed marsh wren, *Evolution,* 18, 252–261, 1964.

Williams, G.C., *Adaptation and Natural Selection: A Critique of Some Current Evolutionary Thought*, Princeton University Press, Princeton, NJ, 1966 (reprinted 1996).

Wilson, E.O., *Sociobiology: The New Synthesis*, Harvard University Press, Cambridge, MA, 1975.

Wynne-Edwards, V., *Animal Dispersion in Relation to Social Behaviour*, Oliver and Boyd Ltd., Edinburgh, 1962.

# 3 How Niche Construction Contributes to Human Gene-Culture Coevolution

*John Odling-Smee*

## CONTENTS



## ABSTRACT

Organisms, through their metabolisms, activities, and choices, define and partly create their own niches, a process called *niche construction*. Niche construction transforms both biotic and abiotic sources of natural selection in environments, and by doing so generates forms of feedback in evolution that change the dynamics of the evolutionary process. In an extended version of evolutionary theory, Odling-Smee, Laland, and Feldman (2003) explicitly combined niche construction with natural selection. This combination has considerable implications for our understanding of evolution in general and human evolution in particular. The biology of niche construction is described first. Then human cultural processes, based largely on the social transmission of information, are introduced. Finally, the significance of human cultural niche construction for both prior human evolution and our contemporary lives is discussed.

## 3.1 INTRODUCTION

The social transmission of information is one of the prerequisites of culture. Human cultural processes could not exist unless a capacity for nongenetic communication had evolved first in diverse social species. But how did it evolve? Why did it evolve, and once it had evolved, what were its eventual consequences for human evolution? How did the unusual capacity for transmitting cultural information via cultural inheritance, when added to the general ability of all organisms to transmit genetic information via genetic inheritance, affect the evolution of our hominid ancestors? How does this same capacity affect our lives today? How might it affect our futures?

These questions cannot be addressed without evolutionary theory, but as soon as we turn to it we encounter a problem. Contemporary evolutionary theory, from hereon the *standard theory*, provides a necessary but still insufficient basis for understanding the evolution of our own species. Standard evolutionary theory tells us a lot about human nature and human behavior, and quite a lot about the likely origins of human cultural processes. However, it is weak, and incomplete, when it comes to demonstrating how human cultural processes interact with human genetic processes.

In part this is because cultural processes are so rare in nature. They are dominant and potent in humans, but in most species they do not even exist. Moreover, even when they are found in other animal species they are seldom potent. So evolutionary theory rarely has to explain how cultural processes work. Standard evolutionary theory also finds it hard to account for the impact that human cultural processes have evidently had on recent human evolution. Lewontin drew attention to one reason. In standard evolutionary theory, "The environment 'poses the problem'; the organisms 'posit solutions'" (Lewontin, 1983, p. 276). Human cultural activities, however, pose problems as well as posit solutions. Humans do this by constantly changing the environments that pose the problems. That does not mean that human cultural processes are outside the scope of biological evolution. But it does mean that some aspects of modern human cultural activity are beyond the reach of standard evolutionary theory.

In this chapter human evolution will therefore be placed in the context of a more comprehensive theoretical approach that my colleagues and I have been working on for some time. We call it *extended evolutionary theory* (Odling-Smee, Laland, and Feldman, 2003), and we claim that it offers a more secure and potentially more useful framework for making sense of both prior human evolution and our contemporary lives.

## 3.2 NICHE CONSTRUCTION

Extended evolutionary theory differs from standard evolutionary theory primarily by adding a new process, called *niche construction*, to the standard processes of evolution. First, I will describe how niche construction affects evolution in general. Then I will apply it to human evolution and discuss how the resulting altered perspective could help us understand human evolution better.

Figure 3.1a summarizes how biological evolution is currently thought to work. According to the standard theory, autonomous natural selection pressures in environments (E) act on populations of diverse organisms (or phenotypes) and influence

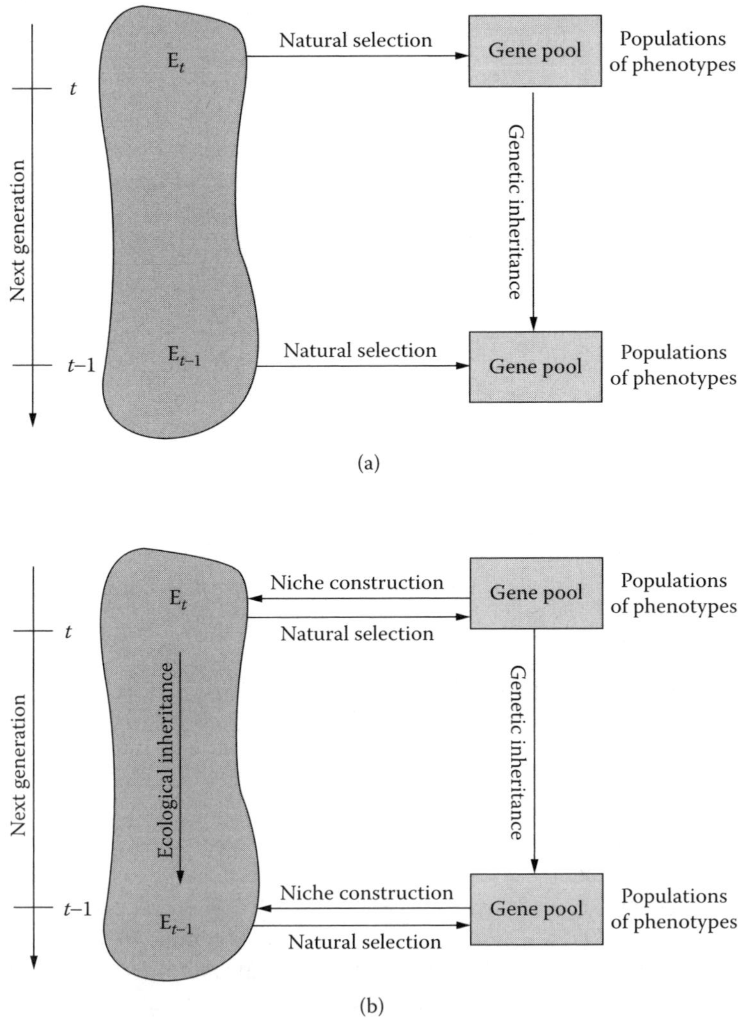


**FIGURE 3.1** (a) Standard evolutionary theory. (b) Extended evolutionary theory.

which individuals survive and reproduce and pass on their genes to the next generation via genetic inheritance. Hence, the adaptations of organisms are assumed to be products of natural selection molding organisms to fit preestablished environmental templates. These templates are dynamic because processes that are independent of organisms frequently change the worlds to which organisms have to adapt. However, the changes that organisms bring about in their own worlds are seldom thought to have evolutionary significance. G. C. Williams (1992, p. 484) is succinct: "Adaptation is always asymmetrical; organisms adapt to their environment, never vice versa."

The difficulty with this standard view is that it largely ignores the feedback in evolution that is caused by the modification of environmental natural selection pressures by the activities of organisms. Organisms, through their metabolisms, behaviors, and choices, partly define, create, and destroy their own environments (Lewontin, 1983; Jones, Lawton, and Shachak, 1994, 1997), and in doing so, they inevitably transform some of the natural selection pressures that act on themselves, and on each other, in their environments. Subsequently, these transformed selection pressures feed back to select for different genotypes in populations from those previously selected (Laland, Odling-Smee, and Feldman, 1996, 1999). It follows that adaptation cannot be just a matter of organisms responding to autonomous natural selection pressures in environments. Sometimes it must also involve organisms responding to natural selection pressures that have been previously transformed either by their own actions or by the actions of their ancestors or by the actions of other organisms in shared ecosystems. We call any organism activity that modifies natural selection *niche-construction.*

Elsewhere, we have given many examples of niche construction from all the kingdoms of life (Odling-Smee, Laland, and Feldman, 2003). Since space is limited I will give only a single nonhuman example here, earthworms. This example has the attraction of being originally described by Darwin (1881).

By burrowing, dragging organic material into the soil and mixing it up with inorganic material and by their casting, earthworms dramatically change both the structure and chemistry of soils. Soils that contain earthworms typically demonstrate enhanced plant yield, less surface litter, more topsoil, more organic carbon, nitrogen, and polysaccharides, as well as better porosity, aeration, and drainage. It follows that earthworms must live in partly self-constructed worlds. At least some of the natural selection pressures that act on earthworms must have been transformed by the prior activities of earthworms.

One evolutionary consequence of these soil changing activities of earthworms was studied by Turner (2000). Turner compared the physiological characteristics of earthworms, in particular the characteristics of their kidneys, with those of other animals, and he made a surprising discovery. Earthworms are equipped with the "wrong" kidneys. Different kinds of kidneys are typically found in different animals, depending on where they live. Turner compared freshwater, marine, and terrestrial kidneys and pointed out that each kind of kidney does a different adaptive job. Animals living in freshwater are in danger of being flooded by excess water, so freshwater kidneys have to excrete surplus water. Animals living in the sea are in danger of being killed by excess salts, so marine kidneys have to get rid of salts. Animals living on land are in danger of drying up, so terrestrial kidneys must prevent desiccation by retaining as much water as possible. The crucial point is that earthworms, which are obviously terrestrial animals, are equipped with very nearly typical freshwater kidneys. What seems to have happened is that earthworms have retained the freshwater physiology of their earlier aquatic ancestors, even though they now live on land: " … earthworms are essentially aquatic oligochaetes, poorly equipped physiologically for life on land. Yet there they are." (Turner, 2000, p. 105).

Yes, they are, but only because they niche construct, and Turner (2000) devotes an entire chapter to demonstrating how they do it. His whole chapter is recommended reading, but I will quote just his conclusion: "Earthworms when they came onto land, seem to have … pursued a strategy of using ATP energy to work against soil weathering. Along the way, they essentially co-opted the soil as an accessory organ of water balance. The advantage of adopting this strategy is clear: it is accomplished much more rapidly than a retooling of internal physiology." (p. 119).

What the earthworms seem to be telling us, contra Williams (1992), is that adaptation is a two-way process. Adaptation cannot be just a matter of natural selection selecting for adaptive traits in responsive organisms, because it is also a matter of organisms transforming at least some of the natural selection pressures they encounter in their local environments to suit themselves, by niche construction.

## 3.3 THE LIMITATIONS OF THE STANDARD THEORY

Innumerable examples of niche construction are known in biology, so how does standard evolutionary theory currently deal with them? Is it really necessary to go beyond the standard theory to incorporate niche construction in evolution?

One person who thinks not is Richard Dawkins. Dawkins (1982) has shown that genes in organisms are not restricted to expressing bodies or phenotypes. They often express what he calls "extended phenotypes" that reach out beyond the bodies of the organisms to affect components of their environments. One of his examples is beavers, which are famous for regulating the water levels in the rivers in which they live by constructing dams. Dawkins argues that beaver dams are extended phenotypes of beaver genes, which have evolved under Darwinian selection in the same way as beaver teeth or tails.

Dawkins is clearly right, except that he does not take his own argument far enough. When beavers build dams they are likely to influence the probability that the particular genotypes that underlie their dam-building behavior will be selected, as Dawkins says. But they are also likely to do more than that. Beaver dams are important components in beaver environments and they are bound to transform the natural selection pressures that are acting on populations of beavers in multiple ways, thereby ensuring that modified natural selection pressures feed back to other beaver genes and other beaver phenotypes too, their teeth and tails for instance, or their feeding or social behaviors, or their susceptibility to predation and disease, and so on. Moreover, many of these self-induced selection pressures are likely to remain in place for as long as a dam remains in place in the beavers' environment. Usually that is considerably longer than the lifetime of any individual beaver, given that beaver dams are constantly being repaired, rebuilt, and replaced by the repetitive niche-constructing activities of successive generations of beavers. Thus, from our perspective, Dawkins overlooks two points. First, extended phenotypes can set up plural feedback cycles in evolution in the form of modified natural selection pressures that feed back, not only to the genes that are responsible for expressing the "extended phenotypes" in the first place, but also to other genes at other genetic loci in either the same or in different populations. Second, some "extended phenotypes" are heritable by succeeding generations of organisms through the medium of an external environment.

In this respect Dawkins is representative. On an *ad hoc* basis niche construction, and the feedback it gives rise to, is cropping up all over the place in standard evolutionary theory today, for instance in habitat selection, frequency and density dependent selection, indirect genetic effects, maternal inheritance, and coevolution. Yet in spite of this, the full ramifications of niche construction are still being neglected. What is missing is a body of theory that sets out to explore niche construction as a general evolutionary process in its own right.

## 3.4 EXTENDED EVOLUTIONARY THEORY

This is why we recently proposed an extension to the standard theory of evolution that explicitly incorporates niche construction (Odling-Smee, Laland, and Feldman, 2003). The basic idea is summarized in Figure 3.1b, which illustrates a scheme in which the evolution of organisms depends on both natural selection and niche construction. Genes are transmitted by ancestral organisms to their descendents by genetic inheritance under the direction of natural selection, exactly as in Figure 3.1a. However, selected habitats, modified habitats, and modified sources of natural selection in those habitats are also passed on by these same organisms to their descendents via an "ecological inheritance," under the direction of niche construction. The net result is that the selective environments encountered by organisms are partly determined by independent sources of natural selection, exactly as described by the standard theory, for instance, by climate, weather, or physical and chemical events. But they are also partly determined by what organisms do, or previously did, to their own and each others' environments by niche construction.

*Ecological inheritance* (Figure 3.1b) is a second proposed new process in evolution. It refers to the inheritance via an external environment of one or more natural selection pressures that have previously been modified by niche construction, and it works very differently from genetic inheritance (Odling-Smee, 1988; Odling-Smee, Laland, and Feldman, 2003). For instance, unlike genetic inheritance, ecological inheritance does not depend on the transmission of discrete "replicators" (Dawkins, 1989). It only depends on the persistence between generations of whatever physical changes are caused by the niche-constructing activities of ancestral organisms in the local environments of their descendents. Another difference is that when organisms inherit naturally selected genes via genetic inheritance, they are inheriting genetically encoded semantic information (Odling-Smee, Laland, and Feldman, 2003). But when organisms inherit modified natural selection pressures via ecological inheritance they do not inherit any information directly. Instead, they only inherit transformed natural selection pressures in their environments that subsequently select for their genes and thereby determine the "content" or "meaning," or in conventional terms the "fitness," of whatever semantic information is encoded in their naturally selected genes. A third difference is that, in sexually reproducing organisms, genetic inheritance is transmitted between generations by only two parents, whereas ecological inheritance is typically transmitted by multiple niche-constructing organisms.

## 3.5 IMPLICATIONS FOR THE HUMAN SCIENCES

We can now return to humans. What are the implications of this extension of evolutionary theory for humans, given that human evolution includes potent cultural processes as well as genetic processes and that human niche construction is so often cultural? I will illustrate this question with one well-known example of human cultural niche construction.

When our pastoralist ancestors first domesticated cattle, and drank milk, they set up some self-induced natural selection pressures in favor of genes that increase lactose absorption in human adults by synthesizing an enzyme, lactase, that breaks down the lactose in milk. Consequently, between 70% and 100% of human adults who come from parts of the world with a long history of dairy farming can continue to drink milk long after their childhood is over. Others, from different areas, cannot (Bodmer and Cavalli-Sforza, 1976; Feldman and Cavalli-Sforza, 1989; Durham, 1991; Holden and Mace, 1997; Hollox et al., 2001; Enattah et al., 2002). This example shows that, through their cultural activities, human phenotypes have sometimes influenced human genetic evolution in the past. Yet, according to standard evolutionary theory, humans should not be able to do this. Phenotypes are not supposed to be able to influence the genetic evolution of their own populations on the basis of culturally acquired traits. Doing that sounds suspiciously like Lamarckism even though it is not. Niche construction, including cultural niche construction, is thoroughly Darwinian because it only modifies orthodox Darwinian selection pressures. Nevertheless, the lactose tolerance example, and others like it, are uncomfortable data for the standard theory to digest, other than on an *ad hoc* basis. The best way to illustrate this point is by considering two of the ways biologists have hitherto tried to understand the role of human cultural processes in human evolution from the perspective of standard evolutionary theory (Laland and Brown, 2002).

The first is sociobiology, originally introduced by E. O. Wilson (1975). Sociobiology is firmly based on standard evolutionary theory and it suffers the consequences. Standard evolutionary theory models only a single general inheritance system in evolution, namely genetic inheritance (Figure 3.1a). Hence, the only way any organisms can possibly contribute to evolutionary descent must be by passing on their genes to their descendents, and sociobiology has to assume that this same limitation applies to humans too. Thus humans, or more accurately human phenotypes, are assigned exactly the same status in biological evolution as the phenotypes of every other species. We become just "survival machines" or "vehicles" for our genes, and our only evolutionary function is differential survival and reproduction relative to natural selection and chance (Dawkins, 1989). That promotes the version of human evolution that is summarized in Figure 3.2a. It is a picture that recognizes the existence of human cultural processes, and it may even recognize the existence of cultural inheritance within human cultures. However, the most that any of these cultural processes can do for human evolution is to influence which individuals are the fittest, and which the least fit, and, therefore, who contributes their genes to human genetic inheritance in each generation. In this scheme human cultural processes have no other power to codirect the course of human evolution.

The limitation of this approach is that it cannot explain how our human ancestors, by inventing agriculture, apparently did manage to contribute to human evolution in another way. They transformed one, and probably more than one, natural selection pressure in their environments by "agri-cultural" niche construction and affected their own genetic evolution as a consequence. For this to have happened humans must be more than just "survival machines" in evolution. To some extent, at least, humans must also be active participants, and even co-directors, in their own evolution.

The first biologists to challenge sociobiology on biological rather than on political grounds were the gene-culture coevolutionary theorists (Cavalli-Sforza and Feldman, 1981; Boyd and Richerson, 1985; Durham, 1991). Like the sociobiologists, these theorists recognized that human culture exists and that it incorporates a separate cultural inheritance system, but unlike the sociobiologists, and in the light of the human lactose tolerance example and others like it, they also realized that human

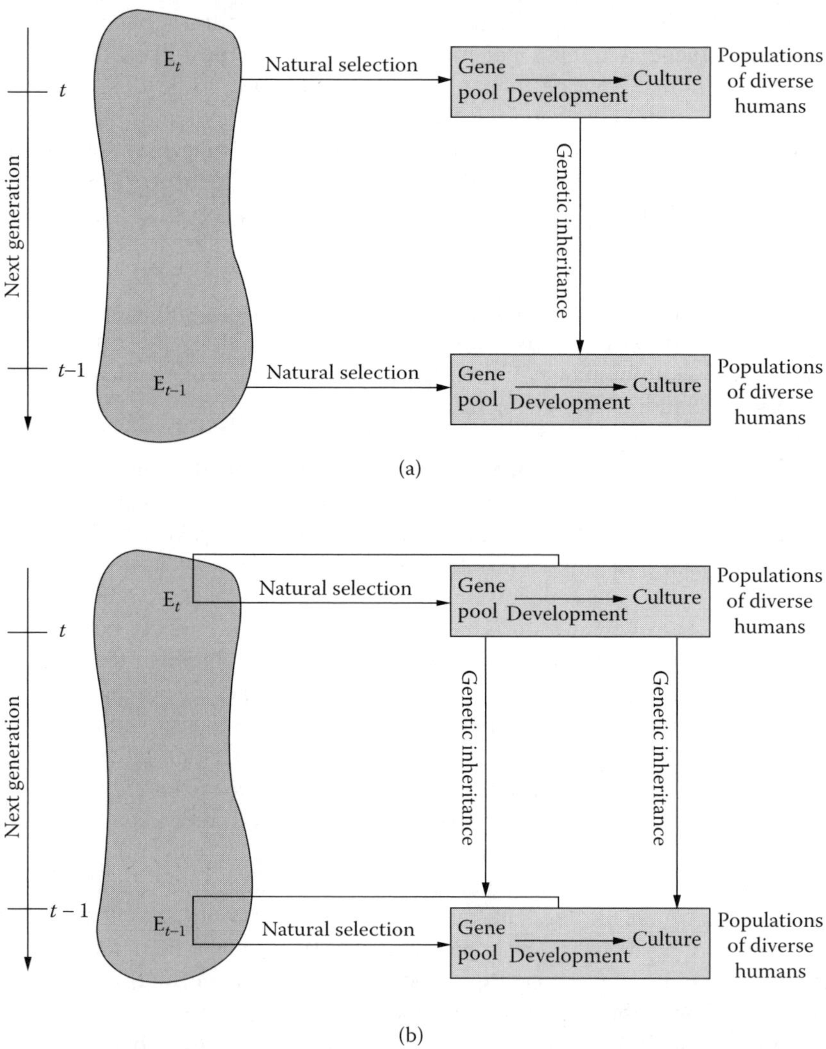


**FIGURE 3.2** (a) Human sociobiology: single inheritance. (b) Gene-culture coevolution: dual inheritance. (c) Gene-culture coevolution: triple inheritance and niche construction.

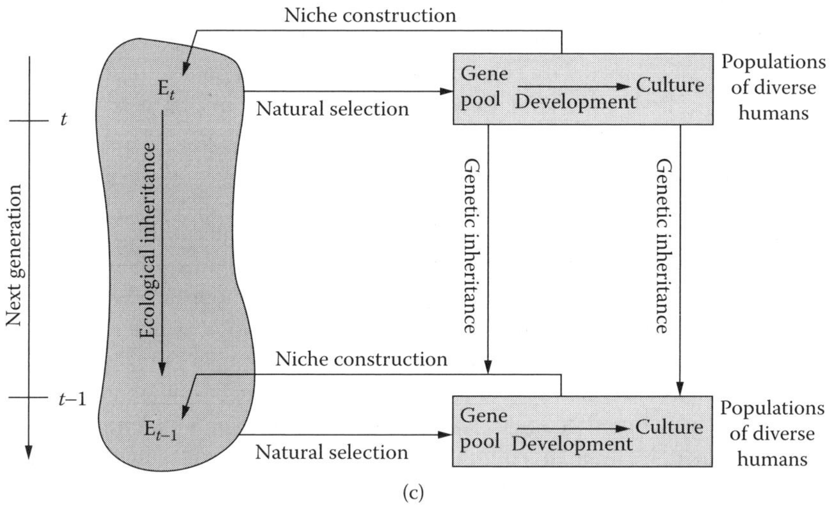


**FIGURE 3.2** (Continued).

cultural inheritance can sometimes lead to the modification of natural selection pressures in human environments in ways that do have consequences for human genetic evolution. This point is symbolized by the feedback arrows in Figure 3.2b.

Gene-culture coevolutionary theory is clearly an advance on sociobiology because it can explain more data, but it still faces two problems. First, it regards the modification of natural selection pressures in environments as an exclusively human ability. It is seen as something that only humans can do, and humans can only do it because of culture. However, as Darwin's earthworms demonstrate, this is not the case. Niche construction is a general phenomenon in evolution. It is something that all organisms do. It does not depend exclusively on human cultural processes (Odling-Smee, Laland, and Feldman, 2003). Second, gene-culture coevolutionary theory did not initially refer either to niche construction or ecological inheritance. However, without these intermediate processes this so-called dual inheritance approach is underequipped to explain how human cultural processes can influence human genetics. Alden Smith (2000, p.32) has criticized it for being "theoretically rich" but "empirically impoverished" because the theory provides insufficient mechanisms to encourage empirical work.

## 3.6 EXTENDED GENE-CULTURE COEVOLUTIONARY THEORY

To counter this criticism we have proposed that, instead of basing gene-culture coevolution on standard evolutionary theory, it should be based on extended evolutionary theory (Laland, Odling-Smee, and Feldman, 2000; Odling-Smee, Laland, and Feldman, 2003). Figure 3.2c illustrates what happens when this is done. Extended evolutionary theory converts gene-culture coevolution from a "dual inheritance" to a "triple inheritance" theory. It now incorporates: (i) genetic inheritance directed by natural selection, (ii) ecological inheritance directed by niche construction, and (iii) cultural inheritance directed by cultural processes. In this scheme cultural inheritance is different from both genetic and ecological inheritance in that it is not a general inheritance system in evolution. It is almost exclusively human. However, it acts by influencing both of the other two inheritance systems, which are general.

Cultural inheritance influences human genetic inheritance by contributing to differential survival and reproduction in human populations in the ways that both sociobiology and the earlier versions gene-culture coevolutionary theory say it does. So here there is broad agreement among all three of the approaches in Figure 3.2. However, cultural inheritance also influences human ecological inheritance by contributing to the transformation of natural selection pressures in human environments through cultural niche construction. Eventually that influences human genetic inheritance again, but by another route. Natural selection pressures that have previously been transformed by human cultural niche construction subsequently select for different human genotypes, in the way that the human lactose tolerance example demonstrates.

We propose that this triple inheritance version of gene culture coevolution is closer to the kind of comprehensive theory that we are ultimately going to need to understand human evolution. It is more complicated than the earlier theories, and it is still incomplete. For instance, it omits human developmental processes, which will eventually have to be included too (Oyama, Griffiths, and Gray, 2001). However, it has the merit of breaking some current logjams by casting some old questions and old arguments in a new light and by suggesting some new ways of answering them. I'll illustrate this claim by focussing on two such questions.

## 3.7 POSSIBLE ORIGINS OF CULTURAL INHERITANCE

Because human cultures depend on the social transmission of information, both within and between generations (Cavalli-Sforza and Feldman, 1981; Boyd and Richerson, 1985; Durham, 1991; Laland, Odling-Smee, and Feldman, 2000), one obvious question is how could the inclusion of niche construction and ecological inheritance in evolutionary theory illuminate our understanding of the processes that convey social information among humans?

Table 3.1 summarizes four principal kinds of niche construction in the form of a $2 \times 2$ table (Odling-Smee, Laland, and Feldman, 2003). Perturbational niche construction occurs when organisms actively change one or more of the components of their external environments at particular places by physically changing them. Relocational niche construction occurs when organisms actively move somewhere else in space and choose, or preferentially bias, the direction and distance they travel and often the time when they travel as well. Inceptive niche construction occurs when organisms either perturb their environments or relocate in such a way that they introduce a novel change in their environments by innovating. Counteractive niche construction occurs when organisms either perturb their environments or relocate in such a way that they either wholly or partly neutralize a novel change in their environments that has been caused by some other agent. The relevant kind

**TABLE 3.1**
**The Principal Kinds of Niche Construction**

| | Perturbation | Relocation |
|---|---|---|
| Inceptive | Organisms initiate a change in their selective environment by physically modifying their surroundings e.g., emission of detritus | Organisms expose themselves to a novel selective environment by moving to or growing into a new place e.g., invasion of a new habitat |
| Counteractive | Organisms counteract a prior change in the environment by physically modifying their surroundings e.g., thermo-regulation of nests | Organisms respond to a change in the environment by moving to or growing into a more suitable place e.g., seasonal migration |

*Source:* Odling-Smee et al., 2003. Reprinted with permission of Princeton University Press.

of niche construction here is perturbational niche construction in both its inceptive and counteractive forms.

We have classified the principal kinds of environmental components that organisms can perturb by niche construction (Odling-Smee, Laland, and Feldman, 2003). They are abiotic, or nonliving environmental components, such as a water resource or the soil; artifacts, such as nest, burrows, trails, or dams, which owe their "design" as well as their improbable thermodynamic status primarily to the actions of the niche-constructing organisms themselves; and other organisms. Other organisms can then be subdivided into either organisms from another species, or conspecifics.

The perturbing of other organisms by niche construction is achieved in two principal ways. Niche-constructing organisms may either perturb other organisms by changing their physical status, for example by killing them or causing them to starve, or conversely by giving them food or physically protecting them. All these kinds of perturbation primarily affect energy and matter relations in populations, and energy and matter flows in ecosystems. For example, the classical predator–prey or host–parasite or competitive or mutualist relationships that are modeled in ecology, and in coevolutionary theory predominantly involve the physical perturbation of some organisms by the niche-constructing activities of other organisms. We have discussed this category of niche construction extensively before (Odling-Smee, Laland, and Feldman, 2003). There is, however, also a second and probably equally important way of perturbing other organisms that we have not previously drawn much attention to. It involves niche-constructing organisms "communicating" with other organisms and perturbing them by doing so. The word *communicate* is used here in its general engineering sense. It includes and goes beyond the more restricted sense in which it is currently used by behavioral ecologists to describe animal communication. It is in this general sense that perturbing other organisms by communicating with them is likely to have played a major role in the origins of diverse information transmission systems in biology, including cultural inheritance in humans.

The minimal requirements for any communication systems are (i) a transmitter, (ii) a receiver, and (iii) a communication channel that is capable of carrying "messages" between them (Shannon, 1948; Shannon and Weaver, 1949). Hence, two kinds of perturbing communication are possible. Niche-constructing organisms may either perturb other organisms by transmitting messages to them or by receiving messages from them.

When an organism niche constructs by transmitting a message, it typically does so by generating a signal, whether iconic, where the signal directly represents X (e.g., a pheromone trail); indexical, where the signal draws attention to X (e.g., an alarm call); or symbolic, where the signal stands in the place of X (e.g., human words; Bickerton, 1995), which the transmitting organism may either broadcast to multiple other organisms or send to a single chosen organism, for example a prospective foe or mate. Transmitting organisms can also transmit either positive signals, which potentially benefit the receiver, or negative signals, which potentially harm the receiver. For example, an animal giving an alarm call in the vicinity of its kin may favorably affect its kin's fitness. Similarly, a human parent may actively teach its offspring a behavior that benefits its offspring. Alternatively, an animal may give a "dishonest signal" to protect its own fitness, or use its so-called Machiavellian intelligence (Byrne and Whiten, 1988) to deceive another animal. The latter may harm the receiver but provide the transmitter with an advantage.

A receiving organism can also niche construct via communication but in a less obvious manner. A receiving organism may initiate a perturbing communication with another organism without waiting for the other organism to transmit a message to it first. To do that the receiving organism need only actively seek or solicit information from another organism, possibly one that it has actively chosen and either at the other organism's expense or not. For example an animal may observe another animal and intercept and translate whatever cues or signals are stemming from the other animal (see Maynard Smith and Harper, 2003, for the distinction between cues and signals). It may then subsequently copy, or partly copy, one or more aspects of the other animal's activities, particularly its niche-constructing activities. For instance a receiver may imitate a new foraging technique that it observes in another animal. Alternatively, it may actively avoid imitation after witnessing a negative consequence arising from the action of another animal. These kinds of receiver-initiated communication are not normally treated as communication by behavioral ecologists, who usually restrict that term to transmitter-initiated messages only. However, there is some connection with the directed social learning described by Coussi-Korbel andand Fragaszy (1995). The phenomenon also appears to be present, although not so described, in other instances of social learning in animals (Heyes and Galef, 1996; Heyes and Huber, 2000).

The immediate consequence of both transmitter- and receiver-initiated communications is likely to be the perturbation of some component in the environment of the niche-constructing organism and possibly of other organisms too. For instance, organisms that transmit messages are likely to modify or manipulate the behaviors of one or more of the other organisms in their environments relative to themselves. Conversely, organisms that actively solicit messages from other organisms are more likely to change their own behaviors first but probably in a way that changes their relationship both to other organisms and one or more components of their environments. For example, if an organism acquires a new foraging technique by copying another organism, it may subsequently gain access to an environmental resource that was hitherto inaccessible to it (e.g., Sherry and Galef, 1984). A longer-term consequence of both these kinds of perturbing niche construction may be the inheritance, via ecological inheritance, of one or more modified natural selection pressures in the local environments of descendent organisms.

Put together, these two kinds of communication via the transmission and reception of messages, may then eventually go further by generating communication networks, or "webs" that connect not just two organisms to each other but multiple organisms to each other by two-way channels of communication, carrying messages to and from many organisms via links or chains of varying lengths. Here, too, I am using the term *network* more broadly than is customary in behavioral ecology. The networks may vary in size from small to very large, and the existence of any such communication network is likely to modify the environments of all the organisms that are contributing to it. This last consequence may seem surprising because, typically, the networks themselves do not appear to exist physically, except in the collective central nervous systems or possibly other systems of all the participating organisms. They are, therefore, deceptively intangible and may seem to possess only virtual reality. Yet they may radically alter the environments of all the organisms that are connected to each other in a shared network.

One possible and even likely consequence of the existence of animal communication networks among conspecifics, for example, is the construction of animal societies. The participating animals are likely to become social as a consequence of their ability to communicate with each other. If they do become social, however, they may subsequently transform large numbers of natural selection pressures in their own and in each others' environments, including their social environments. Depending on whether the semantic information that is carried by a network involves learning or not, the societies the networks construct will be different. If learning is not involved, the resulting societies will lack cultural processes and lack cultural inheritance. Insect societies are the paradigms here (e.g., Holldobler and Wilson, 1994). However, if the information is learned and if socially learned information is transmitted and received by organisms via a shared communication network, the resulting societies should incorporate at least some cultural processes, including horizontal cultural inheritance processes (Cavalli-Sforza and Feldman, 1981). If in addition, the communication network extends across generations by allowing the communication of at least some learned information between generations, then such societies should include vertical cultural inheritance as well (Cavalli-Sforza and Feldman, 1981).

These speculations may or may not be correct. For the moment, however, their principal purpose is to illustrate how the addition of niche construction and ecological inheritance to evolutionary theory could make a difference. It could open up new lines of inquiry about the origins of human social and cultural life and the subsequent course of human gene-culture coevolution (Flack et al., 2006).

## 3.8 BETWEEN PROCESS INTERACTIONS

The second question I want to raise concerns how all the processes shown in Figure 3.2c interact with each other and by doing so drive human gene-culture coevolution. It is not possible to deal with all of them now, so I will narrow down the question by focussing exclusively on the different ways human cultural niche construction affects human cultural processes or human genetics.

Figure 3.3 summarizes the three principal routes via which human cultural niche construction influences either human cultural history or human genetic evolution. The feedback loop at the top of Figure 3.3, route 1, depicts a scenario in which human cultural niche construction transforms a human environment in some way that later feeds back to the same human population and, subsequently, causes a further cultural change in that population. For example, suppose human cultural niche construction changes a human environment by polluting it and the pollution feeds back to influence subsequent cultural processes in the same population in favor of the invention and spread of a new technology, for instance recycling plants to counteract the pollution. In all such cases cultural niche construction is both responsible for causing the problem and for providing the solution. Therefore, route 1 can be colloquially described as a nurture–nurture route. Via route 1 cultural niche construction only affects subsequent cultural processes. It may change human history, or prehistory, but it does not affect human genetics.

In contrast, route 2, the lower and more complicated feedback loop in Figure 3.3, does affect human genetics. It could therefore be described as a nurture–nature route. In this case cultural niche construction changes a human environment as before, but this time, either because there is no cultural response to the change or else an insufficient one, the feedback from the changed environment transforms one or more natural selection pressures in the human environment in a way that does eventually select for different human genes. We have already met one such example, human

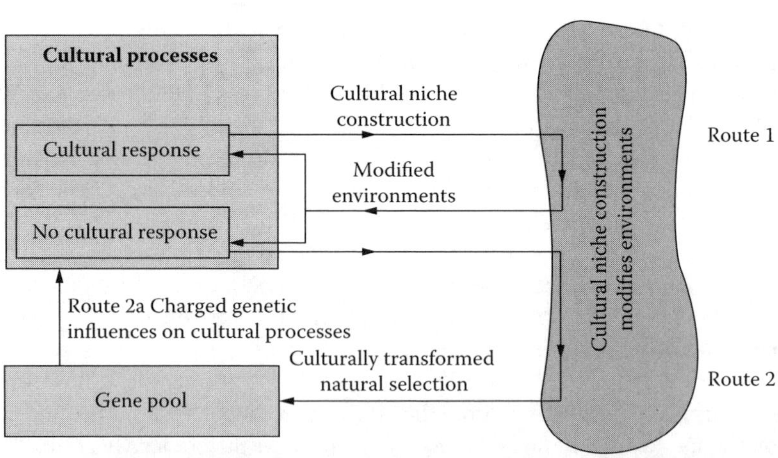


**FIGURE 3.3** Alternative responses to feedback from cultural niche construction.

lactose tolerance in pastoralists. Another comes from Kwa-speaking cultivators in West Africa (Durham, 1991). A genetic change occurred among these people as a consequence of their practice of cutting clearings in tropical rainforests to enable them to grow their crops. The clearings increased the amount of standing water, thereby providing better breeding grounds for malaria-carrying mosquitoes. That increased the prevalence of malaria, which apparently modified natural selection in favor of an increase in the frequency of the hemoglobin S allele because, in heterozygotes, the S allele confers a degree of protection against malaria.

Route 2a (Figure 3.3) is the same as route 2 except that the genotypes selected by culturally modified natural selection underpin the evolution of culture itself. So route 2a might be described as a nurture–nature–nurture route. A candidate example is the evolution of the human brain. Aiello and Wheeler (1995) proposed that our ancestors used their increasingly larger brains to improve their diets in ways that subsequently caused modified selection pressures to select for smaller human guts. That permitted the evolution of still larger brains, because more resources could then be diverted from smaller guts to bigger brains. Aiello and Wheeler also proposed that the resulting enlargement of the human brain probably happened in two episodes, initially when our *homo erectus* ancestors improved their diets through more efficient hunting, and later when our *homo sapiens* ancestors started to cook. Cooking externalizes part of the digestive process, which again permits the digestion of food by smaller guts. Cooking is an obvious case of cultural niche construction, so if Aiello and Wheeler are right it becomes an excellent example of route 2a in action.

The evolution of the human brain takes on added significance because it incorporates the evolution of language (Pinker, 1994; Bickerton, 1995; Deacon, 1997), taking us back to the social transmission of information in human societies via communication networks. By providing much more efficient channels of communication between transmitters (human talkers) and receivers (human listeners), language apparently increased the potency of human cultural processes by orders of magnitude.

Niche construction may have played a more direct role in the evolution of language, too. Deacon suggested that the evolution of language involved a form of feedback from human-altered human environments to human genetics, with consequences that he attributed to "Baldwinian selection" (Deacon, 1997, especially Fig. 11.1 p. 323). Deacon's Baldwinian selection, however, seems to be the same process as niche construction. Certainly it is doing the same job here.

In addition to these primary interactions between genetic and cultural processes, there can also be secondary ones among routes 1, 2, and 2a (Figure 3.3), depending on the details of how any particular human population responds to the consequences of its own cultural niche-constructing activities.

For example, if the Kwa had realized that they were inadvertently intensifying their own selection by malaria because of their agriculture, they might have altered their agricultural practices, for example, by draining the standing water in their local environments. Or if they had had access to modern medicines, they might have used drugs to counteract the malaria. Either of these responses could have converted route 2 to route 1, because the additional counteractive cultural niche construction could have cancelled out the intensifying natural selection by malaria and thus removed the need for a genetic response. It is also possible for the opposite kind of interaction to occur. For example, in modern human societies we do not always recognize the negative consequences of some of our contemporary cultural niche-constructing activities, or if we do, we are reluctant to pay for effective countermeasures. For instance assuming global warming is indeed a negative, technologically amplified consequence of human cultural niche construction, it clearly poses a threat of future climate change (Stainforth et al., 2005) and some changing natural selection pressures. So if we are unwilling to change our cultural practices or to invest in sufficient corrective technology, we may eventually cause route 1 to revert to route 2 by exposing modern human populations to several transformed natural selection pressures that could again select for different human genes. Another topical example is the evolution of antibiotic resistance in bacteria in response to antibiotics (Ewald, 1994; Palumbi, 2001). That too could eventually lead to the selection of different human genes.

## 3.9 CONCLUSION

This chapter illustrates some of the theoretical consequences that adding niche construction and ecological inheritance to evolutionary theory could have for how we understand our own evolution. But it does not exhaust them. Extending evolutionary theory in this manner is likely to raise many other questions, too, about evolution in general and human gene-culture coevolution in particular even with respect to the two topics discussed here.

For example, the existence of powerful language-based communication networks in human societies raises a further question about how human cultural inheritance relates to human ecological inheritance. Is the triple inheritance scheme illustrated in Figure 3.2c a permanent theoretical feature, or may it eventually be possible to revert to a different version of the dual inheritance in Figure 3.2b, one based this time on genetic inheritance and ecological inheritance, in line with Figure 3.1b, instead of genetic inheritance and cultural inheritance, as in Figure 3.2b, by reducing human cultural inheritance to no more than a special component of human ecological inheritance? If so, human cultural inheritance would have to be treated as a very special component of human ecological inheritance because of the intangible properties of social information-communicating networks. Cultural inheritance appears primarily to involve the inheritance of "cultural software" in the form of the social information that supports human cultures. Superficially, it does not look at all like the inheritance of the physical changes in environments that are usually implied by ecological inheritance, except possibly when the information is expressed in the form of an inherited material culture. Cultural inheritance also differs from other inheritance systems in that it is not inherited at birth but mostly only after a human child has developed language skills and starts to listen to other people talking to it or possibly teaching it. Yet, in other respects cultural inheritance is a typical component of ecological inheritance because it functions in the same way. It evidently can transmit modified environments, including modified social environments, from one human generation to the next.

The interactions between human cultural processes and human genetic processes also raise further questions, for example, about the self-induced or anthropogenic changes that are caused by human cultural niche construction. When are anthropogenic changes only potential sources of transformed natural selection pressures for humans, and what can cause them to become actual? Also are there circumstances in which "runaway" cultural niche construction could change our future environments so radically that it would not only transform human natural selection pressures but might also inflict local human extinctions?

Diamond (2005) discusses a number of human populations in which local extinctions did occur. He also indicates how the same thing could happen again, and on larger scales. To stop it happening again, one concept we may have to give up is standard evolutionary theory's attractive, simplifying assumption that natural selection pressures in environments are independent of what organisms do to their environments. The trouble is, they are not.

## ACKNOWLEDGMENTS

I am grateful to Kevin Laland and Robert Aunger for helpful criticisms of an earlier draft of this chapter.

## REFERENCES

Aiello, L.C., and Wheeler, P., The expensive-tissue hypothesis, *Curr. Anthropol.,* 36, 199–221, 1995.

Alden Smith, E., Three styles in the evolutionary analysis of human behavior, In *Adaptation and human behavior: an anthropological perspective,* Cronk, L., Chagnon, N. and Irons, W., Eds., Aldine de Gruyter, New York, 27–46, 2000.

Bickerton, D., *Language and human behavior,* University of Washington Press, Seattle, WA, 1995.

Bodmer, W.F., and Cavalli-Sforza, L.L., *Genetics, evolution and man,* Freeman, San Francisco, 1976.

Boyd, R., and Richerson, P.J., *Culture and the evolutionary process,* University of Chicago Press, Chicago, 1985.

Byrne, R., and Whiten, A., *Machiavellian intelligence: social expertise and the evolution of intellect in monkeys, apes and humans,* Clarendon Press, Oxford, 1988.

Cavalli-Sforza, L.L., and Feldman, M.W., *Cultural transmission and evolution: a quantitative approach,* Princeton University Press, Princeton, NJ, 1981.

Coussi-Korbel, S., and Fragaszy, D.M., On the relation between social dynamics and social learning, *Anim. Behav.* 50, 1441–1453, 1995.

Darwin, C., *The formation of vegetable mold through the action of worms, with observations on their habits,* Murray, London, 1881.

Dawkins, R., *The extended phenotype,* Freeman, Oxford, 1982.

Dawkins, R., *The selfish gene,* 2nd ed., Oxford University Press, Oxford, 1989.

Deacon, T.W., *The symbolic species: the coevolution of language and the brain,* Norton, New York, 1997.

Diamond. J., *Collapse. How societies choose to fail or survive.* Allen Lane, London, 2005.

Durham, W.H., *Coevolution: genes, culture and human diversity,* Stanford University Press, Stanford, CA, 1991.

Enattah, N.S., et al., Identification of a variant associated with adult-type hypolactasia, *Nature Genet.*, 30, 233–237, 2002.

Ewald, P.W., *Evolution of infectious disease*, Oxford University Press, Oxford, 1994.

Feldman, M.W., and Cavalli-Sforza, L.L., On the theory of evolution under genetic and cultural transmission with application to the lactose absorption problem, In *Mathematical evolutionary theory,* Feldman, M.W., Ed., Princeton University Press, Princeton, NJ, 145–173, 1989.

Flack, J.C., Girvan, M., de Waal, F.B.M., and Krakauer, D.C., Policing stabilizes construction of social niches in primates, *Nature*, 439, 426–429, 2006.

Heyes, C., and Galef, B.G., *Social learning in animals: the roots of culture*, Academic Press, New York, 1996.

Heyes, C., and Huber, L., Eds., *The evolution of cognition,* MIT Press, Cambridge, MA, 2000.

Holden, C., and Mace, R., Phylogenetic analysis of the evolution of lactose digestion in adults, *Hum. Biol.* 69, 605–628, 1997.

Holldobler, B., and Wilson, E.O., *Journey to the ants: a story of scientific exploration,* Belknap Press, Cambridge, 1994.

Hollox, E.J., et al., Lactose haplotype diversity in the Old World, *Am. J. Hum. Genet.,* 68, 160–172, 2001.

Jones, C.G., Lawton, J.H., and Shachak, M., Organisms as ecosystem engineers, *Oikos,* 69, 373–386, 1994.

Jones, C.G., Lawton, J.H., and Shachak, M., Positive and negative effects of organisms as physical ecosystem engineers, *Ecol.*, 78, 1946–1957, 1997.

Laland, K.N., and Brown, G.R., *Sense and nonsense. Evolutionary perspectives on human behaviour,* Oxford University Press, Oxford, 2002.

Laland, K.N., Odling-Smee, F.J., and Feldman, M.W., On the evolutionary consequences of niche construction, *J. Evol. Biol.* 9, 293–316, 1996.

Laland, K.N., Odling-Smee, F.J., and Feldman, M.W., Evolutionary consequences of niche construction and their implications for ecology, *Proc. Natl. Acad. Sci. USA*, 96, 10242–10247, 1999.

Laland, K.N., Odling-Smee, F.J., and Feldman, M.W., Niche construction, biological evolution, and cultural change, *Behav. Brain Sci.*, 23, 131–175, 2000.

Lewontin, R.C., Gene, organism, and environment, In *Evolution from molecules to men,* Bendall, D.S., Ed., Cambridge University Press, Cambridge, 1983.

Maynard Smith, J., and Harper, D., *Animal signals,* Oxford University Press, Oxford, 2003.

Odling-Smee, F.J., Niche constructing phenotypes, In *The role of behavior in evolution,* Plotkin, H.C., Ed., MIT Press, Cambridge, MA, 73–132, 1988.

Odling-Smee, F.J., Laland, K.N., and Feldman, M.W., *Niche construction. The neglected process in evolution,* Monographs in Population Biology, 37, Princeton University Press, Princeton, NJ, 2003.

Oyama, S., Griffiths, P.E., and Gray, R.D., *Cycles of contingency: developmental systems and evolution*, MIT Press, Cambridge, MA, 2001.

Palumbi, S.R., *The evolution explosion,* Norton, New York, 2001.

Pinker, S., *The language instinct: the new science of language and mind*, Allen Lane, St Ives, 1994.

Shannon, C., The mathematical theory of communication, *Bell Syst. Tech. J.,* 27, 379–423, 1948.

Shannon, C., and Weaver, W., *The mathematical theory of communication,* Illinois University Press, Urbana, 1949.

Sherry, D.F., and Galef, B.G., Cultural transmission without imitation: milk bottle opening by birds, *Anim. Behav.* 32, 937–938, 1984.
Stainforth, D.A., et al., Uncertainty in predictions of the climate response to rising levels of greenhouse gases, *Nature*, 433, 403–406, 2005.
Turner, J.S., *The extended organism: the physiology of animal-built structures,* Harvard University Press, Cambridge, MA, 2000.
Williams, G.C., Gaia, nature worship, and biocentric fallacies, *Q. Rev. Biol.*, 67, 479–486, 1992.
Wilson, E.O., *Sociobiology: The New Synthesis.* Harvard University Press, Cambridge, MA, 1975.

# 4 State and Value: A Perspective from Behavioral Ecology

*John M. McNamara and Alasdair I. Houston*

## CONTENTS



## ABSTRACT

We view natural selection as producing strategies that approximately maximize an appropriate measure of fitness. Although the trajectories of the evolutionary process may be important, we concentrate on the endpoints of the process. We discuss a state-dependent approach to behavior in which an animal is characterized by state variables such as size, immunocompetence, and energy reserves. State-dependent reproductive value can be used to characterize evolutionary stability. This can be done even when there are intergenerational effects (phenotypic inheritance of traits) such as cultural transmission. We use an example of the coevolution of female mate choice rules and male trait value (the Fisher process) to show that it is possible to characterize the endpoints of evolution when there is both genetic and cultural transmission of information. This approach provides a rigorous justification of intuitive accounts based on value and costs. It also has the advantage over genetic models when there are many traits. We also discuss the implications of state-based models for various topics in the study of behavior, including negotiation and cooperation.

## 4.1 INTRODUCTION

In this paper we review various theoretical issues that arise in analyzing how evolutionary pressures might have shaped the behavior of animals, including humans. We are particularly interested in cases that involve both genetic and cultural inheritance. In such cases, the spread of a gene can be analyzed only by taking account of the correlation between the gene and the cultural trait. We discuss a state-dependent approach that, although it does not deal with this correlation explicitly, is able to handle such cases. In this approach, individuals carrying a particular gene are characterized in terms of their state. State could for example be an individual's wealth or social status, and vertical cultural inheritance is represented by the relationship between the state of parents and the state of their offspring.

The approach that we describe is used in behavioral ecology to characterize the stable endpoints of evolution by natural selection. We begin by describing the concept of an Evolutionarily Stable Strategy (ESS) and how endpoints are characterized. If the best strategy for an individual depends strongly on what other individuals are doing, there can be more than one endpoint. We discuss various cases where there are multiple endpoints. When inheritance is based on both genes and culture, the endpoint specifies both the genetic composition of a population and its social organization. We illustrate this in an example based on mate choice in which female preference for males (genetically determined) and male phenotype (culturally determined) coevolve.

## 4.2 STRATEGIES AS ADAPTATIONS

Much of behavioral ecology is concerned with trying to establish whether observed behavior can be understood in adaptive terms. In other words, is the behavior the result of following a fitness-enhancing behavioral strategy that has been shaped by the action of natural selection? Although the trajectories of the evolutionary process may be important, here we concentrate on the endpoints of the process (this is the usual approach in behavioral ecology, but see, e.g., Enquist et al. (2002)). We can view evolution as selecting between strategies on the basis of an appropriate measure of fitness. Fitness may be independent of the strategies of other population members, in which case endpoints can be characterized as points where fitness is locally maximized (Hammerstein, 1996; Weissing, 1996). In many cases of interest fitness depends on the strategies of other population members, in which case endpoints can be characterized in game theoretic terms (e.g., Hammerstein, 1996). This can usually be done even when in addition to natural selection there is sexual selection (Kokko et al., 2002; McNamara et al., 2003). The approach has the potential to also model the endpoints of cultural evolution, as well as to investigate cases of gene-culture coevolution. The characterization of endpoints can work even if the production of mutations includes an element of intention, as is the case in art and technology (e.g., Ziman, 2000).

It is important to be clear about the distinction between a strategy and the resulting behavior. For example, consider clutch size in birds. The number of eggs laid by a particular bird is not fixed, but depends on the food supply and time of year (Drent and Daan, 1980; Rowe et al., 1994). The rule specifying the dependence of clutch size on conditions is what we refer to as a *strategy*. The observed clutch size in a given year is a consequence of the strategy followed and environmental conditions. Natural selection acts on strategies rather than on clutch size directly.

Another example is provided by the time of day at which a male bird sings to attract a mate. McNamara et al. (1987) regard a strategy as a rule specifying whether a male bird should sing or forage as a function of both time of day and its energy reserves. The strategy they predict involves a threshold level of reserves, above which the bird sings and below which it forages. Despite the fact that this threshold is constant during the first half of the day, a bird following the strategy has a burst of singing at dawn, after which singing drops to a lower level. This burst is a result of the bird starting the day with high reserves, which in turn resulted from the bird ending the previous day with enough reserves to ensure survival even if the night is worse than expected. This example shows that it is not possible to consider a particular time of day in isolation. Selection does not act on singing at dawn *per se*, and the dawn chorus is a consequence of the strategy for regulation of body reserves in the face of the conflicting demands of foraging and singing.

Similar remarks hold for annual routines. Many life history tradeoffs are mediated by state variables such as body size, energy reserves, and state of the immune system. A strategy is now a rule that specifies how activities such as foraging, reproduction, and migration depend on these variables and time of year. Because current behavior typically influences the future values of the state variables, it is often not possible to consider a particular time of year in isolation. Natural selection does not then act on the timing of an activity such as reproduction, rather on the state-dependent rule that organizes behavior over the whole year (Houston and McNamara, 1999; McNamara et al., 2004b). This work serves to emphasize that although the rule is deterministic, it does not follow that behavior is deterministic. In these annual routine models, the rule that specifies if and when to reproduce is the same each year, but fluctuations in the state variables from year to year can result in reproductive behavior varying from year to year. In particular, if reproduction in one year reduces condition in the next year, there may be a tendency to breed in alternate years (McNamara et al., 2004b).

We believe that the state-dependent approach is a particularly valuable tool for understanding the reproductive behavior of humans. In this context, models sometimes assume that parents decide how many offspring to produce before they start to reproduce. It seems unlikely that humans make decisions in this way. We might expect people to decide whether to have another child on the basis of a variety of factors including the current family size, health, and wealth (e.g., Gyimah and Fernando, 2002; Palloni and Rafalimanana, 1999). It is decision rules of this form that natural selection might be expected to act on rather than on the number of children. Models based on sequences of actions and their consequences not only give the parents more flexibility, they also force a detailed consideration of the effect that various actions have on state and how state changes during the production of children. From a specification of these consequences consistent costs emerge (Houston and McNamara, 1999; McNamara and Houston, 1986).

In all the above examples, we have emphasized that natural selection acts on strategies rather than resulting outcomes. The predictions that follow from this view can be dramatically different from those that view selection as acting on outcomes (McNamara et al., 1999).

In some cases natural selection might produce strategies that are so flexible and powerful that the behavior chosen (e.g., clutch size) is the best in all circumstances (food availability, time of year, etc.), but this is a strong assumption. The strategies that animals employ are implemented by neural and hormonal mechanisms whose flexibility may be limited (e.g., McNamara and Houston, 2002; Sih, Bell, and Johnson, 2004; Sih et al., 2004). (For a contrasting view, see Neff and Sherman, 2004.) Thus, in most cases it seems more likely that natural selection shapes strategies so that animals make good decisions in the circumstances that are typically encountered, resulting in individuals that follow the strategy having high average fitness. Behavior is likely to be suboptimal under circumstances that occur rarely in the organism's natural environment. For example, when given the choice some birds prefer to incubate a large model egg to their own egg even though the model egg is much bigger than any that are laid by the species. If within the normal range of egg sizes large eggs result in higher reproductive success, then this preference for a "supernormal stimulus" can be viewed as a side effect of a rule that performs well for the normal range of egg sizes. There has been considerable discussion of the ideas of side effects in the context of human cognitive abilities. Taking an analogy from architecture, Gould and Lewontin (1979) introduced the term *spandrel* for a feature that is a necessary consequence of some other feature of an organism. We note that this definition makes a spandrel a special case of a side effect in which the form of the side effect is determined rather than arbitrary. A spandrel is a "nonadaptive architectural byproduct of definite and necessary form — a structure of predictable size and shape that then becomes available for later and secondary utility" (Gould, 1997). Although the validity of the original architectural example has been questioned (Dennett, 1995; Houston, 1997a; Lauder, 1996), the term has now become established in psychology. Gould (1997) sees spandrels as important for our understanding of the evolution of the human mind: "Mustn't the ever cascading spandrels of the human brain be more weighty than the putative primary adaptations of ancient African hunter–gatherer ancestors in setting the outlines of what we now call "human nature?" For further discussion see Buss et al. (1998), Gould (2002), and Hampton (2004).

As we now illustrate in the context of risk-sensitive foraging, there may be more than one possible explanation for a phenomenon in terms of side effects. Foraging is risk sensitive if preference depends on not just the mean energy gain but also on the variance (McNamara and Houston, 1992; Real and Caraco, 1986). Laboratory experiments on risk sensitivity usually involve the choice between two foraging options whose parameters do not change during the experiment. Various models attempt to explain the observed behavior as an adaptation that minimizes the probability of death from starvation (McNamara and Houston, 1992). Typically these models consider the animal to be in an environment that is exactly like the laboratory procedure in that the options do not change. In contrast, McNamara (1996) finds the optimal strategy for a much richer environment in which options that differ in variability change over time. He argues that some aspects of the observed behavior are side effects of rules for coping with such complex environments. Kacelnik and Bateson (1996) suggest that risk sensitive foraging is not even a consequence of selection for responding to differences in variability but is a side effect of general rules for learning.

In summary, natural selection acts on decision rules (i.e., behavioral strategies). The nature and generality of the decision rules that members of a population use is an empirical question. We do not expect nonhuman animals to make mathematical calculations to find the best course of action; instead we expect them to follow rules that perform well (e.g., McNamara and Houston, 1980). Even though humans are able in principle to calculate optima, they may follow rules (e.g., Hutchinson and Gigerenzer, 2005). The important point is that selection acts on the rules that produce behavior rather than on the behavior itself. What matters is that an individual produces the appropriate behavior for its circumstances. In many cases of interest, this will require that an individual's behavior depends on a range of variables that characterize the individual and its environment, which means that behavior is specified by an appropriate state-dependent rule.

## 4.3 EVOLUTIONARY STABILITY

Maynard Smith (1982) characterized an endpoint of the process of evolution by natural selection as an evolutionarily stable strategy (ESS). The idea of such a strategy is that if almost all individuals adopt the strategy, then no mutant strategy can invade into the population under the action of natural selection. To formalize this idea suppose that almost all population members follow strategy $\pi$ . We can refer to such a strategy as the resident strategy. Consider some rare mutant strategy $\pi'$. Let $\lambda(\pi',\pi)$ be the equilibrium growth rate in the numbers of individuals following strategy $\pi'$ (while this strategy remains rare). We will be more specific as to what we mean by equilibrium below. In this notation $\lambda(\pi,\pi)$ is the growth rate of a mutant strategy that is identical to the resident strategy and is, hence, the growth rate of the resident population as a whole. The mutant strategy will increase in frequency if $\lambda(\pi',\pi) > \lambda(\pi,\pi)$. Thus, a necessary condition for a strategy $\pi^*$ to be an ESS is that

$$\lambda(\pi,\pi^*) \leq \lambda(\pi^*,\pi^*) \qquad \text{for all strategies } \pi. \tag{4.1}$$

We can define the fitness of a strategy to be its equilibrium growth rate $\lambda$ (Metz et al., 1992). As the notation makes clear, the fitness of a mutant strategy depends on the resident strategy. Condition 4.1 says that a necessary condition for $\pi^*$ to be an ESS is that the fitness of mutant strategies (when the resident strategy is $\pi^*$) has a maximum at $\pi^*$ . We can also regard $\lambda(\pi',\pi)$ as the payoff to strategy $\pi'$ against strategy $\pi$ in a game. Condition 4.1 then says that $\pi^*$ is a Nash equilibrium for the game with this payoff. At a Nash equilibrium it is not possible to increase the payoff by adopting a different strategy.

Condition 4.1 may not be sufficient to ensure that a strategy $\pi^*$ is an ESS. A mutant strategy that has fitness equal to the resident strategy can increase by drift when rare. Thus, stability demands that every such equal-fitness strategy must be selected against as numbers following the strategy grow. For details of this second order criterion see Maynard Smith (1982). Of course the stronger condition

$$\lambda(\pi,\pi^*) < \lambda(\pi^*,\pi^*) \qquad \text{for all strategies } \pi \text{ with } \pi \neq \pi^* \tag{4.2}$$

is sufficient to ensure that $\pi^*$ is an ESS. Suppose that the resident strategy is $\pi$. Define a strategy $\hat{\pi}$ to be a best response to this resident strategy if it satisfies

$$\lambda(\hat{\pi},\pi) = \max_{\pi'} \lambda(\pi',\pi)\ , \tag{4.3}$$

where the maximum is taken over all possible mutant strategies, $\pi'$. A mutant that adopts a best response strategy maximizes its rate of growth within the resident $\pi$ population. As can be seen from Condition 4.1 a necessary condition for a strategy $\pi^*$ to be an ESS is that it is a best response to itself.

In some cases there may be more than one best response to a given resident strategy. In other cases, such as the example based on the Fisher process that we describe below, best responses are unique. Suppose that they are unique; that is for every resident population strategy there is a unique best response to that strategy. Denote the best response to resident strategy $\pi$ by $B(\pi)$. Then by Conditions 4.1 and 4.2 we see that the condition

$$B(\pi^*) = \pi^* \tag{4.4}$$

is both necessary and sufficient to ensure that $\pi^*$ is an ESS.

If a population with resident strategy $\pi^*$ is at an ESS then it is stable against invasion by mutant strategies. But this does not mean that the strategy will evolve in the population in the first place. One way to examine this is to see whether the resident strategy will always evolve back to $\pi^*$ if the whole resident strategy is perturbed away from $\pi^*$. If it always returns for small perturbations, then the strategy $\pi^*$ is said to be convergence stable (Eshel, 1983). A convergence stable strategy is surrounded by a basin of attraction. Within this region, the population returns to the equilibrium if it is moved away from it. As our example based on the Fisher process will illustrate, this basin may be the whole space of strategies or some restricted subset. We would expect the stable endpoints of the process of evolution to be both an ESS and convergence stable. Such a strategy is said to be a continuously stable strategy (CSS).

So far we have been characterizing an organism in terms of its phenotype (i.e., strategy). But if a strategy is coded by genes, a rigorous model of evolution by natural selection is given by consideration of the underlying genetics. In the simple case where fitness is lifetime reproductive success (see below) and strategies are coded for at a single locus, the phenotypic approach can be given a rigorous underpinning in terms of genetics (e.g., Charlesworth, 1994). Because in our view it is easier and more intuitive to work with a phenotypic characterization in terms of strategies, this is the approach we adopt.

## 4.4 FITNESS AND REPRODUCTIVE VALUE OF A RARE MUTANT STRATEGY

In the above we introduced the idea of the fitness of a strategy as the equilibrium growth rate in numbers following the strategy when the strategy is rare. The definition of this growth rate depends on circumstances. One of the crucial determinants of fitness is the form of stochasticity that an animal encounters. Environmental stochasticity is stochasticity that affects all population members in the same way. Examples of this form of stochasticity include fluctuations in weather and population density. In contrast demographic stochasticity is stochasticity that affects each individual independently of what happens to other population members given the environmental conditions. For example in a large population, being killed by a predator is probably almost independent of whether others are killed. When there is demographic stochasticity but no environmental stochasticity the fitness of a strategy is the mean equilibrium growth rate in the number of individuals following the strategy, where the mean is an average over demographic stochasticity. When there is environmental stochasticity an adequate fitness measure cannot be obtained simply by averaging over this stochasticity. In particular, the variance (over environmental stochasticity) is important (Gillespie, 1977; Grafen, 1999; Lewontin and Cohen, 1969; McNamara, 1995).

If there is density dependence acting on a population then population numbers cannot grow without limit. If the population is large, there is no environmental stochasticity, and population dynamics are stable, then the population will tend to some equilibrium population size at which population growth rate is 1. If the population is also at an evolutionarily stable equilibrium then no mutant strategy will have growth rate greater than 1. This observation allows us to define an especially simple fitness measure in the special case in which all offspring that survive to independence are the same. (The case in which not all surviving offspring are the same is discussed below.) In this case it can be shown that at evolutionary stability resident population members are not only maximizing their growth rate but are also maximizing their mean lifetime production of surviving offspring (Lifetime Reproductive Success [LRS]; Taylor et al., 1974). Thus, by taking LRS to be our measure of fitness we obtain the same ESS as if we had instead worked with growth rates. Because LRS is usually easier to work with than a growth rate, it is usual to adopt it as the fitness measure in this case (Grafen, 1988). For a discussion of the relationship between growth rate and various fitness measures, see Mylius and Diekmann (1995).

Note that if we consider some group of individuals following a given strategy, then these individuals will be in different states (e.g., energy reserves, age, size) and hence have different abilities to leave descendants that inherit the strategy. Reproductive value measures this relative ability.

When fitness is simply LRS, the reproductive value, $V(x)$, of an animal in state $x$ is the expected number of surviving offspring produced by the animal in the future. If, for example, state is age, then typically reproductive value first rises with increasing age as reproductive maturity is approached (because the older the animal the more likely it is to survive until maturity) and then falls as the organism approaches the end of its reproductive life. For a small bird attempting to survive a harsh winter the most important state variable might be its level of energy reserves. The bird uses its reserves to buffer itself against periods when foraging is poor. Typically the reproductive value, $V(x)$, of a bird with reserves $x$ increases as $x$ increases, but at a decreasing rate. As a simple illustration of how a behavioral ecologist might use this suppose that a bird with reserves $x$ faces a trade off between gaining food and avoiding predation. The derivative $dV/dx$ measures the marginal advantage (in terms of future reproductive success) of gaining energy. $V(x)$ is the value of the animal's life and is hence the cost (in terms of future reproductive success) of being killed. It is easy to show that the maximum risk that an animal should take in getting a unit of food is $(dV/dx)/V$. Because this decreases as $x$ increases, we conclude that it is adaptive for the bird to take fewer risks as its reserves increase. This is no surprise, but the result does depend on the shape of the function $V$, which depends on assumptions about how the bird's energy reserves affect future reproductive success.

Reproductive value depends on future expectations, which depends on the focal animal's own behavior in the future. In the predation risk example reproductive value depends on how plentiful food is likely to be in the future. This depends not only on the environmental food availability but also on the behavior of the focal animal in the future. Because current reproductive value determines the current best action, this means that the best thing to do now depends on what will be done in the future.

The moral here is that value is not absolute; it depends on circumstances. This is illustrated by Houston (1997b), who shows that context-dependent value can arise from errors in decision making. The example involves an animal choosing between foraging options that differ in energy gain and handling time. It is assumed that errors that are costly in terms of rate of energetic gain are less common that errors that are cheap. In turn, the rate of energetic gain depends on the errors that are made. Houston finds the simultaneous solution to these two conditions. As is shown, at this solution it is possible to obtain a set of choice probabilities that violate a form of transitivity.

In the simple case in which the fitness of an invading mutant is LRS, there is a direct connection between the maximization of fitness and the maximization of reproductive value. A strategy maximizes fitness (expected lifetime number of surviving offspring) if and only if the action taken in every possible state maximizes reproductive value given that reproductive value will be maximized in the future (Caswell, 1982; McNamara, 1993; Schaffer, 1981; Taylor et al., 1974). This criterion means that the strategy maximizing fitness can be found using the technique of dynamic programming (Clark and Mangel, 2000; Houston and McNamara, 1999). To give an example, suppose that an animal reproduces annually and must decide its reproductive effort each year. As the effort of an individual of age $n$ increases, the number of surviving offspring from the current breeding attempt, $B_n$, increases, but the probability that the animal will survive to breed next year at age $n + 1$, $S_n$, decreases. Let $V_n$ be the reproductive value of an age $n$ individual, given that future behavior is optimal. Then

$$V_n = \max\{B_n + S_n V_{n+1}\}\ , \tag{4.5}$$

where the maximization is over effort at age $n$. In applying this equation we work backwards in time over the animal's lifetime, at each stage maximizing reproductive value given maximization at all future times. For example, suppose that all animals are incapable of breeding by age 100. Thus $V_{100} = 0$. The reproductive, $V_{99}$, of a 99-year-old can then be found from Equation 4.5 and then the reproductive value of a 98-year-old, and so on.

## 4.5 MAXIMIZATION OF REPRODUCTIVE VALUE AT EVOLUTIONARY STABILITY

In the predation risk example the reproductive value of an animal depends on the food that it will obtain in the future. This not only depends on its own future behavior but also depends on how aggressively other population members will contest items that the animal finds in the future (e.g., Grafen, 1987). This type of dependence is general; the reproductive values of the individuals following a given mutant strategy depend on the resident population strategy. At evolutionary stability the strategy of resident population members maximizes fitness. Thus, when fitness is LRS, at evolutionary stability all resident population members are behaving so as to maximize the measure of reproductive value that is generated by their collective behavior (Houston and McNamara, 1999). In fact, given that best responses are unique, this criterion characterizes ESSs (Figure 4.1).

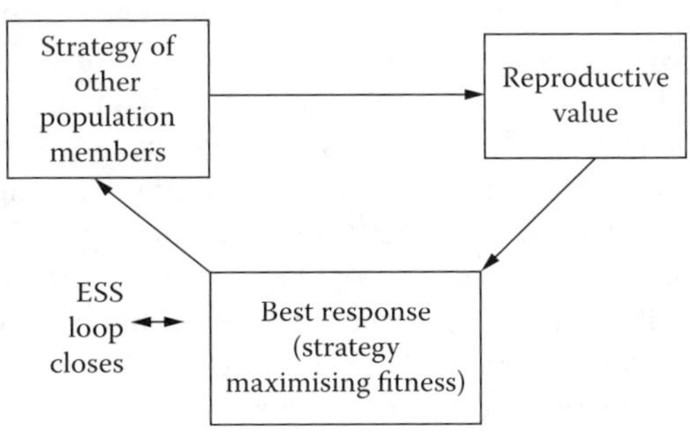


**FIGURE 4.1** The feedback loop. The strategy of population members determines individual reproductive value. The strategy that maximizes this reproductive value maximizes fitness within the population. The population strategy is evolutionarily stable if and only if this fitness maximizing strategy coincides with the strategy of population members.

The feedback loop illustrated in Figure 4.1 requires that reproductive value at evolutionary stability satisfies a consistency condition. This constrains reproductive values and the costs and benefits of actions at evolutionary stability. To illustrate this suppose that population members need a territory to breed. If a territory is occupied an intruder can contest ownership of the territory with the owner. To model such contests between two individuals suppose that each contestant can adopt one of three strategies: Hawk; i.e., fight until injured or opponent withdraws, Dove; i.e., display but withdraw if opponent fights, Bourgeois (respect ownership); i.e., play Hawk if owner, play Dove if intruder. Let $V$ be the increase in reproductive value as a result of gaining a territory and let $C$ be the loss in reproductive value as a result of sustaining an injury. Theory predicts that if $V > C$, then Hawk is the only ESS; if $V < C$, then Bourgeois is an ESS. However, as Grafen (1987) points out, $V$ and $C$ cannot be assigned arbitrary values; their value depends on the behavior of others. Respect for ownership means that territory owners keep their territories, which makes $V$ high. The cost of being injured in a fight is the resulting loss in expected future reproductive success. Thus, if vacant territories are very rare, then this cost is low. Grafen therefore concludes that $V < C$ is unlikely if ownership is respected, and hence we should not expect to find this strategy in nature. Grafen's conclusions are supported by a detailed analysis that explicitly considers a sequence of interactions between different owners and intruders (Mesterton-Gibbons, 1992). For more on the importance of feedback, see Heino et al. (1997), Laland et al. (2000), McNamara et al. (2000), and Houston and McNamara (2002).

If the strength of feedback is sufficiently strong in Figure 4.1 there can be more than one consistent solution. In this way it is possible to have multiple stable social structures. For example Houston and McNamara (1991) consider a situation in which each male in a population encounters a fixed maximum number $N$ of females during the breeding season. Each female encountered is contested by another male drawn at random from the population. In each contest a male must decide whether to play Hawk or Dove. The probability a male wins a contest is greater if the male plays Hawk, but if opponent also plays Hawk the probability the male is killed is also greater. The winner of a contest mates with the female. Each male attempts to maximize the expected number of females that he mates with. For this scenario Houston and McNamara took a strategy $p$ to specify the probability of playing Hawk in a contest. They showed that for sufficiently large $N$ there can be two CSSs. At one CSS all individuals play Hawk. This is stable because, given others are playing Hawk, a male has no chance of obtaining a female unless he fights for her. Thus, Hawk is his best response. At the other CSS individuals play Hawk with a probability $\pi^*$ where $0 < \pi^* < 1$. In such a population a male that always fought would die well before the end of the season and obtain few matings, whereas a male that always played Dove would survive but could have obtained more matings by being more aggressive. The best response in this population is to also play Hawk with probability $\pi^*$.

McNamara (1994) presents a model in which the strategy of a population member determines how much of its available energy to allocate to reproduction at each age. Energy not allocated to reproduction enhances survival. Increasing the allocation to reproduction increases the number of surviving young this year but reduces the chances of survival to breed again next year. In this population individuals compete for food, with older individuals outcompeting younger ones. McNamara shows that there can be two CSSs in this situation. At one CSS all individuals allocate all their energy to reproduction at age 1. Consequently, all individuals die after their first reproductive bout (semelparity). At the other CSS all individuals put some energy into survival and some into reproduction, ensuring that there are older animals in the population and that some individuals manage to breed more than once (iteroparity).

More than one stable endpoint can be found in models of niche construction; see for example Laland et al. (2001).

In summary, reproductive value depends on the behavior of other population members. This feedback between individuals and the population is illustrated in Figure 4.1. At evolutionary stability each population member maximizes its reproductive value, given the behavior of others, so that the loop in Figure 4.1 is closed. There may be a unique strategy that closes the loop and hence a unique ESS. However, if the feedback between the population and individual reproductive value is sufficiently strong, there may be more than one strategy that closes the loop. There will then be more than one social structure that is evolutionarily stable.

## 4.6 INTERGENERATIONAL EFFECTS

Sometimes the phenotype of an individual depends not just on its genotype and environment but also on the phenotype of its mother. This dependence is known as a maternal effect; see Mousseau and Fox (1998) and Reinhold (2002) for reviews. The inheritance of state is an example of an intergeneration effect. Such effects are widespread and are not restricted to the influence of an individual's mother. There can be paternal effects, grandparental effects, etc. (Odling-Smee et al., 2003). Offspring can vary in many components of state such as size, gender, quality, attractiveness, and dominance status. Such attributes are important because they affect the ability of the offspring to produce offspring themselves. Furthermore, if differences between offspring are passed on to grandchildren, these will also differ in their ability to produce offspring, and so on.

To illustrate how intergenerational effects can persist over several generations consider the following simple example based on a hypothetical asexual organism. Individuals of this species reproduce at age 1 and then die, so that those individuals present in one year are the offspring of those present the previous year. Suppose that individuals of a particular genotype of this species are in one of two states — High quality and Low quality. High quality individuals leave 1.5 offspring (on average). Low quality individuals leave 1 offspring (on average). Quality tends to be inherited: offspring have the same quality as their parent with probability $1-p$ and have the opposite quality with probability $p$. This parameter thus specifies the degree to which quality "mutates" between generations. We assume that it lies in the range $0 \leq p < 0.5$. When $p = 0.5$ the quality of offspring are indcpendent of parental quality. As $p$ decreases, the correlation between offspring quality and parental quality increases. For this model we consider the expected number of children, grandchildren, great grandchildren, and so on left by an individual of each type.

The expected number of descendants left after $n$ generations increases for both types of individual (Figure 4.2a). However, the ratio of the expected number of descendants left by a High quality individual to the expected number left by a Low quality individual tends to a limiting value as time, and hence the number of generations, increases (Figure 4.2b). This limiting ratio is defined as the reproductive value of High quality relative to Low quality. As Figure 4.2b shows, the limiting value, and

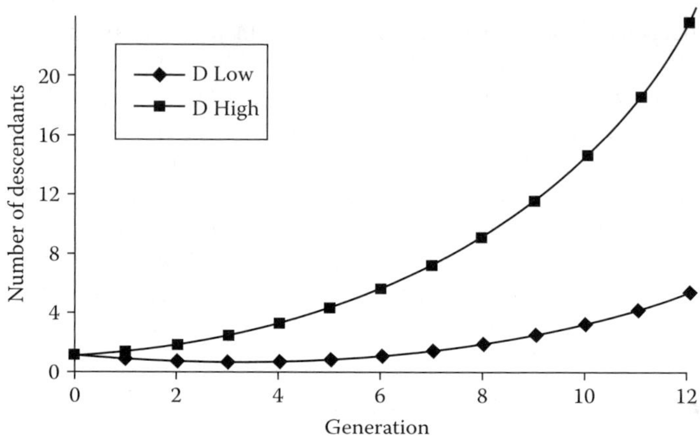


(a)

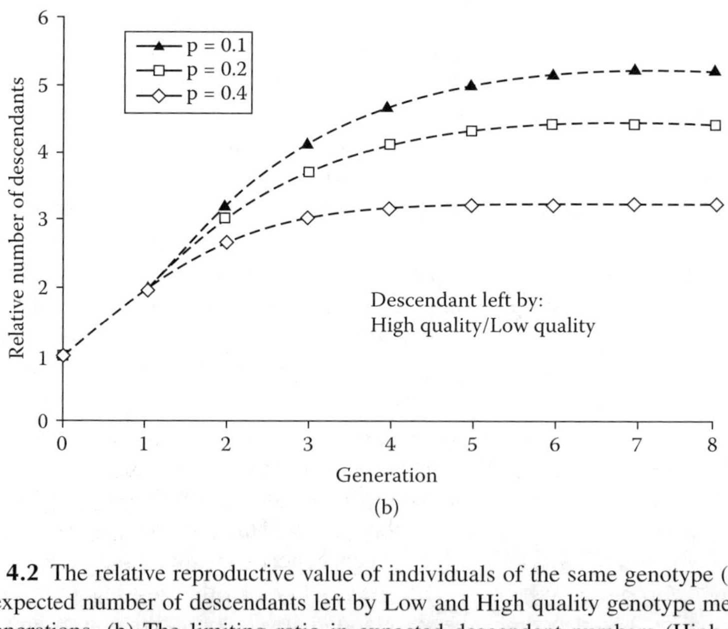


(b)

**FIGURE 4.2** The relative reproductive value of individuals of the same genotype (see text). (a) The expected number of descendants left by Low and High quality genotype members in future generations. (b) The limiting ratio in expected descendant numbers (High to Low), defined as the reproductive value of High quality relative to Low quality for three values of the mutation parameter p.

the rate at which this limit is reached depend on the mutation parameter, $p$. When offspring quality is independent of parental quality ($p = 0.5$), the limit is reached after one generation and is just the ratio of the number of offspring left (i.e., 1.5). When $p$ is low the initial advantage of being High quality in terms of offspring number is multiplied down the generations because the offspring of High quality individuals tend to be High quality and to have more offspring themselves. The smaller $p$ the greater the eventual limiting value and the more generations it takes to approach this limit.

The concept of reproductive value as illustrated by this example can be extended to general state variables and to sexually reproducing species. Suppose that the resident population strategy is $\pi$. We define how the reproductive value of members of this resident population depends on their state. To deal with sexually reproducing species it is necessary to take into account the fact that not all descendants of a given focal individual may inherit the individual's behavioral strategy. Usually the probability that a descendant inherits a specific gene controlling behavior from a given individual is equal to the degree of relatedness of the descendant to that individual. Motivated by this, suppose that an individual is in state $x$ in year 0. Let $n_t(x)$ be the expected number of descendants, weighted by relatedness, left by this individual in year $t$. This number typically depends on the state $x$ of the individual. To look at the dependence on state we compare the focal individual in state $x$ with an individual in some reference state $x_0$. The ratio of descendants left by these individuals, $n_t(x) / n_t(x_0)$, tends to a limit as $t$ increases. Denote the limit by $V_\pi(x)$, i.e.,

$$V_\pi(x) = \lim \frac{n_t(x)}{n_t(x_0)} . \tag{4.6}$$

This is the reproductive value of state $x$ (relative to state $x_0$) in the resident $\pi$ population. In applications of the idea of reproductive value, it is only the magnitude of the reproductive value of one state relative to another that is important. Thus, the choice of reference state $x_0$ does not affect applications.

Now consider the rare mutant strategy $\pi'$ present in the resident $\pi$ population. Let an individual following the mutant strategy be in state $x$ at time 0. Let $n'_t(x)$ be the number of descendants, weighted by relatedness, left by the individual in year $t$. This number changes as $t$ increases. For example, for the case illustrated in Figure 4.2a this number increases over time for both of the initial states $x$ = High and $x$ = Low. The annual proportionate growth in numbers between years $t$ and $t + 1$ is $n'_{t+1}(x) / n'_t(x)$. Provided that numbers do not grow so large that density-dependent effects act, this ratio tends to a limit as $t$ increases; i.e., denoting the limit by $\lambda(\pi', \pi)$ we have

$$\lambda(\pi', \pi) = \lim \frac{n'_{t+1}(x)}{n'_t(x)} . \tag{4.7}$$

Because $n_t'(x) / n_t'(x_0)$ tends to a limit for each $x$ and $x_0$, the limit in Equation 4.7 is independent of $x$. This limit $\lambda(\pi',\pi)$ is the asymptotic growth rate in the numbers of individuals following the mutant $\pi'$ strategy and is defined to be the fitness of the strategy within the resident $\pi$ population.

Both fitness and reproductive value can be defined in terms of matrices as follows. Consider an individual following the rare mutant strategy $\pi'$ present in the resident $\pi$ population. Suppose the individual is in state $x$ in year $t$. Let $a_{yx}$ be the mean number of descendants (discounted by relatedness) left by the individual in state $y$ in year $t+1$. Then the matrix $\mathbf{A}(\pi',\pi) = (a_{yx})$ of these numbers is the projection matrix of the mutant strategy $\pi'$. The fitness of the mutant strategy, $\lambda(\pi',\pi)$, is then the maximum eigenvalue of this nonnegative matrix. In the special case $\pi' = \pi$ the matrix $\mathbf{A}(\pi,\pi)$ is the projection matrix for members of the resident population. The growth rate of the resident population is the maximum eigenvalue $\lambda(\pi',\pi)$ of this matrix, and the components of the corresponding (row) eigenvector are the reproductive values, $V_\pi(x)$, for resident population members (cf., Caswell, 1989).

When there are intergenerational effects, the fitness of a strategy cannot be measured by just counting numbers of offspring. Instead, as the definition makes clear, fitness is the annual growth rate of numbers of descendants following the strategy far into the future. There is no arbitrary point of cutoff in looking at growth, such as just counting grandchildren or even great grandchildren. This is important when there are strong intergenerational effects because maximization of fitness does not necessarily maximize expected numbers of offspring or even grandchildren or great grandchildren. To illustrate this consider again the example of an asexual species in which individuals are either High quality or Low quality. Suppose now that Low quality individuals have only one reproductive action, denoted L(1), open to them. Under this action they leave on average 1 surviving offspring, where each offspring is Low quality with probability 0.8 and High quality with probability 0.2. High quality offspring have two possible reproductive actions. Under action H(1.5), 1.5 offspring survive on average, where each surviving offspring is High quality with probability 0.9 and Low quality with probability 0.1. Under action H(2), 2 offspring survive on average, where each surviving offspring is High quality with probability 0.5 and Low quality with probability 0.5. Thus H(1.5) produces less offspring than H(2) but the offspring tend to be of higher quality. We have emphasized that a strategy is a contingent rule. Here there are just two possible strategies, which we refer to as Strategies 1 and 2. An individual following Strategy 1 takes action L(1) when Low quality and action H(1.5) when High quality. An individual following Strategy 2 takes action L(1) when low quality and action H(2) when high quality. Suppose that in generation 0 there are equal numbers of individuals following the two strategies in a population. For simplicity suppose that all these initial individuals are of High quality. Figure 4.3b illustrates what happens in subsequent generations. After one generation there are 1.5/2 = 0.75 times as many individuals following Strategy 1 as following Strategy 2. The mean quality of the Strategy 1 individuals is, however, higher (Figure 4.3a). As Figure 4.3 shows, the difference in mean quality increases over time and translates into a higher rate of increase in numbers in subsequent

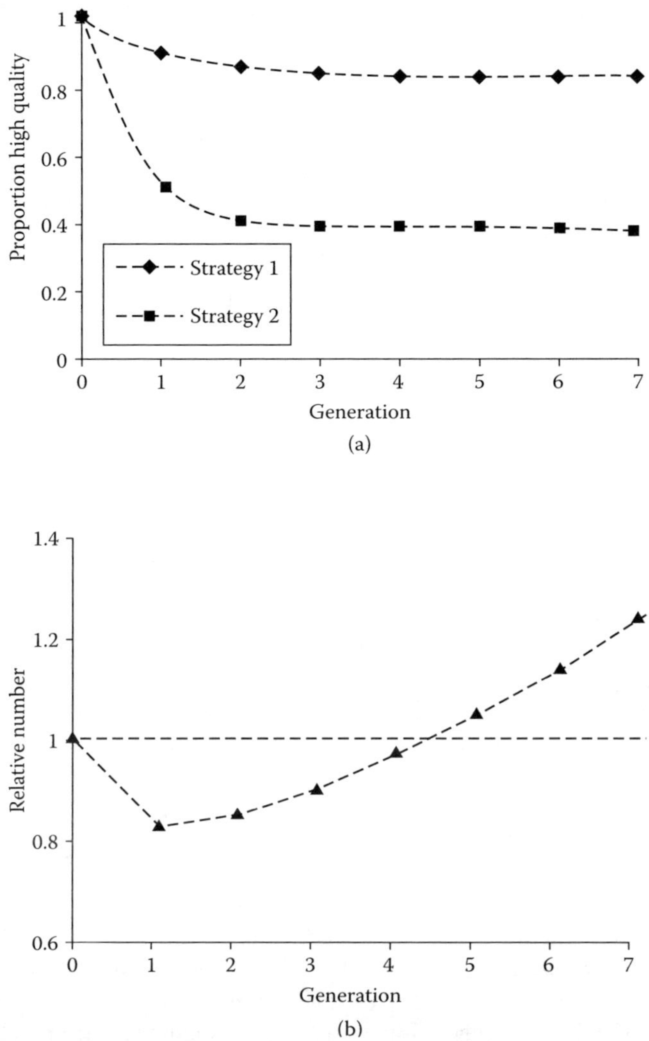


**FIGURE 4.3** Comparison of the fitness of two strategies labeled Strategy 1 and Strategy 2 (see text). Initially there are equal numbers of individuals following both strategies. All these individuals are High quality. (a) The proportion of high quality individuals following each strategy in subsequent generations. (b) Ratio of the numbers of individuals following Strategy 1 to numbers following Strategy 2 in subsequent generations.

generations. But it takes until the fifth generation before numbers following Strategy 1 exceed numbers following Strategy 2. Eventually numbers of individuals following Strategy 1 far exceed those following Strategy 2. For example the ratio of numbers of individuals is 34,851 after 100 generations. Strategy 1 has the higher fitness, but that would not have been apparent by just counting offspring, grandchildren, or even great grandchildren.

Trivers-Willard sex ratio theory (Trivers and Willard, 1973) provides a second example illustrating the importance of counting descendants sufficiently far into the future. This theory is concerned with whether it is better for a high quality female to invest in sons or daughters (Braza, 2004; Brown, 2001; Koziel and Ulijaszek, 2001). Suppose that females are limited in the number of offspring that they can produce, whereas males are not limited. Suppose also that there are maternal effects: high quality females tend to produce high quality offspring. Trivers and Willard considered this situation and argued that if a high quality female produces a daughter, then this daughter is limited in the number of offspring she can produce even though she is high quality. But if a high quality female produces a son, then being high quality he will gain many matings and so produce many offspring. So a high quality female can maximize her number of grandchildren by producing sons. By an analogous argument a low quality female maximizes her number of grandchildren by producing daughters. But these arguments ignore the fact that the offspring of daughters will be high quality. Thus, by producing daughters, a high-quality female gets fewer grandchildren, but these are higher quality than if she invests in sons. Because high quality grandchildren will produce more and higher quality children themselves, this can mean that maximizing grandchildren does not necessarily maximize the number of descendants left far into the future. In particular, it can be better for a high quality female to produce daughters (Leimar, 1996).

In this section we have shown that the presence of intergenerational effects means that the fitness of a strategy is not represented by the number of offspring that are produced. Intergenerational effects can be handled by a state-based approach in which fitness is the long-term growth rate in the number of descendents produced by a strategy.

## 4.7 COMBINING FEEDBACK AND INTERGENERATIONAL EFFECTS

We now give an example based on a particular model of sexual selection with genetically transmitted preferences in females for culturally transmitted traits in males (McNamara et al., 2003). This model illustrates some important general points about evolution with intergenerational effects. The method used also illustrates an approach presented by Taylor (1990) by which the endpoints of evolution can be characterized when intergenerational effects are present.

Males differ in a cultural trait. Sons tend to resemble their fathers as a result of cultural rather than genetic inheritance. Examples include song in some species of birds and wealth, status, or playing a musical instrument in humans. Females can recognize the type of a male when choosing a mate. Female choice is genetically determined and the evolution of this choice is investigated. As female choice rules evolve, the types of males that they mate with will change and, hence, the frequency of male types will change. We seek an endpoint of the process.

In assessing the choice that a female makes, it is not adequate to count the number of offspring that she produces. If there is a preference in the population for a particular type of male, then males of the preferred type will have sons that are in turn preferred (Fisher, 1930). Thus by mating with a preferred male, a female may produce more grandchildren even if her number of offspring produced by mating with such a male is not increased. The number of offspring can even be decreased, but "sexy sons" (Weatherhead and Robertson, 1979) may compensate. As we have said above, even counting grandchildren may not be sufficient to capture the advantage of mating with a preferred male; it may be necessary to count descendants further into the future.

Our model is based on a discrete breeding season of fixed length and nonoverlapping generations, so that all individuals die after their first breeding season. Females mate at most once, whereas males can mate many times. Females encounter males sequentially as a Poisson process of unit rate. On encounter a female either accepts the male as a mate or rejects him. A female cannot return to a male after she has rejected him. Once a female has accepted a male as a mate she stops searching. If by the end of the season she has not mated she does not breed. For simplicity of exposition we start by considering the case where there are only two male types, labeled type 1 and type 2. Males may differ in the number of surviving offspring a female produces after mating with them. Such differences are known as "direct benefits" as opposed to indirect or genetic benefits. Direct benefits could be the result of differences in the quality of male territories or the level of care that males provide. The sex ratio of offspring is 1:1. Transmission of type to sons is not perfect; there is a probability $1 - p$ that a son is of the same type as his father and a probability $p$ that he is the other type ($0 < p < 0.5$). Very choosy females run the risk of failing to obtain a mate. If, for example, the female prefers type 1 males, she will always accept a male of this type if encountered but will only accept a type 2 male toward the end of the season. The degree of preference for type 1 males is specified by the critical time in the season at which she ceases to be choosy. The later this time, the more likely she is to obtain a male of the preferred type, but the more likely she is to not encounter any male after the critical time and so obtain no mate at all.

The behavior of females is genetically determined by a single haploid locus. An allele at this locus codes for which type of male is initially preferred and the critical time in the season after which both types are accepted. Alleles are carried in males but not expressed. Consider a population with a resident allele $\pi$ coding for the mate choice rule. We assume demographic stability so that the frequency of each male type is at equilibrium. We want to know whether a rare mutant allele $\pi'$ spreads within this resident population. Consider copies of the mutant allele that are present at the start of the breeding season in year $t$. At this time, an allele can be in one of three types of individual: a mutant female, a mutant type 1 male, or a mutant type 2 male. In each case for an allele in a given type, we find how many copies there are in the next generation in each class. These numbers are the elements of the projection matrix $\mathbf{A}(\pi',\pi)$ for the mutant allele. In this case it is a $3\times3$ matrix. Because the mutant is rare, we can ignore a male mutant meeting a female mutant. Thus, the number of matings obtained by a mutant male depends on just the mate choice rule of the resident population. Similarly, because of rarity, a mutant female meets only resident males. The probability that she mates with a male of each type (and the probability of not mating) depends both on her mate choice rule and the equilibrium frequency of males of each type in the population. This frequency depends on the resident mate choice rule. The numbers of copies of the mutant allele in each of the three types left at the beginning of year $t + 1$ is then determined by these mating frequencies together with the number and type of surviving offspring from each class of mating. The fitness of the mutant strategy is its equilibrium growth rate $\lambda(\pi', \pi)$. As explained above this is the maximum eigenvalue of the projection matrix $\mathbf{A}(\pi', \pi)$. Criteria for a strategy to be an ESS are as usual given by Conditions 4.1 and 4.2.

So far we have described the problem in terms of the transmission of information (genetic for the female choice rule, cultural for male type). It is possible, however, to characterize an endpoint in terms of phenotypes and values. Let the resident mate choice rule be $\pi$. We defined reproductive value within this resident population as a relative measure of the ability to leave descendants far into the future (Equation 4.6). For convenience we can use the female as the reference state in this definition. The reproductive value of the female is then just 1, and the reproductive values, $V_\pi(1)$ and $V_\pi(2)$ of a type 1 male and a type 2 male, respectively, are the number of descendants (discounted by relatedness) left by each male far into the future relative to a female.

Consider now a rare mutant allele $\pi'$ for female choice. We can define a payoff, $W(\pi',\pi)$ to this female as the expected value of the offspring left next year by the female. Here in assigning value to offspring we take the value of daughters to be 1, the value of type 1 sons to be $V_\pi(1)$, and the value of type 2 sons to be $V_\pi(2)$. Then it can be shown (McNamara et al., 2003) that

$$W(\pi', \pi) < W(\pi, \pi) \quad \text{if and only if } \lambda(\pi', \pi) < \lambda(\pi', \pi), \tag{4.8}$$

i.e., the resident population is stable against invasion by the mutant allele if and only if the payoff to a mutant female is less than the payoff to a resident female. Condition 4.8 allows us to replace the criterion for invasion based on asymptotic growth rates by one based on the phenotypic concepts of payoffs and value. For the problem considered here best responses are unique. Thus, the criterion

$$W(\pi, \pi^*) < W(\pi^*, \pi^*) \ \text{ for all } \ \pi \neq \pi^* \tag{4.9}$$

is both necessary and sufficient for $\pi^*$ to be an ESS. Alternatively, defining the phenotypic best response $B_W(\pi)$ to strategy $\pi$ by

$$W(B_W(\pi), \pi) = \max_{\pi'} W(\pi', \pi), \tag{4.10}$$

where the maximum is over mutant strategies $\pi'$, we see that a necessary and sufficient condition for $\pi^*$ to be an ESS is that

$$B_W(\pi^*) = \pi^*. \tag{4.11}$$

We can also characterize evolutionary stability in terms of a suitable measure of attractiveness. This measure will again be a relative measure. To define it we set the attractiveness of a type 2 male equal to 1 (without loss of generality) and define the attractiveness of a type 1 male as follows. Let a population have resident mate choice strategy $\pi$. Consider a female within this population. We assume that the descendants of this female follow the resident mate choice strategy and look at the expected number of these descendants (weighted by relatedness) present at some specified time far in the future. This expected number depends on the type of male that the female mates with. We define the attractiveness of a type 1, $r(\pi)$ , as the ratio of the descendant number that results from mating with a type 1 male to the number that results from mating with a type 2 male. This measure of attractiveness depends on two factors, the number of offspring of each type that results from a mating together with the attractiveness of male offspring of each type to resident females. It follows that attractiveness depends on the resident population strategy. From the definition of reproductive value it can be shown that a mate choice strategy $\pi'$ maximizes $W(\pi', \pi)$ if and only if it maximizes the expected attractiveness of the male chosen. Here the attractiveness is $r(\pi)$ if the female mates with a type 1 male, 1 if she mates with a type 2 male, and 0 if she fails to mate at all. Thus, by Equation 4.11 an ESS can be characterized as the mate choice strategy that maximizes the expected attractiveness of the male chosen, where attractiveness is that generated by the strategy itself.

We now look at some results from the model when the two types of male are essentially equivalent; i.e., the direct benefit is the same so the female leaves exactly the same number of offspring regardless of which type of male she mates with. Figure 4.4 shows the effect of increasing the preference for type 1 males in the resident population. The degree of preference is represented by the switch time $\pi$. Before this time resident females mate only if a type 1 male is encountered. After $\pi$ a female mates with the first male encountered. If the female has not chosen a male by time $T$ she has left it too late and does not mate at all. Thus, the switch time $\pi$ lies in the range $0 \leq \pi \leq T$. Because females encounter males at unit rate during search, the season length $T$ is just the mean number of males encountered if the female searches for the whole breeding season.

Preference for type 1 males is increased by increasing the switch time $\pi$. As $\pi$ increases, the probability of mating with a type 1 male becomes larger. Because type tends to be inherited by sons, the stable proportion of type 1 males in the population increases as female preference for this type of male increases (Figure 4.4a). Because resident females prefer type 1 males, the sons of type 1 males will tend to get more matings than the sons of type 2 males. Thus the attractiveness $r(\pi)$ of type 1 males is greater than that of type 2 males (taken to be equal to 1). As preference for type 1 males increases in the resident population so does the attractiveness of type 1 males (Figure 4.4b). Figure 4.4c shows the best response $B_W(\pi)$ to the resident mate choice strategy $\pi$. Not surprisingly, because attractiveness of type 1 males increases with the preference of resident females for this type, so does the preference for type 1 males under the best response strategy. An ESS is given by Equation 4.11. As can be seen when the fidelity of cultural transmission is low (low resemblance between fathers and sons), the only solution of this equation is $\pi^* = 0$. This ESS

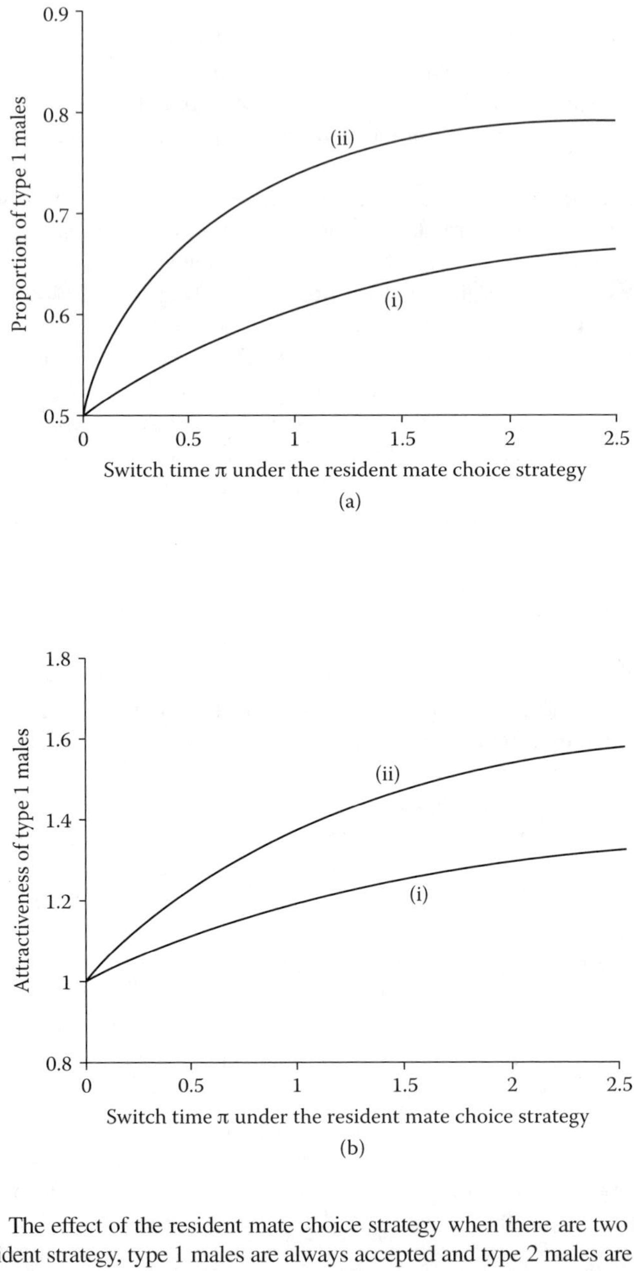


**FIGURE 4.4** The effect of the resident mate choice strategy when there are two types of males. Under the resident strategy, type 1 males are always accepted and type 2 males are rejected before the switch time $\pi$. (a) The proportion of males that are type 1. (b) The attractiveness of type 1 males relative to type 2 males, $r(\pi)$. (c) The best response $B(\pi)$ to the resident strategy. For case (ii), arrows mark the three solutions of the equation $B_w(\pi^*) = \pi^*$, and it is also indicated whether or not each ESS is continuously stable. The probability that a son is the same type as his father is $1 - p$. Cases are (i) low fidelity of cultural transmission, $p = 0.25$; (ii) high fidelity, $p = 0.1$. $T = 2.5$ throughout.

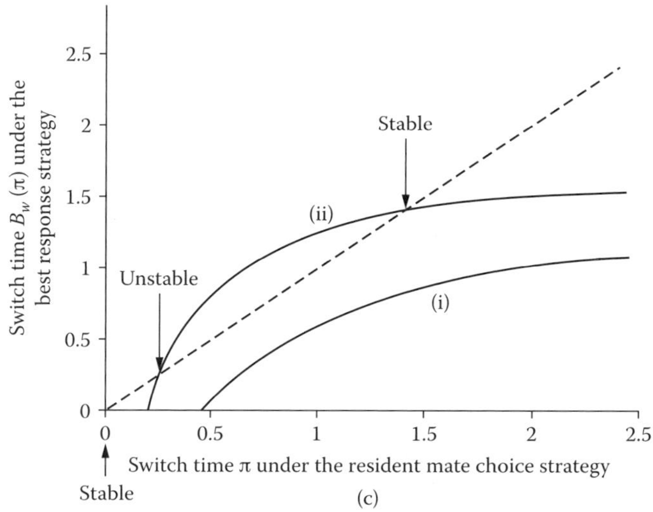


**FIGURE 4.4** (Continued).

is convergence stable and so is a CSS. At this CSS there is no preference, with females mating with the first male encountered. In contrast, when fidelity of cultural transmission is high there are three solutions of Equation 4.11. The middle solution is not convergence stable, and if a population with this mate choice strategy is perturbed it will evolve to either the no preference solution ($\pi^* = 0$) or the solution with a strong preference for type 1 males. These two alternatives are both convergence stable and so are CSSs. Note that the CSS with preference for type 1 males will not evolve if the original resident population strategy is one of low preference because the middle ESS forms an invasion barrier. Of course, once there is sufficient preference (the barrier is crossed) preference will increase until the CSS with preference is reached (cf., Laland, 1994).

The model is based on optimal choice by the female over a breeding season of length $T$. As $T$ increases, a female meets more males during search and can afford to be more choosy, and this means that the range of parameters for which there is a stable preference increases (cf., McNamara et al., 2003). Figure 4.5 shows that at the CSS with preference, the relative attractiveness of the type 1 males increases as the length $T$ of the breeding season increases.

In this model, the number of offspring left and the fidelity of inheritance of type is the same for both male types. Thus, whenever there is a CSS with preference for type 1 males, there will also exist a corresponding CSS with preference for type 2 males.

We now look at a case with many types of male, with each type differing in the direct benefit to females and hence the number of offspring that result from a mating. (This sort of case is hard to handle with explicit genetic models.) We rank males from good to poor on the basis of the direct benefit that they provide. Mutation tends to produce a move toward high quality ("biased mutation," cf., Iwasa et al., 1991; Pomiankowski et al., 1991). It seems obvious that females should prefer good males, and as Figure 4.6

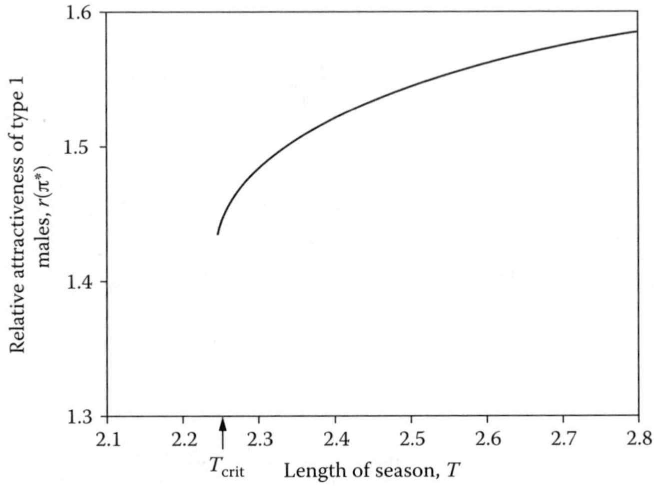


**FIGURE 4.5** The effect of the length of the breeding season $T$ on the attractiveness of type 1 males relative to type 2 males at the continuously stable ESS with preference for type 1 males. The season length $T$ represents the expected number of males encountered if a female searches for the whole breeding season. For $T < T_{crit}$ no preference is the only ESS. For $T > T_{crit}$ a second continuously stable ESS is separated from the no-preference ESS by an invasion barrier. $p = 0.15$.

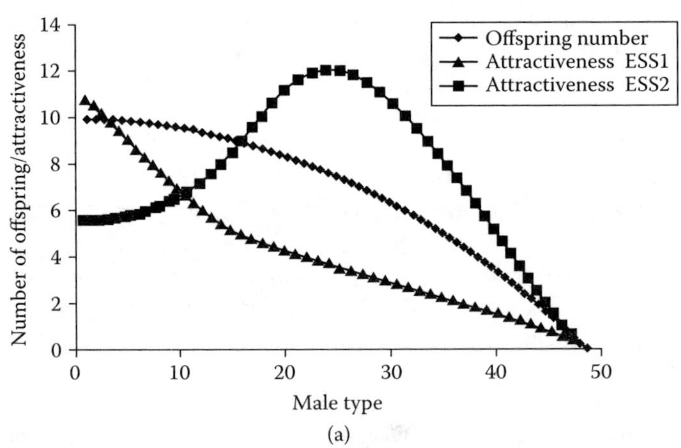


**FIGURE 4.6** Example with 50 types of male, ranked so that males of type 1 provide the most direct benefits (most offspring) and males of type 50 fewest direct benefits (fewest offspring). Two ESSs, referred to as ESS1 and ESS2, are illustrated. (a) For each male type: the number of surviving offspring left by a female if she mates with a male of this type and the attractiveness of a male of this type at each ESS. (b) The two ESS mate choice strategies, each specified by a sequence of switch times. (c) The proportion of males of each type at each of the ESSs. $T = 4$.

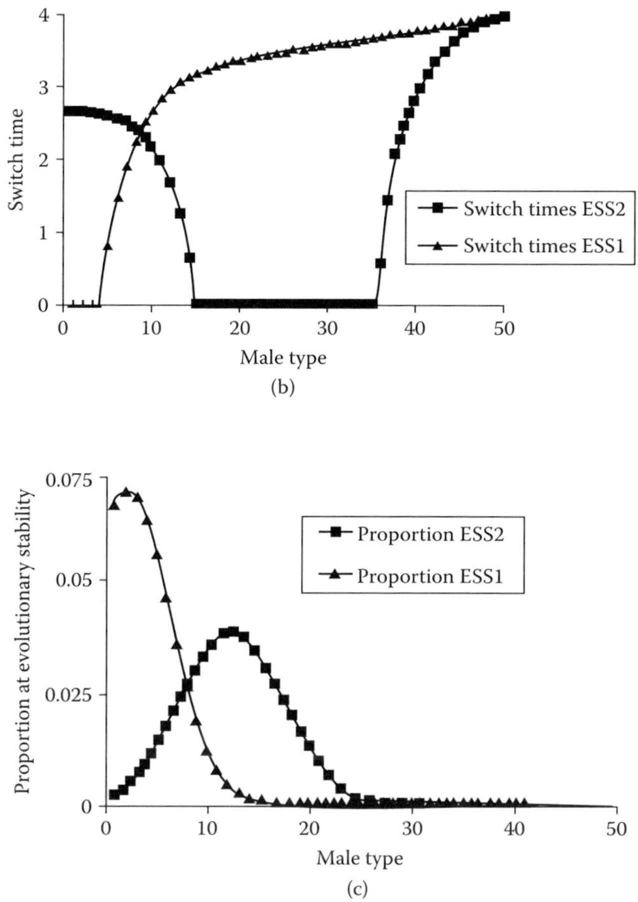


**FIGURE 4.6** (Continued).

shows, such a solution is indeed possible (ESS1). But the figure shows that it is also possible to find an ESS at which females prefer intermediate males and tend to be reluctant to mate with either very good or very poor males (ESS2). This illustrates the sexy sons effect: at this endpoint females prefer to mate with certain males even though this results in fewer offspring. The disadvantage of a small number of offspring is compensated for by the fact that the offspring (and their descendants) are attractive.

We have illustrated a general phenotypic approach to intergenerational effects, including cultural transmission, in which preference determines value and attractiveness emerges in a self-consistent way. This approach makes it easy to look at cases with many phenotypes. It has been claimed that intergenerational effects can only be handled using a genetic approach and that a phenotypic approach will not be valid. For example, Kirkpatrick (1985) says,

"At any evolutionary equilibrium, the forces acting on the genes must equilibrate within each generation. The sexy son hypothesis, however, is based on the premise that a loss in fitness to females caused by mate choice can be reconstituted in future generations by their male descendants. Unfortunately for this argument, there is no mechanism by which Mendelian inheritance can recoup fitness deficits from earlier generations" (pp. 801–802). and "Female choice may evolve in response to immediate, material forces that affect female survival or fecundity, but not to any force that can be attributed to the fitness of descendants" (p. 802).

If a phenotypic model does not include state then Kirkpatrick's claim is correct. But if state is included, then effects that last for more than one generation into the future, including the effects of cultural transmission, can be represented by accounting for not just the number of offspring but also their state.

## 4.8 DISCUSSION

We have presented a general and powerful framework for identifying the endpoints of evolution under selection. The framework can handle sequences of decisions as well as single actions. A central feature of our approach is the dependence of reproductive value on state. Individuals may differ in state in a variety of ways; some state variables (e.g., energy reserves) may change relatively quickly, whereas others may change quite slowly or be fixed (e.g., differences in quality or adult size). There are several ways in which differences in state can be important.

### 4.8.1 State-Dependent Decisions

Because the consequences of performing an action may depend on an animal's state, it is advantageous for an animal to make state-dependent decisions. For example, consider an animal that is foraging in an environment in which it is at risk of being killed by a predator. If its energy reserves fall to zero then it dies of starvation. McNamara and Houston (1990) investigate the optimal strategy when the animal can choose between two foraging patches that differ in terms of energetic gain and danger of predation. Optimal choice of foraging area depends on state; the option with better intake rate but greater risk is chosen when reserves are low, whereas the safer option with the lower intake rate is chosen when reserves are high. This effect is general (McNamara, 1990) and means that an animal's choices may change as a result of changes in its state. McNamara and Houston (1990) show that the optimal state-dependent strategy achieves a much higher survival probability than the best state-independent strategy. Schuck-Paim et al. (2004) present data from starlings that shows that such state-dependent decisions can result in behavior that appears to be irrational.

### 4.8.2 Phenotypic Plasticity

It is clearly important for an animal to adjust its behavior to the environment in which it finds itself. The ability of an organism to determine its phenotype according to its environment is known as phenotypic plasticity. Organisms are often particularly plastic during development. Because the environment can be regarded as an aspect of the animal's state, phenotypic plasticity is an example of state-dependent decision making. How an organism's phenotype depends on its environment is known as its norm of reaction. In our terms the norm of reaction is a state-dependent strategy. Not all the norms of reaction observed in nature will be adaptive. Houston and

McNamara, 1992 point out that in general the optimal norm of reaction cannot be found by taking the optimal behavior for each environment in isolation. Fitness is determined by strategies (i.e., contingent rules) not by actions, so that the best action to take in one state may depend on the actions that would have been taken were the animal in another state (Houston and McNamara, 1992).

### 4.8.3 Intergenerational Effects

As we have demonstrated, intergenerational effects are handled naturally by a state-dependent approach. Many people have more or less independently developed life history models that include intergenerational effects (e.g., Barro and Becker, 1989; Grafen, 1998; Grafen, 2000; Lee, 2003; McNamara and Houston, 1996; Robson and Kaplan, 2003; Rogers, 1990, 1995). Lee (2003) and Robson and Kaplan (2003) focus on the transfer of energy from parents to offspring. In this case a constraint based on the conservation of energy is appropriate. A similar constraint applies in the case of wealth (Rogers, 1995) but does not necessarily hold if status or skill is being transferred (Kaplan et al., 1995; McNamara and Houston, 1996). Models based on niche construction (Odling-Smee et al., 2003) also deal with intergenerational effects.

Intergenerational models have implications for how wealth is inherited. Rogers (1995) assumes that wealth is allocated as equally as possible to all offspring. Mace (1998) looks at the evolution of both the spacing of births and the allocation of wealth. We have emphasized that if an evolutionary approach to intergenerational effects is adopted, it is not adequate to assess fitness by counting offspring, and it may even not be adequate to count grandchildren.

### 4.8.4 Side Effects

Mulder (1998) defines the demographic transition as a " ... radical decline in number of offspring that parents produce, despite overall increases in the availability of resources." After the transition, the positive correlation between wealth and fertility is destroyed. A fundamental issue is whether behavior after the transition is adaptive given the current conditions or is a side effect of the previously adaptive rules being followed in a new environment (Grafen, 1998, 2000; Mulder, 1998). We suspect that state-dependent rules will have the potential to produce a wide range of side effects, and hence the demographic transition may not be the product of a currently adaptive strategy.

### 4.8.5 Negotiation

By interacting with each other, individuals can gain information about each other's state; there is no need to do this if all individuals are the same. We now outline some contexts in which such interactions can have important implications for the outcome of evolutionary games.

Consider a male and female that have mated. How hard should they work to raise the resulting young? The probability that the young survive to become independent is likely to increase as the total effort of the parents increases. The time and energy that a parent devotes to the young will limit that parent's reproductive success from other reproductive opportunities. There are various reasons for this. Caring may mean that an individual misses mating opportunities or suffers a reduction in growth or survival (for a review, see Houston et al., 2005). So the benefit of care to a parent depends on the effort of both parents, whereas the cost to a parent depends on just its own effort. Thus as Trivers (1972) and Maynard Smith (1977) point out, a conflict of interest will be present. This view replaced the earlier view of a cooperative enterprise. The cooperative view remained in anthropology for longer — but is now being replaced by an approach based on conflict, e.g., Bird (1999).

The standard approach in behavioral ecology is that adopted by Houston and Davies (1985), who argued that parental effort should be the stable outcome of a game. They found the equilibrium on the assumption that there was no interaction between the parents. This approach makes sense if all parents are the same; there is then no reason to respond to the effort of the other parent because its effort can be known in advance. In contrast, if parents differ in quality then it becomes important for parents to interact.

McNamara et al. (1999) consider a model in which there is a cost-free period of negotiation that produces the final effort of each parent. Rather than looking at stable levels of effort, it is now necessary to look at stable negotiation rules. McNamara et al. show that the level of effort that results from adopting the stable negotiation rule is less than the Houston–Davies effort.

### 4.8.6 Effect of Variability

The standard analytic approach to evolutionary games assumes either that all individuals are the same (monomorphic population) or that a polymorphism is maintained because each type present has the same fitness. Various results have shown that the presence of additional variability in a population (i.e., variability not maintained because variants have equal fitness) can increase the stable level of cooperation (e.g., Crowley and Sargent, 1996; Fishman et al., 2001; McNamara et al., 2004a; Page and Nowak, 2001; Sherratt and Roberts, 2001). To see why this is so, consider a game between two individuals like the Prisoner's Dilemma game in which there is a tension between trying to obtain the benefit that results if both cooperate while avoiding the undesirable payoff that is obtained as a result of cooperating when the other player defects (i.e., does not cooperate). If there is only a single play of this game, then the stable outcome is for both players to defect. If there is a fixed number of plays that is known to both players, then by working backwards from the last play it can be shown that defecting on all plays is the stable outcome. In many games where this sort of argument holds, people are more cooperative than predicted (e.g., Camerer, 2003; Colman, 2003). As Luce and Raiffa (1957) point out, this may be associated with variability in the level of cooperation. If people expect opponents to be drawn from a population in which individuals differ in their level of cooperation, then it can make sense to start an interaction by being cooperative because if the opponent is cooperative the benefit from prolonged cooperation by both players will more than compensate for cooperating for one play against an opponent that defects.

## ACKNOWLEDGMENTS

We thank Kevin Laland, Sean Rands, and Jonathan Wells for comments and the Leverhulme Trust for support.

## REFERENCES


Barro, R.J., and Becker, G.S., Fertility choice in a model of economic-growth, *Econometrica*, 57, 481–501, 1989.

Bird, R., Cooperation and conflict: the behavioral ecology of the sexual division of labor, *Evol. Anthropol.*, 8, 65–75, 1999.

Braza, F., Human prenatal investment affected by maternal age and parity, *Hum. Ecol.*, 32, 163–175, 2004.

Brown, G.R., Sex-biased investment in nonhuman primates: can Trivers & Willard's theory be tested, *Anim. Behav.*, 61, 683–694, 2001.

Buss, D.M., et al., Adaptations, exaptations, and spandrels, *Am. Psychol.*, 53, 533–548, 1998.

Camerer, C., *Behavioral game theory*, Princeton University Press, Princeton, NJ, 2003.

Caswell, H., Optimal life histories and the maximization of reproductive value — a general theorem for complex life-cycles, *Ecology*, 63, 1218–1222, 1982.

Caswell, H., *Matrix population models*, Sinauer, Sunderland, MA, 1989.

Charlesworth, B., *Evolution in age-structured populations*, Cambridge University Press, Cambridge, 1994.

Clark, C.W., and Mangel, M., *Dynamic state variable models in ecology*, Oxford University Press, Oxford, 2000.

Colman, A.M., Cooperation, psychological game theory, and the limitations of rationality in social interaction, *Behav. Brain Sci.*, 26, 139–198, 2003.

Crowley, P.H., and Sargent, R.C., Whence tit-for-tat? *Evol. Ecol.*, 10, 499–516, 1996.

Dennett, D.C., *Darwin's dangerous idea*, Simon & Schuster, New York, 1995.

Drent, R.H., and Daan, S., The prudent parent: energetic adjustments in avian breeding, *Ardea*, 68, 225–252, 1980.

Enquist M., et al., Spectacular phenomena and limits to rationality in genetic and cultural evolution, *Philos. Trans. Roy. Soc. Lond. Series B*, 357, 1585–1594, 2002.

Eshel, I., Evolutionary and continuous stability, *J. Theor. Biol.*, 103, 99–111, 1983.

Fisher, R.A., The genetical theory of natural selection, Oxford University Press, Oxford, 1930.

Fishman, M.A., Lotem, A., and Stone, L., Heterogeneity stabilizes reciprocal altruism interactions, *J. Theor. Biol.*, 209, 87–95, 2001.

Gillespie, J.H., Natural selection for variances in offspring numbers: a new evolutionary principle, *Am. Nat.*, 111, 1010–1014, 1977.

Gould, S.J., The exaptive excellence of spandrels as a term and prototype, *Proc. Natl. Acad. Sci.* USA, 94, 10750–10755, 1997.

Gould, S.J., *The structure of evolutionary theory*, Harvard University Press, Cambridge MA, 2002.

Gould, S.J., and Lewontin, R.C., The spandrels of San Marco and the panglossian paradigm: a critique of the adaptationist programme. *Proc. Roy. Soc. Lond.*, B 205, 581–598, 1979.

Grafen, A., The logic of divisively asymmetric contests — respect for ownership and the Desperado effect, *Anim. Behav.*, 35, 462–467, 1987.

Grafen, A., On the use of data on lifetime reproductive success, In *Reproductive success*, Clutton-Brock, T.H., Ed., Chicago University Press, Chicago, 1988.

Grafen, A., Fertility and labour supply in Femina economica, *J. Theor. Biol.*, 194, 429–455, 1998.

Grafen, A., Formal Darwinism, the individual-as-maximizing-agent analogy and bet-hedging, *Proc. Roy. Soc. Lond. Series B*, 266, 799–803, 1999.

Grafen, A., A biological approach to economics through fertility, *Econ. Lett.*, 66, 241–248, 2000.

Gyimah, S.O., and Fernando, R., The effects of infant deaths on the risk of subsequent births: a comparative analysis of DHS data from Ghana and Kenya, *Soc. Biol.*, 49, 44–57, 2002.

Hammerstein, P., Darwinian adaptation, population genetics and the streetcar theory of evolution, *J. Math. Biol.*, 34, 511–532, 1996.

Hampton, S.J., Domain mismatches, scruffy engineering, exaptations and spandrels, *Theor. Psychol.*, 14, 147–166, 2004.

Heino, M., Metz, J.A.J., and Kaitala, V., Evolution of mixed maturation strategies in semelparous life histories: the crucial role of dimensionality of feedback environment, *Phil. Trans. Roy. Soc. Lond. Series B*, 352, 1647–1655, 1997.

Houston, A.I., Are the spandrels of San Marco really panglossian pendentives? *Trend. Ecol. Evol.*, 12, 125–125, 1997a.

Houston, A.I., Natural selection and context-dependent values, *Proc. Roy. Soc. Lond. Series B*, 264, 1539–1541, 1997b.

Houston, A.I., and Davies, N.B., The evolution of cooperation and life history in the dunnock Prunella modularis, In *Behavioural ecology*, Sibly, R.M., and Smith, R.H., Eds., Blackwell Scientific Publications, Oxford, 471–487, 1985.

Houston, A.I., and McNamara, J.M., Evolutionarily stable strategies in the repeated hawk dove game, *Behav. Ecol.*, 2, 219–227, 1991.

Houston, A.I., and McNamara, J.M., Phenotypic plasticity as a state-dependent life-history decision, *Evol. Ecol.*, 6, 243–253, 1992.

Houston, A.I., and McNamara, J.M., *Models of adaptive behavior*, Cambridge University Press, Cambridge, 1999.

Houston, A.I., and McNamara, J.M., A self-consistent approach to paternity and parental effort, *Philos. Trans. Roy. Soc. Lond. Series B*, 357, 351–362, 2002.

Houston, A.I., Szekely, T., and McNamara, J.M., Conflict between parents over care, *Trend. Ecol. Evol.*, 20, 33–38, 2005.

Hutchinson, J.M.C., and Gigerenzer, G., Simple heuristics and rules of thumb: where psychologists and behavioral biologists might meet. *Behav. Process.*, 69, 97–124, 2005.

Iwasa, Y., Pomiankowski, A., and Nee, S., The evolution of costly mate preferences II. The "handicap" principle, *Evolution*, 45, 1431–1442, 1991.

Kacelnik, A., and Bateson, M., Risky theories — the effects of variance on foraging decisions, *Am. Zool.*, 36, 402–434, 1996.

Kaplan, S.H., et al., Fertility and fitness among Albuquerque men: a competitive labour market theory, In *Human Reproductive Decisions: Biological and Social Perspectives*, Dunbar, R.I.M., Ed., Macmillan Press, Basingstoke, United Kingdom, 1995.

Kirkpatrick, M., Evolution of female choice and male parental investment in polygynous species: the demise of the "sexy son", *Am. Nat.*, 125, 788–810, 1985.

Kokko, H., et al., The sexual selection continuum, *Proc. Roy. Soc. Lond. Series B*, 269, 1331–1340, 2002

Koziel, S., and Ulijaszek, S.J., Waiting for Trivers and Willard: do the rich really favor sons? *Am. J. Phys. Anthropol.*, 115, 71–79, 2001.

Laland, K.N., On the evolutionary consequences of sexual imprinting, *Evolution*, 48, 477–489, 1994.

Laland, K.N., Odling-Smee, J., and Feldman, M.W., Niche construction, biological evolution, and cultural change, *Behav. Brain Sci.*, 23:131–146, 2000.

Laland, K.N., Odling-Smee. J., and Feldman, M.W., Cultural niche construction and human evolution, *J. Evol. Biol.*, 14, 22–33, 2001.

Lauder, G.V., The argument from design, In *Adaptation*, Rose, M.R., and Lauder, G.V., Eds., Academic Press, San Diego, 55–91, 1996.

Lee, R.D., Rethinking the evolutionary theory of aging: transfers, not births, shape social species, *Proc. Natl. Acad. Sci. USA*, 100, 9637–9642, 2003.

Leimar, O., Life-history analysis of the Trivers and Willard sex-ratio problem, *Behav. Ecol.*, 7, 316–325, 1996.

Lewontin, R.C., and Cohen, D., On population growth in a randomly varying environment, *Proc. Natl. Acad. Sci. USA*, 62, 1056–1060, 1969.

Luce, R.D., and Raiffa, H., *Games and decisions,* John Wiley, New York, 1957.

Mace, R., The coevolution of human fertility and wealth inheritance strategies, *Philos. Trans. Roy. Soc. Lond. Series B*, 353, 389–397, 1998.

Maynard Smith, J., Parental investment: a prospective analysis, *Anim. Behav.*, 25, 1–9, 1977.

Maynard Smith, J., Evolution and the theory of games, Cambridge University Press, Cambridge, 1982.

McNamara, J.M., The policy which maximizes long-term survival of an animal faced with the risks of starvation and predation, *Adv. Appl. Probability*, 22, 295–308, 1990.

McNamara, J.M., State-dependent life-history equations, *Acta Biotheor.*, 41, 165–174, 1993.

McNamara, J.M., Multiple stable age distributions in a population at evolutionary stability, *J. Theor. Biol.*, 169, 349–354, 1994.

McNamara, J.M., Implicit frequency dependence and kin selection in fluctuating environments, *Evol. Ecol.*, 9, 185–203, 1995.

McNamara, J.M., Risk-prone behavior under rules which have evolved in a changing environment, *Am. Zool.*, 36, 484–495, 1996.

McNamara, J.M., Barta, Z., and Houston, A.I., Variation in behavior promotes cooperation in the Prisoner's Dilemma game, *Nature*, 428, 745–748, 2004a.

McNamara, J.M., Gasson, C.E., and Houston, A.I., Incorporating rules for responding into evolutionary games, *Nature,* 401, 368–371, 1999.

McNamara, J.M., and Houston, A.I., The application of statistical decision theory to animal behavior, *J. Theor. Biol.*, 85, 673–690, 1980.

McNamara, J.M., and Houston, A.I., The common currency for behavioral decisions, *Am. Nat.*, 127, 358–378, 1986

McNamara, J.M., and Houston, A.I., Starvation and predation in a patchy environment, In *Living in a patchy environment*, Shorrocks, B., and Swingland, I.R., Eds., Oxford University Press, Oxford, 1990.

McNamara, J.M., and Houston, A.I., Risk-sensitive foraging — a review of the theory, *Bull. Math. Biol.*, 54, 355–378, 1992.

McNamara, J.M., and Houston, A.I., State-dependent life histories, *Nature*, 380, 215–221, 1996.

McNamara, J.M., and Houston, A.I., Credible threats and promises, *Philos. Trans. Roy. Soc. Lond. Series B*, 357, 1607–1616, 2002.

McNamara, J.M., et al., Quantifying male attractiveness, *Proc. Roy. Soc. Lond. Series B*, 270, 1925–1932, 2003.

McNamara, J.M., Mace, R.H., and Houston, A.I, Optimal daily routines of singing and foraging in a bird singing to attract a mate, *Behav. Ecol. Sociobiol.*, 20, 399–405, 1987.

McNamara, J.M., et al., A dynamic game-theoretic model of parental care, *J. Theor. Biol.*, 205, 605–623, 2000.

McNamara, J.M., et al., The effects of background mortality on optimal reproduction in a seasonal environment, *Theor. Popul. Biol.*, 65, 361–372, 2004b.

Mesterton-Gibbons, M., Ecotypic variation in the asymmetric Hawk-Dove game — when is bourgeois an evolutionarily stable strategy? *Evol. Ecol.*, 6, 198–222, 1992.

Metz, J.A.J., Nisbet, R.M., and Geritz, S.A.H., How should we define fitness for general ecological scenarios? *Trend. Ecol. Evol.*, 7,198–202, 1992.

Mousseau, T.A., and Fox, C.W., The adaptive significance of maternal effects, *Trend. Ecol. Evol.,* 13:403–407, 1998.

Mulder, M.B., The demographic transition: are we any closer to an evolutionary explanation? *Trend. Ecol. Evol.*, 13, 266–270, 1998.

Mylius, S.D., and Diekmann, O., On evolutionarily stable life histories, optimization and the need to be specific about density dependence, *Oikos*, 74, 218–224, 1995.

Neff, B.D., and Sherman, P.W., Behavioral syndromes versus Darwinian algorithms, *Trend. Ecol. Evol.,* 19, 621–622, 2004.

Odling-Smee, F.J., Laland, K.N., and Feldman, M.W., *Niche construction: the neglected process in evolution*, Princeton University Press, Princeton, NJ, 2003.

Page, K.M., and Nowak, M.A., A generalized adaptive dynamics framework can describe the evolutionary Ultimatum Game, *J. Theor. Biol.*, 209, 173–179, 2001.

Palloni, A., and Rafalimanana, H., The effects of infant mortality on fertility revisited: new evidence from Latin America, *Demography,* 36, 41–58, 1999.

Pomiankowski, A., Iwasa, Y., and Nee, S., The evolution of costly mate preferences: I. Fisher and biased mutation, *Evolution*, 45, 1422–1430, 1991.

Real, L., and Caraco, T., Risk and foraging in stochastic environments, *Annu. Rev. Ecol. Systemat.*, 17, 371–390, 1986.

Reinhold, K., Maternal effects and the evolution of behavioral and morphological characters: A literature review indicates the importance of extended maternal care, *J. Heredity,* 93, 400–405, 2002.

Robson, A.J., and Kaplan, H.S., The evolution of human life expectancy and intelligence in hunter-gatherer economies, *Am. Econ. Rev.*, 93, 150–169, 2003.

Rogers, A.R., Evolutionary economics of human reproduction, *Ethol. Sociobiol.*, 11, 479–495, 1990.

Rogers, A.R., For love or money: the evolution of reproductive and material motivation, In *Human reproductive decisions: Biological and Social Perspectives*, Dunbar, R.I.M., Ed., Macmillan Press, Basingstoke, United Kingdom, 1995.

Rowe, L., Ludwig, D., and Schluter, D., Time, condition, and the seasonal decline of avian clutch size, *Am. Nat.*, 143, 698–772, 1994.

Schaffer, W.M., On reproductive value and fitness, *Ecology,* 62, 1683–1685, 1981.

Schuck-Paim, C., Pompillo, L., and Kacelnik, A., State-dependent decisions cause apparent violations of rationality in animal choice, *PLOS Biol.*, 2, 2305–2315, 2004.

Sherratt, T.N., and Roberts, G., The importance of phenotypic defectors in stabilizing reciprocal altruism, *Behav. Ecol.,* 12, 313–317, 2001.

Sih, A., Bell, A., and Johnson, J.C., Behavioral syndromes: an ecological and evolutionary overview, *Trend. Ecol. Evol.*, 19, 372–378, 2004.

Sih, A., et al., Behavioral syndromes: an integrative overview, *Q. Rev. Biol.*, 79, 241–277, 2004.

Taylor, H.M., et al., Natural selection of life history attributes: an analytical approach, *Theor. Popul. Biol.*, 5, 104–122, 1974.

Taylor, P.D., Allele-frequency change in a class-structured population, *Am. Nat.* 135, 95–106, 1990.

Trivers, R.L., Parental investment and sexual selection, In *Sexual selection and the descent of man*, Campbell, B., Ed., Aldine, Chicago, 136–179, 1972.

Trivers, R.L., and Willard, D.E., Natural-selection of parental ability to vary sex-ratio of offspring, *Science*, 179, 90–92, 1973.

Weatherhead, P.J., and Robertson, R.J., Offspring quality and the polygyny threshold: "the sexy son hypothesis," *Am. Nat.*, 113, 349–356, 1979.

Weissing, F.J., Genetic versus phenotypic models of selection: Can genetics be neglected in a long-term perspective? *J. Math. Biol.*, 34, 533–555, 1996.

Ziman, J., Ed., Technological innovation as an evolutionary process, Cambridge University Press, Cambridge, 2000.

# 5 An Agnostic View of Memes

*Robert Aunger*

## CONTENTS



## ABSTRACT

A number or scholars have recently suggested that culture evolves based on the replication of units of information called *memes* (by analogy to genes as the units of genetic inheritance). I argue that culture may evolve, but need not be based on a replication process, and that what we know about communication makes it difficult to believe that most social learning involves the replication of information. It is also the case that in humans much of what is interesting about cultural evolution does not involve the interpersonal communication of information. Instead, much of what is cultural depends on the evolution of artifacts — in a word, technology. Technology not only makes certain kinds of communication possible (e.g., mass media, e-mail) but also serves as the primary repository of cultural information (libraries, the WWW). Thus, even if memes do exist, they may play a relatively limited role in explaining the dynamics of cultural change in modern societies.

It has long been noted that there is a tendency for the beliefs and values of one generation to resemble those of the previous generation. It is widely thought that this is due to a kind of inheritance process, whereby the values of a group are passed from one time to the next as a way of preserving their "culture." There are (at least) three causal mechanisms that can explain the recurrence of cultural traits (Aunger, in press). Recurrence takes place through 1) strong individual learning biases, 2) population-level normalizing effects on what is adopted, and 3) replicator-based inheritance. Evolutionary psychologists (e.g., Tooby and Cosmides, 1992) tend to believe that the first mode is most likely; cultural evolutionists (e.g., Henrich and Boyd, 1998) emphasize the second possibility: that conformist transmission, or the tendency to do what others are doing, is responsible for intergenerational cultural similarity. The clear position that the field of memetics takes — the point that separates memetics from all the other positions on cultural evolution — is that social learning primarily involves the replication of particles of information, called *memes*

(Dawkins, 1976; Dennett, 1995; Blackmore, 1999; cf. Gil-White, 2005). Thus, memetics is associated with the third possible means of accounting for cultural continuity. The central hypothesis of memetics is the claim that cultural replicators (memes) are duplicated through the process of social learning, particularly imitation. My strategy in this chapter will simply be to work out what this implies, to take the meme-as-cultural-replicator position as far as it will go — but no further. I do not assume (unlike the dominant memeticists, Dennett, 1997; Blackmore, 1999; and Distin, 2004) that memes exist. In fact, *meme*, as it is currently used, is a concept I consider too diffuse to be subject to scientific testing, much less to support the claim that memes have played a crucial role in the evolution of large brains and human society (Blackmore, 1999).

The several schools of thought noted above all agree that cultural change is an evolutionary process (see Laland and Brown, 2002 for an overview). What they disagree about is the mechanism. What most people don't realize — including distinguished figures in evolutionary theory, such as Richard Dawkins (1976) — is that evolution does not depend on replication. A Darwinian process is defined by descent with modification (Lewontin, 1970). It so happens that the quintessential, prototypical evolutionary process, the origin and regeneration of species, is based on the inheritance of genes, which are replicators. But for cultural change to be an evolutionary process, it is not necessary to hypothesize that memes exist. Cultural evolution could be based on a blending inheritance rule, such as averaging the values exhibited by multiple cultural "parents" (Boyd and Richerson, 1985). Blending inheritance has been proposed as a model of biological evolution as well: Galton (1898) proposed (before Mendel's law was rediscovered) that an individual's genetic makeup was a combination of contributions from parents, grandparents, and other ancestors. This process of convergence on the mean would still be Darwinian because it exhibits descent with modification, but it would not preserve the identity of particular bits of information through transmission.

Most observers agree that cultural information is inherited through a process of social learning. Social learning in humans often involves communication, verbal or nonverbal. But our best current models of communication (Sperber and Wilson, 1995; Grice, 1989) are consistent with any of the three models of cultural change outlined above. We, therefore, cannot just study the way people talk to each other to discriminate between the evolutionary psychological, cultural selectionist, and memetic approaches to cultural evolution because communication models are not specific about how the brain treats information it acquires from social exchange. We have to know more about how communication works — how what I have in my head comes to resemble what you have in your head — before we can argue that the process is one in which a particular bit of information gets replicated, and hence can confirm that memes exist.

To determine if memes exist, it is also necessary to have a firm idea of what replication involves. Replication can be defined as a special relationship between a source and a copy such that four conditions hold (Aunger, 2002):

- Causation: The source must play an active role in bringing about the conditions that lead to a copy being made.
- Similarity: The source and copy must resemble each other in relevant respects.

- Information Transfer: What makes the copy similar to the source must be derived from the source.
- Duplication: The source and copy must coexist for some time.

In the Dawkins/Blackmore/Dennett tradition, cultural information can be present in any behavior or artifact, as well as in minds, because people can extract that information from any of these forms and copy it for themselves (e.g., through a generalized notion of imitation, *in sensu* Blackmore, 1999). In standard memetics, then, replication is considered to be a much more general process than defined above. This allows memes to be instantiated in behavior, artifacts, or ideas.

Other memeticists argue that memes are restricted to particular kinds of physical forms. For Distin (2004), for example, memes are restricted to representational objects. Thus, some artifacts (such as books and CDs) can house memes, but nonrepresentational artifacts such as everyday objects (spoons, cars) cannot. Some behavior can be symbolic, as when it is ritualized (e.g., dance) or when two lights flashing on a hilltop mean the British are invading (as in the Paul Revere myth), and thus embody memes, whereas everyday behavior cannot. In this case, memes can exist in many forms, but they are all symbolic forms, so this is a more restricted manifestation than standard memetics because most behavior is not symbolic and only a few kinds of artifacts are. The same basic problem remains, however: that memes must be able to replicate from any given symbolic form to any other. Thus, memes can move from a mind to a CD and back again.

It is true that any replicator can be thought of as information. Of course, information must take a form or have a material instantiation. But in many cases, there can be multiple representations of a given bit of information. For example, it is possible for a gene sequencer to store the information about a particular gene in a computer, which is then reconstituted using a different source of DNA by another machine. So is it legitimate to think of a gene being replicated in this fashion? What about the following example. Say a piece of paper on which a phrase is typed is scanned, then the phrase is converted to a character sequence by character recognition software, and then stored in computer memory later to spoken aloud by another piece of software so that someone can hear the phrase and store it in their brain. Do the piece of paper and the brain have the "same" information? And has the "same" meme gone from a piece of paper to a brain — is this an instance of replication?

These are central questions for memetics. Dawkins, Blackmore, Dennett, and Distin answer "yes" to both questions. For all of these scholars, the crucial fact is that symbolic information has a protean ability to take multiple forms without its essential nature being changed. The fancy software, lab equipment, and human activity involved in producing the gene copy or meme copy in these cases can be thought of as very complex replication-assistance machines. So a meme with the content the "hello" can be in the representational form of words on paper, a spoken phrase, or the idea in someone's brain. Identifying a *particular* meme (a token of a given type) only requires being specific about its informational content and the form in which it is represented.

I argue that there are dangers in taking this "pure information" route. Why is it not possible to argue that memes exist only in brains and that these other forms are signals of various kinds (even written text, or digital memory)? Even if the information can be considered to be the same, only in certain contexts does the same information have the ability to instigate the process through which its content is duplicated. For example, only if the physical chain of amino acids is present inside a cell, with everything that implies, can a DNA molecule reproduce itself. The same is true of other replicators: the protein replicators, prions, require contact with other prions inside bodies; computer viruses only work when present in computers using a specific operating system. The process by which a meme replicates itself can become very complex indeed, involving transfers to artifacts of various kinds (as in the example above) for varying lengths of time prior to finding its proper form again in another brain. A pattern of ink on paper can be the source material, the template, for multiple replications. However, it cannot cause itself to be used as a template. Genes have this unique ability only when inside organisms: they can produce proteins which then read off the gene sequence and construct a copy of that gene.

In essence, my argument is that a replicator can take many forms, but only one form is active; the other forms are inert. Thus, given the criteria for "true" replication above, which includes an ability to play a causal role in the duplication of information, it turns out that replication is a sufficiently complex operation that it cannot proceed on multiple substrates in multifarious contexts as supposed by most memeticists. As a result, memes cannot be produced in the form of behavior and ideas and in artifacts like books or TV shows (contra Dawkins, 1976; Dennett, 1995; Blackmore, 1999).

If memes could exist in brains, in speech, and in artifacts, they would be the superheros of the replicator world, able to transform themselves into any shape or form at will, like Proteus of Greek mythology. Instead, memes must be confined to one physical substrate, just as their brethren, the biological replicators genes and prions, are. I thus argue that only one substrate can be associated with memes.

So which substrate is the proper "home" of memes? Artifacts do not satisfy the conditions of replication because they cannot transfer information directly from one to another; behavior fails the replication test because it does not duplicate (an imitated behavior usually occurs after the model behavior has been displayed, so there is a sequence of one behavior, then another). The only remaining candidate is that memes are mental objects.

If memes can only take one form and its physical substrate is tied to brains, then how can it be that memes make their way from inside one brain to another? Brains are not in direct physical contact with one another, so memes have to take some intermediary form to "jump the gap" between people — something like a spoken phrase. I propose that meme replication solves the gap problem through a process I call *instigation*. Memes as brain states produce motor behavior that produces signals, such as spoken language, which on being heard initiate a complex process in the receiving individual's brain that culminates in a new brain state that qualifies as being the same as the memetic state that produced the signal in the first place. The role of signals in this view is not to bring along the information that will be incorporated into the meme copy, but rather to set in motion the kind of local transformation that will result in the copy being produced. The signals themselves are not memes, nor need they contain representations of memes (i.e., the same information as the meme that produced it). Instead, the signal contains the information necessary for the signal receiver's brain to reconstruct the source meme *in situ*. This may be some trigger that relies on innate (and hence shared) structures in the receiving brain that can be depended on to do the job of reproducing the relevant memetic information from that trigger. Thus, the content of the signal does not necessarily map in any way onto the content of the meme. I thus define meme replication as a process in which such neuronal states are duplicated through a process instigated by signals (Aunger, 2002).

This perspective has a number of advantages. First, it adheres to my argument that memes can only be in one place — people's heads — not anywhere else. Second, it sidesteps Sperber's (2000) complaint that communication is too imprecise a process to result in replication and is consistent with the "poverty of the stimulus" postulate in communication theory (that the amount of information contained in messages is insufficient to define the inferences made on the basis of those messages). Signals are not memes, so the content of communication can be different from the consequences of it. As a result (third), it keeps memetic evolution Darwinian because what happens to signals (Lamarkian influences on memetic phenotypes) has no necessary influence on the "germ-line" of meme copies in brains. The information content of offspring memes is not affected by anything that happens to degrade or modify signals because they are built out of local materials in the receiving person's brain. The signal's message can just be "Go to work to produce meme X!" — nothing more. Fourth, it keeps the account of memetic "generations" clean: from brain to brain constitutes one generation, no matter how complex the intermediary process. Fifth, it keeps the evolutionary lineages clear: replications of a meme can be traced by "counting heads," not artifacts or behavior or anything else. So memes can persist, form long lineages, and explain cultural similarity over time. In effect, *all* the hurdles I believe meme theory must jump to explain cultural transmission are leapt with this one conceptual move: seeing signals as instigators of meme reconstruction rather than as alternative embodiments of memes themselves.

But what is the likelihood that information replicates in brains? The brain dynamicist school (e.g., Freeman, 1999; McCrone, 1999) believes that it is basically senseless to break the brain into bits. In their lingo, borrowed from complexity theory, the whole cortex flexes when stimulated by new data, then eventually settles down into a state of relaxation that represents the individual's refined response to some situation. The idea of active memes is antithetical to such a view because the meme argument requires that some parts of the brain be able to function as individualized, independent units of information storage, transmission, and replication.

For brain dynamicists, it is also difficult to see why information should be duplicated in the brain because they think of the brain as being involved in the production of a single, complex representation of everything you know or feel or think, and that huge complex changes every second. But if information is in any sense localized in the brain — if you can say that particular areas play important roles in the retrieval and representation of, say nouns or the ability to recognize faces (as brain scans nowadays seem to suggest) — then the role of particular nodes in the storage of that information for future use must be guaranteed. But a single node is a degenerate source — neurons die all the time. Backups become necessary, and that means more than one copy of the same information, which presumably occurs through a process of replication (just what memes are supposed to do) would have utility. Other eminent brain scientists are quite happy to think of the brain as composed of individuated circuits (Fuster, 2002 or Mountcastle, 1998), more in concert with the position of a memeticist.

The replication model I have suggested for memes does not prove that memes exist; it just gives them a *way* to exist. The model may make sense, but that does not force the objects of the model to be real, to be things-in-the-world. Given the current state of ignorance concerning how brains store and manipulate information, an agnostic position on the existence of memes is the only scientifically responsible position to take. My objective has simply been to make a diligent attempt to take what is a very poorly understood concept more scientific and, in particular, susceptible to empirical testing. I am content to challenge people more competent than myself to settle the matter one way or another.

I think it is quite possible that the brain does not work in the way necessary to be a storehouse and replicator of memes. In this case, then, I think we have to admit that cultural evolution proceeds by a mechanism other than replication. If the model I have sketched here — or a similar one consistent with the constraints I have developed for any model of meme replication — is not empirically validated, then the idea that memes exist must be abandoned because alternative approaches to meme replication have already been eliminated. That is, we will no longer have any reason to refer to memes in our explanations of sociocultural phenomena. I suggest this is a contribution to science since it would eliminate one hypothesis for how culture works. We will then be left with the interesting prospect of having to develop a new evolutionary theory of cultural change that does not depend on the replication of bits of information. Culture then would not be like biology in important respects. This conclusion, it seems to me, may present its own problems because gene-culture coevolutionary models would then be attempting to simultaneously describe one process that conforms to the replicator dynamic (genetic evolution), while the other does not (although Henrich and Boyd, 2002 have shown that a strong attractor-style transmission process works very much like a replicator process in terms of population dynamics).

Is this theory rich enough to explain the complexities of human culture? Probably not. Memes are just parasites on the normal system of thinking and communicating. Memes cannot explain big topics like the evolution of big brains or human culture (contrary to Blackmore, 1999) because not all thinking replicates information, nor does all communication. So, memes are the result of a potentially important (from an evolutionary point of view), but also a potentially empty, subset of cultural processes. Memetics, if it explains anything at all, can only explain how a parasite arose on the system for thinking and exchanging information in social species with relatively big brains. It is not, and cannot be, a general theory of culture.

In the end, then, it seems that worrying about the mechanism by which culture evolves may not be central to the explanation of how or whether culture differs from other evolutionary processes. It is also not central to the question that concerns (at least implicitly) most memeticists: how human culture differs from the culture of other animal species. So the issue of whether culture is underwritten by memes may be peripheral. Certainly, human culture has a very different evolutionary trajectory than nonhuman culture. No doubt any explanation of this difference must include reference to the greater ability of humans to communicate specialized information (i.e., language) and to learn socially (i.e., imitation). Either of these might be sufficient to explain why human culture is able to accumulate more rapidly than cultural traits in other animals. It also seems likely that the ability to store information in the environment, in the form of artifacts, is also crucial (Donald, 1991; Clark, 1998; Aunger, 2002).

There is a lot more happening in culture than just the replication of information. Explaining human culture requires attention to the independent dynamics of technology, not just changes in beliefs and values, because technology increasingly has the power to transform the physical environment in ways that influence both human culture and biology (Aunger, 2002).

What roles do artifacts play in cultural evolution? Certainly books and other representational artifacts play a crucial role in the evolutionary dynamics of modern culture. Other memeticists argue that information is still a meme when on paper or computer screen, even though it is not capable of initiating action independently. I argue, on the contrary, that the pattern of writing on paper is a "signal template" — a stable substrate from which signals can be created (Aunger, 2002). For example, photons can glance off a piece of paper, becoming "informed" about the content of the writing on it in a way that is interpretable by human brains as a signal; this sensory signal in turn, once inside the brain in the form of interneuronal signals, instigates meme replication by transforming the state of some neurons into a state similar to that of an area in the signal sender's brain. In this way, artifacts can serve as way stations for information content related to memes but do not serve as storehouses nor as creators of memes themselves, which are restricted to their natural place inside brains. Of course, artifacts have roles as stores of information (e.g., books) and as channels of information (the World Wide Web), whether or not the information involved is memetic. So artifacts are guaranteed members of the cultural evolutionary team whereas memes are only possible players.

Still, if memes exist, they have to be the cause of *some* important phenomenon — just like every other replicator operating in the world. Prions cause mad cow disease, computer viruses wreak havoc with the Internet, genes make us. If we root out memes from their hiding places, we will begin to understand how cultural change works, why people cling to certain beliefs that seem not to do them any good (because they're infected by brain parasites with their own interests), and have yet another example of a replicator to compare with those we already know about. This makes the meme project worth working on. I cannot establish that anything associated with this system are memes, but it is still fascinating to think that someday, we might be able to tell whether real people in real situations are in fact spreading memes around when they interact. Memes will only be proven to exist when (if?) the lab-based (rather than theoretical) neuroscientists become interested in this question.

## REFERENCES

Aunger, R., Three roads to cultural recurrence, In *The epidemiology of ideas*, Sperber, D., Ed., Open Court, Chicago, Forthcoming.

Aunger, R., *The electric meme: a new theory of how we think and communicate*, The Free Press, New York, 2002.

Blackmore, S., *The meme machine,* Oxford University Press, Oxford, 1999.

Boyd, R., and Richerson, P.J., *Culture and the evolutionary process*, University of Chicago Press, Chicago, 1985.

Boyd, R., and Richerson, P.J., Memes: universal acid or a better mouse trap? In *Darwinizing culture: the status of memetics as a science*, Aunger, R., Ed., Oxford University Press, Oxford, 143–162, 2000.

Clark, A., *Being there: putting brain, body, and world together again*, MIT Press, Cambridge, MA 1998.

Dawkins, R., *The selfish gene,* Oxford University Press, Oxford, 1976.

Dennett, D., *Darwin's dangerous idea: Evolution and the meanings of life*, Simon and Schuster, New York, 1995.

Distin, K., *The selfish meme: a critical reassessment,* Cambridge University Press, Cambridge, 2004.

Donald, M., *Origins of the modern mind: three stages in the evolution of cognition and culture,* Harvard University Press, Cambridge, 1991.

Freeman, W.J., *How brains make up their minds,* Weidenfeld and Nicolson, London, 1999.

Fuster, J.M., *Cortex and mind: unifying cognition,* Oxford University Press, Oxford, 2002.

Galton, F., A diagram of heredity, *Nature,* 57, 293, 1898.

Gil-White, F.J., Common misunderstandings of memes (and genes): the promise and the limits of the genetic analogy to cultural transmission processes, In *Perspectives on imitation: from mirror neurons to memes*, Hurley, S., and Chater, N., Eds., MIT Press, Cambridge, MA, 2005.

Grice, H.P., *Studies in the way of words,* Harvard University Press, Cambridge, MA, 1989.

Henrich, J., and Boyd, R., On modelling cognition and culture: why replicators are not necessary for cultural evolution, *J. Cognition Cult.,* 2, 87–112, 2002.

Henrich, J., and Boyd, R., The evolution of conformist transmission and between-group differences, *Evol. Hum. Behav.*, 19, 215–242, 1998.

Henrich, J., Boyd, R., and Richerson, P.J., Five common mistakes in cultural evolution, In *The epidemiology of ideas*, Sperber, D., ed., Open Court, Chicago, Forthcoming.

Laland, K., and Brown, G., *Sense and nonsense,* Oxford University Press, Oxford, 2002.

Lewontin, R.C., The units of selection, *Annu. Rev. Ecol. System.,* 1, 1–18, 1970.

McCrone, J., *Going inside: a tour round a single moment of consciousness,* Faber and Faber, London, 1999.

Mountcastle, V., *Perceptual neuroscience: the cerebral cortex*, Harvard University Press, Cambridge, MA, 1998.

Pinker, S., *How the Mind Works,* Penguin, London, 1997.

Richerson, P., and Boyd, R., *Not by genes alone,* University of Chicago Press, Chicago, 2004.

Sperber, D., *Explaining culture: a naturalistic approach,* Blackwell, Cambridge, MA, 1996. Sperber, D., An objection to the memetic approach to culture, in *Darwinizing Culture: The status of memetics as a science*, Aunger, R., Ed., Oxford University Press, Oxford, 163–74, 2000.

Sperber, D., and Wilson, D., *Relevance: communication and cognition*, 2nd ed., Blackwell, Oxford, 1995.

Tooby, J., and Cosmides, L., The psychological foundations of culture, in *The adapted mind,* Barkow, J.H., Cosmides, L., and Tooby, J., Eds., Oxford University Press, Oxford, 19–136, 1992.

# 6 Biological Ends and Human Social Information Transmission

*Jonathan Wells and Simon Strickland*

## CONTENTS



## ABSTRACT

Evolutionary models of human behavior conventionally use genetic replication as the currency by which to assess fitness. Recent developments in evolutionary theory have also led to focusing on the "gene's-eye" view of the organism, whereby individual genes are considered the unit on which natural selection ultimately acts via phenotype. However, organisms can also transmit nongenetic information to future generations. Whereas genomes are dismantled between each successive generation, such that successive descendants contain progressively fewer of the original ancestor's genes, nongenetic information need not degrade over time in the same way. We argue that nongenetic information is more successful than genetic information at conveying the uniqueness of each human. We suggest that individuals can transmit both their genes and their reputation to future generations, such that they can maximize fitness in both genetic and nongenetic currencies. We also consider the significance of the human tendency to project the "soul" into the future. This approach sees humans investing in competing ends, often to the detriment of the maximization of genetic fitness. Focusing on the phenomenon of personal identity allows biologists to address agency and purposive behavior, issues that have traditionally been difficult to incorporate into evolutionary biological approaches to human behavior.

## 6.1 INTRODUCTION

The function of social information transmission in humans is complex, due to its central role within the phenomenon of culture. In nonhuman species, behavior is generally assumed by biologists to be adaptive (Krebs and Davies, 1978). From a biological perspective, it might, therefore, be tempting to assume that all social information ultimately increases genetic fitness as defined by Hamilton (1964). This hypothesis can then be tested empirically (Laland and Kendall, 2003). However, with the application of dual-inheritance theory, it became clear that human cultural transmission can conflict with, rather than enhance, the maximization of fitness (Cavalli-Sforza and Feldman, 1981; Durham, 1991). If we define genetic replication as merely *one particular* currency in terms of which biological fitness may be assessed, the aim of this chapter is to identify other potential currencies. Broadly speaking, therefore, we are seeking to identify the different biological ends toward which human social information transmission is directed.

Traditionally, evolutionary biologists have been concerned with the transmission of genes over time. Genes are assumed to spread in the gene pool because they underlie traits that are adaptive. Over time a population of organisms comes to take on the characteristics of those individuals who breed (Dobhzansky and Boesiger, 1983), and the consolidation of adaptive traits manifests as adaptation. More recently, theorists have paid greater attention to the transmission of other biological phenomena over time. Of particular interest to those studying human behavior are the transmission first of nongenetic or cultural information (Lumsden and Wilson, 1981; Cavalli-Sforza and Feldman, 1981; Boyd and Richerson, 1985; Durham, 1991), and second of environmental niches that organisms create or maintain and then transmit to subsequent generations (Odling-Smee, Laland, and Feldman, 2003). These different types of biological information are all important when considering social behavior.

The very flexibility of complex social behavior makes it difficult to model from an evolutionary perspective. Adaptations, as discussed above, are passed on via genetic inheritance to future populations where, given a similar environment, they are likely to prove adaptive once more. However, complex behavior is less constrained by the genome and allows one of various options to be selected by the organism. Such behavior may either not be heritable (and hence not subject to natural selection) or may not be adaptive in the present environment (Caro and Borgerhoff Mulder, 1987).

Biologists can therefore identify as maladaptive behaviors that decrease genetic fitness (e.g., Laland, 1996; Laland and Williams, 1998), but the *biological ends* of apparently maladaptive behavior have been relatively ignored. Human behavior is not characterized by inflexible physical laws, but nor is it random. Successful biological models should generate testable predictions. The prediction that all human behavior is genetic-fitness-maximizing or adaptive (Alexander, 1980) appears to be invalid, and alternative predictions are required. This chapter, therefore, builds on recent work in evolutionary theory to improve understanding of the alternative fitness currencies of human social information transmission.

## 6.2 GENE-BASED MODELS OF BEHAVIOR

Evolutionary theory has undergone steady refinement in recent decades, incorporating, for example, the concept of inclusive fitness, genetic conflict theory, game theory, and group selection theory. These have in turn allowed the development of diverse and sophisticated approaches for the study of human behavior, as reviewed in Chapter 1. Although addressing different aspects of behavior, the majority of these approaches ultimately remain focused on the gene, which may be considered to be continually subject to natural selection because the environment is always changing (van Valen, 1973). Thus, many biologists seek evidence for either adaptations (e.g., sociobiology, evolutionary psychology) or adaptive behavior (e.g., human behavioral ecology), each of which enhances genetic fitness, or did so in an earlier environment.

Individual organisms are not driven by the goal of fitness, but by a range of proximate goals such as feeding, surviving, and mating. In humans, proximate goals have been proposed to include wealth and status. However, exclusively gene-based models of behavior still consider such proximate goals as subsidiary correlates of the ultimate fitness currency of reproductive success. For example, human behavioral ecologists test the hypothesis that the acquisition of power or wealth increases genetic fitness (Henrich and Gil-White, 2001).

When seeking to explain inconsistency between entirely gene-based models and observed human behavior, biologists have on occasion argued (1) that behavior may merely be neutral for reproductive success, such that genes underlying "unexpected" behavior are not selected out of the gene pool (Hinde, 1987); (2) that some human behavioral traits may be byproducts arising from selection of other traits (Gould and Lewontin, 1979); or (3) that humans are now in a different environment from that in which they evolved, and that it is, therefore, not surprising if not all behavior is fitness maximizing (Tooby and Cosmides, 1990). We argue that these explanations are unsatisfactory because many aspects of human behavior clearly decrease fitness, and if they are directed at new contemporary ends then these ends still require elucidation.

In humans, phenomena such as power, prestige, and wealth involve symbolism in the broad sense that these characteristics are represented or reflected by varieties of behavioral and cultural differentiation. Thus, we maintain that to understand the ends pursued by contemporary humans, it is necessary to incorporate symbolism into biological models.

## 6.3 SYMBOLIC REPRESENTATION

Social information transmission involves communication and interpretation. However, information can be stored and processed in different ways, and a major difference between humans and other species is the fundamental role of symbolic representation in our own species' behavior. Humans live in a dual world, comprising both the physical world that humans share with all species but also a symbolic world that is inhabited through symbolic representation — a virtual world constructed through shared understandings (Goldschmidt, 1990) that are achieved through the social transmission of information. A biological understanding of symbolic representation is therefore key to elucidating both human communication and human behavior more generally.

Symbolism arises merely from a particular way of classifying information, rather than differences in information content per se (Deacon, 1997). In nonhuman species, social communication involves the transmission of information with either no referential communication or a relatively simple referential relationship between the sign and the signified, although studies of primate, cetacean, and avian forms of communication often suggest a high level of sophistication. In contrast, symbolic representation among humans is based on a more complex semantic relationship and requires a shared, socially constructed reference against which information is evaluated. Despite this greater complexity, symbolism is an extremely efficient way to transmit complex information and facilitates the capacity for individuals to gain from others' experience without recourse to trial-and-error learning.

Symbolic phenomena are no less important or "real" to humans than physical phenomena, and they induce the pursuit of constructed goals, such as earning status, analogous to and sometimes correlated with the physiological goals that are attributed to all organisms by evolutionary biologists, such as obtaining food.

For over a century, the discipline of social anthropology has considered how the human social order is symbolically expressed. Humans categorize information symbolically and express it through symbolic language systems (Lowenthal, 1985). By using the past to structure the present, humans avoid the need to keep constructing the symbolic world from new experience, which would jeopardize the extent to which individual constructed worlds coincided (Lowenthal, 1985). Because differences between and within individuals are not necessarily "givens" of the physical world, humans make artificial demarcations and symbolically express them. For instance, different social roles are occupied by undergoing rites of passage (van Gennep, 1960), while the political and religious systems are expressed through ritual (Lewellen, 1983).

It is important not to equate the symbolic world with a purely relativist viewpoint. First, healthy humans generally share common physical senses and psychological capacities, so all individuals must construct symbolic worlds in the same way. Second, such worlds are primarily constructed socially, with each community broadly existing in a shared world. Thus, while subjective experience inevitably contributes to how individuals process information, our common cognitive capacities combined with our intensive social interaction ensure a high degree of unity in symbolic reification. Third, symbolic worlds converge due to the constraining effects of norms and conventions, which themselves influence how individuals process information. However, such constraints are also challenged by creativity, which reshapes and reforms social structures.

To incorporate symbolic phenomena within biological evolutionary models, it is necessary to use consistent terminology. The physical, chemical and biological sciences are in general characterized by consilience, whereby the models of each level (e.g., chemistry) are consistent with those of the level below (e.g., physics; Wilson, 1998). To improve consilience between biological and social anthropological models of behavior, it is necessary to address from a biological perspective the issue of how symbolic information is generated and processed. We first summarize the theoretical approaches used to address nongenetic information and then discuss how existing evolutionary models based on such theory have failed adequately to incorporate symbolic information.

## 6.4 REPLICATORS, DUAL INHERITANCE THEORY, AND MEMETICS

Genes may be considered as replicating units of biological information that part-construct a "survival machine" to bring about through reproduction their renewed presence in future generations (Dawkins, 1976). A succession of small changes in DNA structure, in combination with selection for those conferring advantages in gene function, is sufficient to generate modification of the genome over time, and the process has been conceptualized using the analogy of a blind watchmaker, a purposeless process that nevertheless "designs" for success over time (Dawkins, 1986).

Evolution can therefore be considered an algorithmic process, a logical process operating on its substrate (in this case, genes) and generating an outcome (selective survival; Dennett, 1995). Any trait that increases the spread of replicators must be preferentially selected by the algorithm, and the genome of an organism may be conceptualized as a "library" of information about the nature of selective pressures in previous generations (Plotkin, 1994). Such a metaphor must be used with great caution, for genes do not "control" development of the organism but constantly interact with environmental factors during the ontogenetic process (Oyama, 1985). Thus, organisms are "open systems" (Bertalanffy, 1967) in which genes only convey information through being embedded within self-organizing developmental systems (Oyama, 1985). Furthermore, the ancestors of humans have inhabited many different environments during their evolutionary history, and whether contemporary genes produce similar effects to those of ancestral genes depends on the stability of the environment over the same period. Nevertheless, the genome inevitably reflects the experience of past generations, and adaptations can broadly be considered as "knowledge" of past environments — for example, the bipedal phenotype of contemporary humans expressing genetic knowledge of past savannah habitats in which much hominid evolution occurred.

There is no reason the algorithm of natural selection might not be found operating on other substrates besides genes, and the notion of Darwinian selection has been used to model several biological phenomena, including development of the immune system and the acquisition of knowledge. Many authors have modeled the transmission of nongenetic information over time (see Table 1, Introduction). Recent workers in this area have tended to favor Dawkins's term the *meme* and have developed the approach known as *memetics* (see Chapter 1).

Although there is much that is valuable to evolutionary modelers in memetics, there remain several problems with the theory in its current form. First, there are many varieties of nongenetic information that are stored and transmitted in different ways. The use of a single term is problematic, as it must either ignore some types of information or force apparent consistency where none exists. Second, where nongenetic information lacks the long-term fidelity characteristic of genes, the application of replicator theory can have little value. Third, nongenetic information may be transmitted not only from one to many individuals, but also from many to one. This complicates the distinction between what is transmitted and what is stored and which of these entities is itself the meme (see Aunger this volume).

In view of these problems, we have preferred to employ a simple distinction between genetic and nongenetic information. It will be clear later in the chapter that we are concerned with one specific type of nongenetic information (symbolic information), and our arguments apply only to contexts in which symbolism is used. However, the main point we seek to make here is that symbolic phenomena merely represent a type of nongenetic information and can be incorporated in biological models such as dual-inheritance and memetic models.

Early dual-inheritance (also known as gene-culture coevolution) models considered how the transmission of nongenetic information might enhance the replication of genes (Lumsden and Wilson, 1981) with "culture held on a genetic leash." More recent models have suggested that social information transmission may be contradictory to the maximization of genetic fitness (Cavalli-Sforza and Feldman, 1981; Boyd and Richerson, 1985; Durham, 1991; Richerson and Boyd, 2004). Memetic models have proved useful at addressing the capacity of information to replicate in viral fashion (Lynch, 1996), whereas other models have been successful at modeling the transmission of cultural behaviors alongside genetic replication (Durham, 1991; Feldman and Laland, 1996). Collectively, this work has addressed many symbolic phenomena, including the effects of beliefs, conventions, and traditions on biological outcomes.

Dual-inheritance theory is now an established part of the evolutionary biology toolkit and has contributed to various issues such as the spread of genes for lactose tolerance in relation to subsistence patterns (Durham, 1991) and the relationship between subsistence practices and genes underlying resistance for malaria (Odling-Smee, this volume). The area that dual-inheritance theory addresses least successfully, and which might benefit from more comprehensive memetic models, is our understanding of immediate or short-term behavioral variability.

However, what both these approaches fail to incorporate is the concept of *agency*. Both dual-inheritance and memetic models treat the individual as a relatively passive entity, subjected to the influence of genetic and nongenetic information. Thus, there remains poor understanding of what drives much human behavior, in particular where it is, in genetic terms, nonfitness maximizing. The concept of agency is particularly important for understanding the pursuit of goals that are expressed symbolically.

## 6.5 AGENCY

"Agents act, and agency is the capability, the power, to be the source and originator of acts" (Rappaport and Overing, 2000). Ideas about personhood and volition date back to ancient Greek philosophy and were then further developed in the 18th century along the lines of free will, choice, and intentionality (Dobres and Robb, 2000). Weber differentiated agency from "mere animal behavior" on the basis of characteristics such as consciousness, reflection, intention, purpose, and meaning, each of which is in part socially constructed. In contrast, Durkheim emphasized the importance of the context in which action takes place, highlighting for example socially derived constraints or even coercions (Rappaport and Overing, 2000). Leach (1977) suggested that the essence of being human is to resent the domination of existing structures and to oppose them through creative forces such as defiance, reinterpretation, or imagination. It is this tension between social constructions and individual creativity that makes agency a fundamentally social entity.

Agency is now exploited as a flexible concept with many interpretations. Among the contexts where it is applied are the construction of the individual as a psychological entity, the reproduction and also the challenging of power structures, material creativity and the acquisition of technological skill, and goal-oriented behavior (Dobres and Robb, 2000). Such applications can on occasion be contradictory, and the concept remains controversial. Among problematic issues are whether the actual consequences of behavior were intended, whether agency is expressed by individuals or collectively by groups, and whether the material world is created by free-acting individuals or whether material culture rather constructs not only the world within which individuals act but also the people themselves (Dobres and Robb, 2000). Thus, the idealized notion of free-acting individuals is countered by the reality of people also being structured by institutions and beliefs beyond their conscious awareness or direct control (Bourdieu, 1977).

Agency is relevant to many areas of human behavior that are of importance for biologists, for example gender dynamics, the variability of material culture, the continual process of negotiation that characterizes relationships between individuals and institutions, and the emergence of power imbalances (Dobres and Robb, 2000). What distinguishes agency from mere action is the social significance of the activity (Dobres and Robb, 2000) and the interactions between action, space, and time whereby actions "carry the past into the future" (Barrett, 2000). Human agency is about the pursuit of constructed goals by constructed selves or constructed groups, and hence has a clear basis in symbolic representation. In the next section, we aim to set symbolic information within the broader context of biological information.

## 6.6 BIOLOGICAL INFORMATION PROCESSING

Maynard Smith (1999) has argued that "information enters biology when there is a difference between genotype and phenotype." Information enters complex organisms in numerous ways, including cytoplasm, chromosomes, physical niches, experience, and behavior. The emergence of information processing in evolution, through increasingly complex systems, has been described by Bonner (1980).

All incoming information requires a reference against which it can be interpreted. In lineages of organisms that have become more complex in evolution, the role of the genome as a reference has steadily decreased, and instead experience gathered during the organism's lifetime has increasingly come to dominate. In animals, plastic responses are controlled not just by factors such as allele sensitivity and gene regulation, but also learning. For example, trial-and-error learning allows the organism to store nongenetic information gained from experience in its contemporary environment in addition to genome-located "knowledge" derived from ancestral organisms (Plotkin, 1994). With the development of social learning, information can further pass between different individuals of the species. For example, emulative learning allows the organism's acquisition of experience to be enhanced by others' activities, while mimicry allows the more direct acquisition of information (Plotkin, 2002). Sociality therefore benefits both organisms, which can use experience gained by others, and the transmitted information itself, which can replicate between as well as within individuals.

Thus, with the advent of animals, particularly those with complex cognition, the genome increasingly provides the capacity to acquire and process supplementary information, rather than the information itself. Genes are expressed in learning and may set aptitudes and constraints, but they do not determine what is learned nor how animals behave.

In humans, learned symbolic information is relatively detached from the genome, with the reference constructed from recent or current social consensus rather than the legacy of natural selection of ancestral genes. It is this detachment and hence flexibility that makes symbolism difficult to incorporate within conventional biological models. Symbolism represents a shortcut in information processing, allowing organisms to process and store large amounts of knowledge that would require time and risk to obtain through individual experience (Deacon, 1997). It represents a "reordering of referential relationships" whereby simple learning set generalizations are replaced by complex logical or categorical generalizations. In this way, organisms can know something they have not explicitly learned (Deacon, 1997). However, although symbolism simplifies some aspects of information acquisition, the capacity to symbolize is very complex (Deacon, 1997). Although not restricted exclusively to humans, only in our species does symbolism play a dominant role in information processing.

Information theorists have argued that information is equivalent to eliminating uncertainty. "Signal information" eliminates uncertainty as to what is the signal against a background of "noise"; and "semantic information" eliminates uncertainty as to the message (Lyons, 1977). In turn, to accommodate environmental uncertainty natural selection has favored the processing of information and the evolution of information processing capacities.

The key to understanding the role of symbolic information in biology therefore lies in appreciating its role in addressing uncertainty. Because of its inherent flexibility, learned symbolic information carries potential costs as well as benefits (see Nettle, this volume). The flexibility potentiated by symbolism is assumed to be favored in more changeable environments, in particular highly social environments where the variable behavior of conspecifics requires rapid information processing. Far more so than the physical environment, the social environment can change rapidly over fractions of an individual's life-span, and genome-based knowledge from previous generations is of greatly reduced value as a reference. Conversely, excessive flexibility would forfeit the advantages gained by prior individual or collective experience. Individuals therefore operate under tension between two opposing forces. On the one hand, tradition and conformity can reduce uncertainty on the basis of past experience though they can also act as vehicles for social control (Boyer, 1990). Symbolism contributes to such stabilization mechanisms through rituals, conventions, and norms. On the other hand, creativity can reduce uncertainty by resolving contemporary problems but can also challenge controlling systems. The concept of *bricolage*, proposed by the social anthropologist Lévi-Strauss (1966), elegantly expresses how humans can generate novelty from existing information and experience.

What distinguishes humans from other organisms is that, through the symbolic medium of conscious thought, they have developed a remarkably strong ability to predict, and project strategy into, the longer-term future. Such predictions attempt to address the hazards of uncertainty over long time periods. Some have argued that human modes of thought can be differentiated as analytical and experiential, both of which encode reality in symbols or images (though with contrasting levels of abstraction) and that the experiential mode, in which decision making relies more on affect and emotion than on logical reason, has in practice proven to be a more efficient way to handle uncertainty (Slovic et al., 2004). Although many issues relating to human consciousness and cognitive processes remain poorly understood, we argue that the role of symbolism in projecting strategy into the future is a critical component of the agency and purposive behavior of human individuals.

Human agency can, therefore, be seen as merely one particular way organisms attempt to predict and control their circumstances, the main difference from other species being the timescale over which goals are represented in symbolic terms. Agents compete among each other, for both physical and symbolic resources. However, because agency is generated by both genetic and nongenetic factors, it may not be directed solely to the maximization of genetic fitness. We explore this issue by considering the different relationships that the organism can have with genetic and symbolic information.

## 6.7 INFORMATION IN REPLICATORS

Dawkins (1976) emphasized the value to evolutionary theorists of focusing on individual genes rather than the organism or vehicle that temporarily houses them. Consistency with this approach has likewise led memetic modelers to emphasize the selfish nature of the meme (Brodie, 1996; Lynch, 1996; Blackmore, 1999). Rather than considering how people acquire ideas, for example, it is possible to consider how ideas acquire people by being fecund and readily propagated (Lynch, 1996). To comprehend human behavior, however, it is necessary to develop a more sophisticated model of the relationship between replicating units of information and the organism. One such model is provided by gene-culture coevolutionary theory, which considers genetic and nongenetic units simultaneously without assuming that the former control the dynamics of the latter (Feldman and Cavalli-Sforza, 1976; Boyd and Richerson, 1981). However, this model has limited applicability to the large proportion of human behavior that varies over short time periods, and other approaches are required.

Critics have complained that Dawkins's vehicle is little more than a robot, constructed by genes to carry out their aims, and many researchers have favored the alternative term *interactor* (Hull, 1980). No single gene would be viable were it not located at any given time point in an aggregation of genes, either as a gamete or an organism. Each organism consists of a temporary aggregation of genes in a "gene-team" and develops through a complex interactive relationship with the environment though the process of ontogenesis.

Like genetic information, nongenetic information may also require a store in the organism. To remain viable over the evolutionary timescale, biological information must either confer success on its host organism or spread rapidly between organisms. A continuum of strategies exists, at the extremes of which are symbiosis and parasitism. Just as physical organisms (and hence genes) may be viable parasites, so may units of nongenetic information, as evidenced by the phenomena of diffusion and fashion. The evolution of parasitic nongenetic information is reflected by the evolution of filtering mechanisms that protect the organism from unwanted information while favoring beneficial information (Dennett, 1995). Such filtering mechanisms may be considered to be part of the "architecture of the human mind" as modeled by evolutionary psychologists and are evidenced by psychological studies such as those demonstrating the tendency to distort memories that challenge self-esteem (Forer, 1949).

An example of symbiotic nongenetic information is breast feeding in chimpanzees. This behavior must be learned, the information being socially transmitted between successive generations of female chimpanzees (Budd and Smith, 1943; Wyatt and Vevey, 1935). Thus, chimpanzee breast feeding behavior and chimpanzee genes have a symbiotic relationship, each requiring the other for long-term viability. The value of breast feeding being noninstinctive is assumed to lie in the capacity of this behavior to incorporate cumulative experience and to adapt to local and immediate circumstances (Wells, in press).

Previous memetic models have tended simplistically to treat memes as a single category of information and to focus on "copying" as the primary mode of transmission (Dawkins, 1976; Brodie, 1996, Blackmore, 1999). Instead, it would be more appropriate to consider that nongenetic information exists in a range of forms, with varying value to the organism, and is able to enter organisms by a variety of modes including trial-and-error learning, mimicry, social learning, and socialization. We argue that a critical component of nongenetic information central to human behavior is the concept of the self. Incorporation of this concept counters the tendency of other evolutionary models to portray the individual as a passive entity at the mercy of circulating information units. Most important for our argument, the self can achieve a relationship with the organism that is impossible for any one gene, or indeed the entire genome, to attain.

## 6.8 CONSTRUCTION OF THE SELF

The sense of self in each human mind appears to impose unity on the individual's behavior. Even primitive organisms must have a sense of self to avoid self-harm and pursue self-interested goals, but the complex conscious self of humans has diverse properties and cannot be reduced to a simple construct or function (Olson, 1998). Dennett's (1993) argument that there is no "Cartesian Theater where it all comes together" is consistent with neuroscientific research acknowledging the illusory nature of the human self. There is no definitive self, rather multiple levels of self-consciousness. Nevertheless, returning to themes explored above, we suggest that the self-concept plays a critical role in reducing uncertainty by seeming to maintain continuity in the mind over time while also potentiating agency by which beneficial goals can be pursued.

Human consciousness has been proposed to have evolved in response to selection for the capacity to predict the behavior of conspecifics (Humphrey, 1986), as well as other species' behavior and natural physical phenomena (Liebenberg, 1990). Anthropologists have found the self–other distinction to be common to all healthy members of human societies (Mauss, 1938; Brown, 1991) although it is equally clear that the sense of self varies considerably between societies (Moore, 1994). Self-orientation allows coherent interaction in the social environment, whereas self-motivation allows the pursuit of self-interested goals. Identity is "Janus faced" in being simultaneously subjective, internally manufactured through individual experience (including in social contexts), and "objective" in the sense of being externally imposed (Barth, 1969). Thus, sociality is fundamental to the conscious self.

We focus on two components of identity of importance to modeling human behavior. One component can be conceptualized as the psyche or "*soul*," a construct that each mind develops in a way that brings temporal coherence and stability to the individual's experience. We adopt this term rather than that of *self-concept* (which is widely accepted by psychologists) to facilitate our efforts to explore the social anthropological literature but regard the two terms as broadly synonymous. As we discuss below, much human behavior is influenced by the widespread, but not universal, tendency to project this identity beyond the temporal constraints of the physical body. Another component of identity can be conceptualized as the public persona, which participates in social interactions. This construct is explicitly initiated through the universal human practice of naming, providing a vehicle onto which external identity is projected. We now consider how these components of identity can be considered to compete with genes and exploit the organism for alternative ends.

## 6.9 DIFFERENTIAL REPLICATORY ABILITIES OF GENETIC AND NONGENETIC INFORMATION

Neo-Darwinian evolutionary theory considers the differential success of replicators in a total replicator pool over time. The process of evolution, by selecting genes that promote survival and reproductive success, effectively "designs" gene-teams and hence organisms to maximize reproductive success. The circular relationship between breeding success and evolutionary success means that natural selection engineers organisms as maximizers of breeding, and hence genetic fitness. Given such "retrospective design," from a teleological perspective organisms acquire the "purpose of breeding" and genes acquire the "purpose of replicating."

There is nothing special about the notion of maximizing the fitness of genes as opposed to other replicators (Dawkins, 1976), but those working on coevolutionary models have offered no definition of the organism's *nongenetic fitness*. We propose that the organism can be considered as an entity that can selectively direct its resources toward both genetic and nongenetic fitness by transmitting both genetic and nongenetic information over time.

When genes are transmitted over time, two important effects occur. First, each individual passes on only half her or his genes to each member of the next generation. The human genome project has estimated the number of genes in humans at around 40,000. After about 16 generations, the direct descendants of a human may contain not a single one of that ancestor's genes by direct inheritance. Second, each time an offspring is produced the parental genes are mixed up. The consequence of these effects is that even where genes are successfully transmitted over many generations, they never meet the same selection of genes in any new gene-team. The process of sexual reproduction continually dismantles individual gene-teams.

There is no process of sexual reproduction in the replication of nongenetic information, so it need not necessarily break down over time. The name of any individual is as valid a label denoting that person today as it was millennia ago. The significance of this for evolutionary theory is that genes and nongenetic information may offer different potential returns for a given investment. For any nonhuman species, this approach may be less relevant because the pertinent nongenetic information is symbolic and only viable through self-consciousness and learning. But for our species, the replication of symbolic information is more successful than that of our genes at preserving our uniqueness, our individuality. With the evolution of consciousness came both the awareness of death and the urge to overcome it through immortality (Bauman, 1992). Human literature dating from the Epic of Gilgamesh to the present day is testimony of our powerful desire to overcome death. Although the earliest works are rarely by identifiable authors, the written word and even robust oral traditions can be potent means to ensure that individuals are remembered by future generations. This tendency is, we argue, fundamentally distinct from the tendency to reproduce.

Thus a view of the organism based on nongenetic information is very different to the conventional gene's-eye view of the organism. Dawkins (1976) emphasized that the gene's-eye view leads to us considering the organism as a vehicle designed by a collaborating team of genes that cooperate on account of their shared interest in replicating. The actual organism itself is incidental, and the genes, not the organism, replicate over time. Our novel view, in contrast, can consider how a quantity that summarizes the complete organism replicates over time. Identity comprises a unique piece of information that, in certain forms, need not degrade over time. It is the information that describes the effect of putting together that particular team of genes and creating a unique individual character for a finite time period. As with genes, this information is not the same as the physical organism itself, which remains doomed to death and decay.

Through the combination of their genes and their symbolic identity, humans are able to invest in posterity in several different ways. A finite quantity of resources can be directed toward competing ends: (1) replication of genes (offspring), (2) replication of internal identity (projecting the soul into an afterlife), and (3) replication of external identity (reputation). In addition, humans can elect not to invest resources in long-term ends but rather consume them in the present, which may be termed *hedonism*.

## 6.10 THE SOUL AND AFTERLIFE STATUS

Various factors predispose humans, given certain conditions, to project the concept of the soul beyond death. Freud (1915) argued that all individuals find it impossible to imagine their own death, as they inevitably find themselves still present witnessing their demise. Likewise, when an organism dies, its physical body remains though lacking life. All humans are, therefore, confronted through experience by the fact that life seems distinct from the physical body. Relatively few societies are concerned with where the soul comes from at the beginning of life, whereas most societies are concerned with where it goes at death.

Early anthropologists assumed the notion of a soul persisting in the afterlife to be a human universal. The more extensive ethnographic record now available has identified, in a handful of cases, societies with no apparent concept of an afterlife. These societies tend to be those with little focus on the future, i.e., egalitarian societies with immediate return subsistence patterns (Woodburn, 1982). Similarly, in Western societies the notion of an afterlife conflicts with scientific models of death. In the vast majority of societies, however, belief in an afterlife is present. It can be assumed that all normal humans develop the notion of a central self-concept, distinct from the physical body, and that the majority of social environments contribute to the concept being projected beyond the individual's life span. Boyer (1994) has referred to this predisposition to such beliefs as "the naturalness of religious ideas." A central tenet of social anthropology is that it is necessary, to understand human behavior in diverse societies, to approach that behavior "from the inside" using local terminology and constructs. Thus, scientific models of death offer minimal scope for addressing afterlife belief, whereas anthropological models paradoxically allow the empirical investigation and testing of hypotheses that are essential for scientific knowledge.

The concept of the soul varies between societies, as does the nature of the virtual world constructed by a given society. Social anthropologists have often observed how the soul construct is exploited for social purposes. From a social perspective, it is not death itself but the knowledge of it that is problematic (Elias, 1985). Death is a tragedy for the survivors, who lose part of the social order. The idea of another world is not just to soften the threat of death, the dead are also exploited to provide structure to lives in this world. The ethnographic record illustrates a wide range of ways in which the dead contribute to the structure of the world of the living.

Death is the ultimate rite of passage and is seen as a kind of transitional experience in almost all societies (Huntingdon and Metcalfe, 1979). The deceased leave this world and pass to the next, undergoing rites of separation and incorporation. These rites involve not only the deceased but also the living, for society must reassert itself and be reborn in the absence of the deceased. In some societies, life is considered a limited resource with souls being reborn cyclically in successive generations, and there is a widespread association between death and fertility (Bloch and Parry, 1982).

As well as death sometimes being exploited to structure the self in new individuals, the dead may also serve to maintain the vitality of the wider social order. Descent defined in terms of dead ancestors often provides structure for contemporary groups and families, examples including the unilineal descent group societies of

Africa and the Merina of Madagascar, who use the membership of tomb groups to manufacture quasikinship groups (Bloch, 1971). Many anthropologists have noted how the death of an individual in high office invokes the fundamental ceremonies by which the social order is maintained (Huntingdon and Metcalfe, 1979).

Ancestors can, therefore, continue to play a role in the lives of the living. Often, this role is active as well as structural, for the ancestors may be perceived as representing an important source of power for use by the living. Indeed, the fundamental relationship between religion and power is well known to anthropologists, who have built on the pioneering writings of Marx, Weber, and Durkeim. Kopytoff (1971) for example has argued that many African societies do not distinguish the elders from the ancestors, but rather conceive of a continuous flow of power crossing the boundary between life and death where each generation is subordinate to the next senior generation. Goody (1962) also suggested that the ancestors may deserve veneration because it is through their deaths that important material property passes on to subsequent generations.

These examples provide a background against which to understand how it is that humans frequently regard themselves as maintaining relationships with nonphysical individuals, whether gods, spirits, or ancestors. Commitment to belief in these ideas can play an important role in daily life; and maintenance of social relationships with the characters represented can require physical effort just as is the case for relationships with living people. Among the Kwaio of the Solomon Islands, each generation maintains over time a symbolic world in which spirits play a central social role (Keesing, 1982). Similarly, Vitebsky (1993) has described how the Sora of India frequently conduct dialogues with the dead to manage the uncertainty in their world and handle the psychological process of bereavement. Conversely, it is very rare, though not unknown, for living humans to be treated as deities (Littlewood, 1998).

In ecological terms, many humans invest a proportion of their total resources in ends related to symbolic nonphysical characters, in some cases including the pursuit of afterlife status. Firth (1965) explicitly investigated the economic cost of religious behavior among the Tikopia of Polynesia, but concluded in that society that it was relatively low. Archaeologists have likewise considered the high energy costs of large-scale religious architecture such as temples and shrines. The greater the hierarchy of status in the afterlife, the more individuals attempt to invest in future status during their time in this life. However, in many societies behavior in this life has little consequence for status in the next life, and current investment is therefore minimal.

The soul concept need not contradict genetic fitness. Following extensive research on computer simulations of cooperation, Axelrod (1984) concluded that cooperative behavior flourishes if "the future matters," i.e., if potential interactions are valued as well as current and past interactions. The projecting of human souls beyond death to an afterlife is an indication of the value placed on the future, which may then favor increased levels of cooperation in this life. Thus, afterlife status may be seen as a phenomenon related to both genetic and nongenetic ends, with these ends overlapping to varying degrees. The hypothesis that societies with religious belief have higher mean genetic fitness than those without such beliefs is distinct from the hypothesis that within a given society, individuals who prioritize symbolic ends may have reduced genetic fitness.

## 6.11 REPUTATION AND FAME

In addition to their internal identity, the soul, humans also transmit their external identity within and between populations and over time. External identity is socially constructed through socialization processes beginning from birth. Central to the process is the naming of the individual, an activity that is universal in human societies (Brown, 1991). As the individual develops this identity, her or his activities and behavior project the identity back into the social environment, creating a further construct — reputation.

The anthropological literature demonstrates that naming fulfils two related functions — it distinguishes individuality but also expresses conformity (Alford, 1988). Thus, although names can express what is different about each individual, they may also emphasize what individuals have in common. Where they perform one of these tasks better than the other, alternative means of achieving that second end are used. For example, scarification and ritual clothing are alternative ways of indicating shared status, while nicknames or surnames can distinguish among those who share a name or can mark changes in social position. Names are therefore linked to both groups and individuals, and the identity of both categories may be strenuously perpetuated over time. For example the reputation of the Cossacks may spread, just as may that of an individual soldier.

Reputation is an abstract piece of information that, unlike genes themselves, is not permanently restricted within a single gene team. The information can replicate on a vast scale among peers and descendants and can be achieved without the decay that characterizes genomic transmission. Reputation can be expressed in diverse ways — originally by a name, then by portraits, statues, epitaphs, and literature, and now by diverse audiovisual media. On a wider scale, reputation can evolve into fame. Fame is undoubtedly a rather Western concept, thriving where individuality is already stressed and where the population is sufficiently large to allow wide-scale spread of the information. In "The frenzy of renown" Braudy (1986) has summarized the changing relationship between the individual and the audience in Western civilization in response to changes in demographics, historical events, and technological advances.

Human behavior tends broadly to be consistent with the proximate goals that lead to reproductive success — for example prestige, wealth, and honor may all aid in maximizing the quality and quantity of mating partners and hence, ultimately, offspring. Broadly, therefore, human genes and reputation both underlie genetic fitness. However, we argue that there are circumstances in which reproductive success is clearly irrelevant, in which case reputation is the only remaining currency in which investment may reap a payoff. Such a scenario is given where death is imminent. In contrast to nonhuman animal species, humans often confront certain death with composure and dignity. Death may be transcended through achieving the status of hero or saint. Individuals may earn themselves heroic status by courting death without actually dying and may leave behind fecund nongenetic information when genetic replication is not viable by dying in a memorable and honorable way. Such risk-taking behavior could be said to trade off genetic fitness against reputation. In some societies, such as ancient Greece, a "good" death was one where individuals died in early adulthood, crystallizing them in their prime (Humphreys, 1981), suggesting a very high value on reputation.

Many aspects of culture may be exploited to further reputation. For example, art fulfils many functions and may play a biological role in sexual selection, as argued by Miller (2000). However, although gene-based theories may explain the origins of art, other models are required fully to explain its manifestation in the present. The idea that art acts as a vehicle for the reputation of the artist may be particularly relevant to Western society, where individualism is most developed. Nevertheless, Dissanayake (1988) has previously identified the central role of art in "making things special," suggesting that art is, at the very least, ideally placed to further the reputation of the artist. Art crystallizes information in time and space, invoking the illusion of permanent survival and is thus the ideal medium for preserving identity extracorporeally.

Where art, in particular architecture, is directed to symbolic ends, the evolutionary process of niche construction may also be said to be directed toward symbolic as well as genetic fitness. Monumental architecture, common in many complex societies, has been proposed to symbolize power through violation of the principle of economy of effort that characterizes the majority of human behavior (Trigger, 1990). Material culture endures over time in a way that ritual cannot, but may act as a vehicle not only for power itself but also the powerful. The stamping of the king's name on every brick in ancient Mesopotamian architecture (Trigger, 1990) illustrates the ability of niche construction to convey symbolic information including reputation as well as selective pressures to future generations. This example also illustrates that human nongenetic fitness may comprise both creative and manipulative elements: successful agents can exploit others to further their own symbolic fitness.

Focusing on the flow of nongenetic information therefore offers an enhanced evolutionary model with which to consider the peculiarly human phenomena such as art and music and prestige and achievement in general in which it is an individual's reputation rather than genes that is the prime beneficiary. Biologists have seldom considered reputation as a motivating force in human behavior although archaeologists have discussed the role of monumental architecture in transmitting prospective memories into the future (Holtorf, 1996). Instead, cultural achievement is typically interpreted by biologists as status seeking, with status in turn linked to mate attraction and hence reproductive fitness (Miller, 2000).

It is certainly not the case that investing in reputation is inconsistent with genetic fitness. Anthropologists increasingly appreciate the role of human institutions as environments in which individuals participate in complex game conventions (Skyrms, 1996; Bowles and Gintis, 1998). The functioning of such institutions depends both on expectations about the range of strategies that will be used and the continuity of interactive behavior over time. Reputation clearly plays a critical role in such relationships. Nowak and Sigmund (1998) have used game theory to explore how maintenance of a cooperative reputation increases the likelihood of participating in mutually beneficial interactions even in one-off interactions with strangers. Thus, as in the case of the soul, reputation should be seen as a construct that can be exploited for both genetic and nongenetic ends.

## 6.12 ENDS AND GOALS

In this chapter, we have explored the proposition that human social information transmission has multiple ends, including genetic fitness, the preservation of identity, and the manufacture of reputation. These ends may be considered generic goals of human behavior, and it is worth considering their potential role in evolutionary theory.

Biologists distinguish between the proximate causes of behavior (how things happen) with the ultimate causes (why they happen; Mayr 1982). The logic of the evolutionary process has then led to the formulation of an ultimate goal of behavior, the maximization of genetic fitness, to which proximate goals (survival, obtaining resources, mating, parenting) are considered subservient. If we argue that there are other ends besides genetic fitness that can be pursued, then logic suggests we need to define other ultimate goals of human behavior. Thus, human behavior could be said to be directed toward two ultimate goals, the transmission of both genetic and nongenetic information over time. As we have discussed above, much nongenetic information transmission is directed toward the maximization of genetic fitness, for example, the process of niche construction. Nevertheless, individuals may also pursue nongenetic ends, in particular symbolic ends, to the detriment of genetic fitness. Examples include a celibate priest investing in the afterlife, and a warrior investing in reputation by risking his life in seeking to prove his courage on the battlefield.

The inclusive fitness approach might be invoked to argue that both the priest and the warrior are still investing in genetic fitness. The priest might contribute to the fitness of his genes by investing in his kin. The warrior, if he survives, might use his reputation to win a higher quality mate; his strategy could be considered high risk but with high potential rewards. We suggest that such arguments, although not irrelevant, are an incomplete explanation for the behavior of such individuals. Humans as agents plan actions and pursue symbolic goals with rationality distinct from the "goals" of their genes. More generally, some individuals elect not to breed, while others who cannot breed nevertheless pursue alternative goals. Thus, the pursuit of nongenetic fitness should be considered a significant component of human behavior, as illustrated in Figure 6.1.

## 6.13 SUMMARY

Human identity is a major issue addressed by the discipline of social anthropology and is clearly one of the distinguishing features of our species' biology. Paradoxically, although biologists have addressed many salient human traits such as bipedal walking, encephalization, and long life span as well as behavioral traits such as learning, language, and cooperation, they have rarely considered identity and agency in biological models.

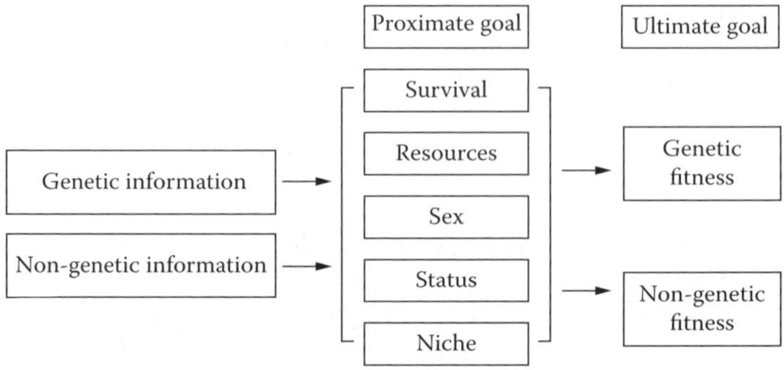


**FIGURE 6.1** A dual fitness model of human behavior. In this model, the organism receives both genetic and nongenetic information. All proximate goals are considered to underlie two ultimate goals, genetic and nongenetic fitness, and the organism can allocate resources between these competing ends. Nongenetic fitness incorporates outcomes such as reputation that cannot be expressed in the currency of genes. Niche construction is considered a proximate goal, with material effects that persist beyond the lifetime of the organism to the potential benefit of both genetic and nongenetic fitness.

We argue that in order for our species to have reached the present era with an ever-growing population, genetic replication has dominated, and remains fundamental to, human behavior. However, symbolic ends have also been pursued for millennia. Teleologically, "ideational evolution" could broadly be considered a form of risk identification and management. Much anthropological insight has been gained into how environmental and other threats are perceived and addressed in largely nonliterate societies, with numerous studies focusing on the role of social structures or conventions in providing a model for understanding the human world. Human identity could be said to represent a strategy for reducing uncertainty, which benefits genetic fitness, but it can also be directed toward other fitnesses. Symbolic identities can replicate, or appear to replicate, for their own benefit, and may play a self-serving role in driving human behavior including social information transmission and niche construction. Elucidating such nongenetic ends should generate a new series of testable hypotheses concerning human behavior.

Whereas Monod (1972) argued that human creativity has been driven by a "profound disquiet" ultimately serving survival as a social animal, the philosopher Popper (1972) argued a more robustly nonutilitarian case. Mallory famously remarked that he attempted Everest "because it is there." There is plentiful evidence that some are attracted by the struggle against a challenge of their own choice and the pleasure of success. Such instances test the creative capacity to handle risks despite biological constraints rather than in pursuit of genetic fitness. Much nonutilitarian behavior involves taking risks for outcomes, which nevertheless may have far-reaching social consequences. That one may gain pleasure from success adds a hedonistic and sometimes individualistic dimension to this type of behavior.

This approach takes us beyond that of Rosenberg (1985), who argued rather despairingly that "if biology is a thoroughly nonintentional science … then [it] can hope to shed no systematic light on, still less absorb, the science of human behavior"; in other words that intentionality is recalcitrant to analysis or definition in behavioral or neurophysioloigical terms and that there are no "laws" about human behavior under its intentional description. As we have argued, an understanding of human agency is critical to comprehensive biological models of human behavior and yet is absent from the major existing approaches including sociobiology, evolutionary psychology, gene-culture coevolution, human behavioral ecology, and memetics. Embracing the concept of agency will allow a new program of hypothesis testing and potentiate a new scientific understanding of human behavior. We predict that such an approach is likely to find identity central to those aspects of human behavior that have proven most difficult to explain using conventional evolutionary biological models.

## ACKNOWLEDGMENTS

We gratefully acknowledge helpful comments by Kevin Laland, John Odling-Smee, and Henry Plotkin.

## REFERENCES

Alexander, R.D., *Darwinism and human affairs*, Pitman Publishing Ltd, London, 1980.

Alford, R.D., *Naming and identity: a cross-cultural study of personal naming practices*, Hraf Press, New Haven, CT, 1988.

Axelrod, R., *The evolution of cooperation*, Basic Books, New York, 1984.

Barrett, J.C., A thesis on agency, In *Agency in archaeology*, Dobres, M.A., and Robb, J.E., Eds., Routledge, London, 61–68, 2000.

Barth, F., *Ethnic groups and boundaries*, Little Brown, Boston, 1969.

Bauman, Z., *Mortality, immortality and other life strategies*, Polity Press, Cambridge, 1992.

Bertalanffy, L. von., *Robots, men and minds: psychology in the modern world*, George Braziller, New York, 1967.

Blackmore, S., *The meme machine*, Oxford University Press, Oxford, 1999.

Bloch, M., *Placing the dead: tombs, ancestral villages, and kinship organisation* in Madagascar, Seminar Press, London, 1971.

Bloch, M., and Parry, J., Introduction: death and the regeneration of life, In *Death and the regeneration of life*, Bloch, M., and Parry, J.P., Eds., Cambridge University Press, Cambridge, 1–44, 1982.

Bonner, J.T., *The evolution of culture in animals*, Princeton University Press, Princeton, NJ, 1980.

Bourdieu, P., *Outline of a theory of practice*, transl. Nice, R., Cambridge University Press, Cambridge, 1977.

Bowles, S., and Gintis, H., *Recasting egalitarianism: new rules for communities, states and markets*, Verso, London, 1998.

Boyd, R., and Richerson, P.J., *Culture and the evolutionary process*, University of Chicago Press, Chicago, 1985.

Boyer, P., *Tradition as truth and communication: a cognitive description of traditional discourse*, Cambridge University Press, Cambridge, 1990.

Boyer, P., *The naturalness of religious ideas: a cognitive theory of religion*, University of California Press, Berkeley, CA, 1994.

Braudy, L., *The frenzy of renown: fame and its history*, Oxford University Press, New York, 1986.

Brodie, R., *Viruses of the mind*, Integral Press, Seattle, WA, 1996.
Brown, D.E., *Human universals,* McGraw-Hill Inc., New York, 1991.
Budd, A., and Smith, L.G., On the birth and upbringing of the female chimpanzee Jacqueline, *Proc. Zool. Soc. Lond. A*, 1–20, 1943.
Caro, T.M., and Borgerhoff Mulder, M., The problem of adaptation in the study of human behaviour, *Ethol. Sociobiol.*, 8, 61–72, 1987.
Cavalli-Sforza, L.L, and Feldman, M.W., *Cultural transmission and evolution: a quantitative approach*, Princeton University Press, Princeton, NJ, 1981.
Dawkins, R., *The selfish gene*, Oxford University Press, Oxford, 1976.
Dawkins, R., *The blind watchmaker*, Longmans, London, 1986.
Deacon, T.W., *The symbolic species: the co-evolution of language and the human brain*, Penguin, London, 1997.
Dennett, D.C., *Consciousness explained*, Penguin, London, 1993.
Dennett, D.C., *Darwin's dangerous idea*, Penguin, London, 1995.
Dissanayake, E., *What is art for?* University of Washington Press, Seattle, WA, 1988.
Dobres, M.-A., and Robb, J.E., Agency in archaeology: paradigm or platitude? In *Agency in archaeology*, Dobres, M.A., and Robb, J.E., Eds., Routledge, London, 3–17, 2000.
Dobzhansky, T., and Boesinger, E., *Human culture: a moment in evolution*, Columbia University Press, New York, 1983.
Durham, W.H., *Coevolution: genes, culture and human diversity*, Stanford University Press, Stanford, 1991.
Elias, N., *The loneliness of the dying*, Transl. Japhcott, E., Basil Blackwell, New York, 1985.
Feldman, M.W., and Cavalli-Sforza, L.L., Cultural and biological evolutionary processes, selection for a trait under complex transmission, *Theor. Popul. Biol.* 9, 238–259, 1976.
Feldman, M.W., and Laland, K.N., Gene-culture co-evolutionary theory, *Trend. Ecol. Evol.*, 11, 453–457, 1996.
Firth, R., *Primitive Polynesian economy*, Routledge and Kegan Paul, London, 1965.
Forer, B.R., The fallacy of personal validation: a classroom demonstration of gullibility, *J. Abnormal Psychol.* 44, 118–123, 1949.
Freud, S., *Thoughts for the times on war and death*, Standard Edition, Vol. 14, p. 289, 1915, edited by J. Strachey et al., Hogart Press and the Institute of Psychoanalysis, London, 1953–1974.
Goldschmidt, W., *The human career: the self in the symbolic world*, Basil Blackwell Inc., Cambridge, MA, 1990.
Goody, J., *Death, property and the ancestors*, Tavistock, London, 1962.
Gould, S.J., and Lewontin, R.C., The spandrels of San Marco and the Panglossian paradigm: a critique of the adaptationist paradigm. *Proc. R. Soc. Lond. B*, 205, 581–598, 1979.
Hamilton, W.D., The genetical evolution of social behaviour (I and II). *J. Theor. Biol.*, 7, 1–16, 17–52, 1964.
Henrich, J., and Gil-White, F.J., The evolution of prestige, *Evol. Hum. Behav.* 22, 165–196, 2001.
Hinde, R.A., *Individuals, relationships and culture*, Cambridge University Press, Cambridge, 1987.
Holtorf, C.J., Towards a chronology of megaliths: understanding monumental time and cultural memory, *J. Eur. Archaeol.*, 4, 119–152, 1996.
Hull, D.L., Individuality and selection, *Annu. Rev. Ecol. Systemat.* 11, 311–332, 1980.
Humphrey, N., *The inner eye*, Faber and Faber, London, 1986.
Humphreys, S.C., *Death and time*, Academic Press, London, 1981.
Huntingdon, R., and Metcalf, P., *Celebrations of death: the anthropology of mortuary ritual*, Cambridge University Press, Cambridge, 1979.

Keesing, R., *Kwaio religion: the living and the dead in a Solomon Island society*, Columbia University Press, New York, 1982.

Kopytoff, I., Ancestors as elders in Africa, *Africa*, 41, 129-142, 1971.

Krebs, J.R., and Davies, N.B., *Behavioural ecology*, Blackwell Scientific, Oxford, 1978.

Laland, K.N., Is social learning always locally adaptive? *Anim. Behav.*, 52, 637–640, 1996.

Laland, K.N., and Kendal, J.R., What the models say about animal social learning, In *Traditions in nonhuman primates: models and evidence*, Fragaszy, D., and Perry, S., Eds, Chicago University Press, Chicago, 33–55, 2003.

Laland, K.N., and Williams, K., Social transmission of maladaptive information in the guppy, *Behav. Ecol.*, 9, 493–499, 1998.

Leach, E.R., *Custom, law and terrorist violence*, Edinburgh University Press, Edinburgh, 1977.

Lévi-Strauss, C., *The savage mind*, Weidenfeld and Nicolson, London, 1966.

Lewellen, T.C., *Political anthropology: an introduction*, Bergin & Garvey, South Hadley, MA, 1983.

Liebenberg, L., *The art of tracking*, David Philip, Claremont, South Africa, 1990.

Littlewood, R., Living gods: in partial defence of Euhemeus, *Anthr. Today*, 14, 6–14, 1998.

Lowenthal, D., *The past is a foreign country*, Cambridge University Press, Cambridge, 1985.

Lumsden, C.J., and Wilson, E.O., *Genes, mind and culture: the co-evolutionary process*, Harvard University Press, Cambridge, MA, 1981.

Lynch, A., *Thought contagion: how beliefs spread through society*, Basic Books, New York, 1996.

Lyons, J., *Semantics*, Cambridge University Press, Cambridge, 1977.

Mauss, M., A category of the human mind: the notion of person; the notion of self, Reprinted in *The category of the person,* Carrithers, M., Collins, S., and Lukes, S., Eds., Cambridge University Press, Cambridge, 1938.

Maynard Smith, J., The idea of information in biology. *Q. Rev. Biol.*, 74, 395–400, 1999.

Mayr, E., *The growth of biological thought*, Harvard University Press/Belknap Press, Cambridge, 1982.

Miller, G.F., *The mating mind*, Heineman, London, 2000.

Monod, J., *Chance and necessity: an essay on the natural philosophy of modern biology*, Collins, London, 1972.

Moore, H., *A passion for difference: essays in anthropology and gender*, Polity Press, Cambridge, 1994.

Nowak, M.A., and Sigmund, K., Evolution of direct reciprocity by image scoring, *Nature*, 393, 573–577, 1998.

Odling-Smee, F.J., Laland, K.N., and Feldman, M.W., *Niche construction: the neglected process in evolution*, Princeton University Press, Princeton, NJ, 2003.

Olson, E.T., There is no problem of the self, *J. Conscious. Studies*, 5, 645–657, 1998.

Oyama, S., *The ontogeny of information: developmental systems and evolution,* 2nd ed., Duke University Press, Durham, NC, 2000.

Plotkin, H.C., *Darwin machines and the nature of knowledge*, Penguin, London, 1994.

Plotkin, H.C., *The imagined world made real*, Allen Lane, London, 2002.

Popper, K.R., Of clouds and clocks, In *Learning, development and culture*, Plotkin, H.C., Ed., John Wiley & Sons, Ltd., Chichester, United Kingdom, 109–119, 1982.

Rapaport, N., and Overing, J., *Social and cultural anthropology: the key concepts*, Routledge, London, 2000.

Richerson, P.J., and Boyd, R., *Not by genes alone: how culture transformed human evolution*, Chicago University Press, Chicago, 2004.

Rosenberg, A. *The structure of biological science*, Cambridge University Press, Cambridge, 1985.

Skyrms, B., *Evolution of the social contract*, Cambridge University Press, Cambridge, 1996.

Slovic, P. et al., Risk as analysis and risk as feelings: some thoughts about affect, reason, risk, and rationality, *Risk Anal.* 24: 311–322, 2004.

Tooby, J., and Cosmides, L., The past explains the present: emotional adaptations and the structure of ancestral environments, *Ethol. Sociobiol.* 11, 375–424, 1990.

Trigger, B.G., Monumental architecture: a thermodynamic explanation of symbolic behaviour, *World Archaeol.*, 22, 119–132, 1990.

Van Gennep, A., *Rites of passage*, Routledge & Kegan Paul, London, 1960.

Van Valen, L., A new evolutionary law, *Evol. Theor.*, 1, 1–30, 1973.

Vitebsky, P., *Dialogues with the dead. The discussion of mortality among the Sora of eastern India*, Cambridge University Press, Cambridge, 1993.

Wells, J.C.K., The role of cultural factors in human breastfeeding: adaptation or biopower? *J. Hum. Ecol.*, Forthcoming.

Wilson, E.O., *Consilience*, Little, Brown & Co, London, 1998.

Woodburn, J.C., Social dimensions of death in four African hunting and gathering societies, In *Death and the regeneration of life*, Bloch, M., and Parry, J.P., Eds., Cambridge University Press, Cambridge, 187–210, 1982.

Wyatt, J.M., and Vevers, G.M., On the birth of a chimpanzee recently born in the society's gardens. *Proc. Zool. Soc. Lond.*, i, 195–197, 1935.

# 7 The Significance of Socially Transmitted Information for Nutrition and Health in the Great Ape Clade

*Andrew Whiten*

## CONTENTS



## ABSTRACT

Major progress has recently been made in identifying an extensive range of traditions in great apes, suggesting they share with humans a distinctively heavy reliance on socially acquired information. This implies that selection pressures for the levels of social information transmission required have been operating for long periods of evolution encompassing our common ancestry. Yet, the adaptive significance of this approach to information acquisition remains to be explicitly studied. Here I examine three case studies where we can begin to examine the significance of ape traditions for fundamental biological outcomes concerning nutrition and health. First, there is evidence from West Africa that a regional tradition of nut cracking permits chimpanzees to survive a seasonal bottleneck in availability of their preferred food, fruit. Second, female chimpanzees have been found to rely on tool-based insectivory more than males, apparently compensating for constraints on their vertebrate hunting capacity and accordingly prioritizing social learning of the skill earlier in development. Third, African apes exploit a variety of plant products for medicinal functions, and new experimental evidence suggests a role for social information in the acquisition process. Together, such discoveries begin to map the biological significance of patterns of social information transmission that characterize the great ape clade.

## 7.1 INTRODUCTION

In humans, the totality of social information transmission that we identify with "culture" plays a massive part in defining the nature of the species: so much so that one perspective on the relation between culture and biology is simply to say that culture must be considered a major, intrinsic *part* of human biology. The human brain is fundamentally adapted to operate in a cultural environment that includes such phenomena as a local language that is socially learned, and without such cultural transmission an organism recognizably human will simply not develop.

But it is also common to think about culture and biology in a second, different way that involves separating and contrasting them. If in this way we distinguish biological phenomena that are not of necessity cultural (such basic functions as eating and reproducing), it becomes possible to ask how what is biological in this sense interacts with social information transmission in a particular species, a repeated theme in this volume. A specific form of this perspective focuses on the interplay between cultural and genetic evolutionary processes, as exemplified in Odling-Smee's contribution to the present volume.

Whichever of these perspectives is focused upon, it is evident that social information transmission represents such a gigantic and complex phenomenon in the human case that our species stands apart from the rest of the natural world in this respect. Human culture has certainly evolved to this extremely distinctive state. Yet, it must have done so from more modest beginnings. In general terms the nature of these origins is not particularly mysterious, for social information transmission — in the sense of social learning, which is the meaning I assume for this chapter — is widespread not only in primates but in the animal kingdom more generally (Heyes & Galef, 1996; Avital & Jablonka, 2000; Fragaszy & Perry, 2003; Galef & Heyes, 2004). In this chapter my interest is in such evolutionary origins, but I shall focus on the most recent ancestry shared with nonhuman species, which provided the relatively sophisticated platform from which our distinctive cultural biology arose. Devoting most space to our closest relative, the chimpanzee, but also considering relevant data on other great apes, my rationale is to identify aspects of social information transmission that humans and the other apes share and that are thus likely to reflect the nature and significance of social transmission in our "concestors" ([common ancestors] Dawkins, 2004) of around 6 million years ago (for characteristics shared with chimpanzees) or longer (about 14 million years for features shared with orangutans; Waddell & Penny, 1996).

An analysis guided by this rationale has, in fact, been completed relatively recently (Whiten, Horner, & Marshall-Pescini, 2003), reviewing the nature of cultural phenomena themselves, rather than their adaptive significance, the topic of the present paper. Whiten et al. noted that although many students of animal behavior tend to treat the terms *tradition* and *culture* as approximate synonyms (a *cultural behavior*, like a tradition, being defined as one that is shared by a population of animals through processes of social transmission; Slater, 1986; de Waal, 2001), others emphasize there is more to human culture than the existence of tradition defined in this way (Galef, 1992). Accepting the latter perspective, Whiten et al. (2003) identified 10 aspects of how human culture appears to go beyond the mere existence of tradition and proceeded to consider how chimpanzees showed some evidence of each of these. These aspects are more rudimentary than their human equivalents of course, but typically more complex than what has been documented in other species. These 10 elements include aspects of population-level patterning of multiple traditions, social learning processes such as imitation and teaching, and the contents of behaviors transmitted, such as tool use or courtship techniques. It is not appropriate here to recount all the details of this recent analysis, but readers requiring more comprehensive information as background to the present paper are encouraged to consult it.

Here I take the second perspective on the relationship between culture and biology alluded to above and ask what we know of the adaptive significance of social information transmission for fundamental aspects of biology, notably nutrition and health, in the great apes. Relatively little attention appears to have been paid to this in the literature to date, most effort having been devoted to establishing the more basic issue of what cultural phenomena exist. Reviewing the literature, I pick out three topics for close examination: (1) the use of socially transmitted information to access food items crucial to coping with seasonal feeding bottlenecks, (2) sex-specific niche construction through cultural variations in tool use, and (3) socially learned aspects of self-medicative behavior. The relevance of each of these to the evolution of human cultural biology will be discussed. Before examining each of these three sets of data, however, a general account must be offered of what has been discovered in recent years about the cultural lives of the great apes.

## 7.2 CULTURE IN THE GREAT APES

Half a century ago, next to nothing was known of the behavior of our closest living relatives in the wild. In the intervening period, however, a handful of longitudinal studies have each logged decades of detailed records of habituated chimpanzees. As the results of these efforts accumulated in the literature, workers began to tabulate the occurrence of behavior patterns across the different African sites, beginning to identify those that varied in ways that suggested they represent local traditions (Goodall, 1986; McGrew, 1992; Boesch, 1996). By 1992, McGrew was able to tabulate 19 forms of *material culture* (tool use) that each appeared to show discontinuous distributions across a dozen study sites (McGrew, 1992).

These important advances suffered a number of methodological difficulties, however (Whiten, 2000). The three key ones that led on to the next stage were that (i) the published literature does not necessarily record the occurrence of all behavior patterns that have been observed at a site — each scientist has her or his own priorities in publishing and this may well exclude reporting a catalogue of all behaviors habitual at their site; (ii) even where behaviors are reported to occur, there may be no data concerning how common they are in the community concerned; and (iii) perhaps most important of all, the absence of behavior patterns that exist elsewhere is often low priority for publication!

Noting these problems, I was fortunate in gaining the agreement of the research directors of the principal long-term study sites to collaborate for the first time in a new and systematic approach not directly dependent on the published literature. Instead, it required these scientists to pool the relevant information. In Phase 1, each worker suggested a list of candidate cultural variants — behavior patterns either common at their site yet suspected not to be universal or, conversely, known to exist elsewhere yet absent at their site. These behaviors were assembled into a catalogue with the crucial issues of category splitting, lumping, and definitions agreed between all parties. This phase of the project generated a list of 65 candidate cultural behavior patterns.

In Phase 2, the research director and colleagues at each site coded each of the 65 behavior patterns as (1) Customary — occurs in all or most able-bodied members of at least one age–sex class (e.g., adult males); (2) Habitual — not customary but seen repeatedly in several individuals, consistent with some degree of social transmission; (3) Present — has been seen, but has not achieved habitual levels; (4) Absent but with a likely environmental explanation (e.g., the key raw materials or species are not available); (5) Absent with no such explanation apparent; or (6) Uncertain through lack of appropriate scrutiny.

The criterion for a behavior pattern counting as a cultural variant was that it occurs customarily or habitually in at least one community studied yet is absent without apparent environmental explanation in at least one other community, a criterion met by 39 behaviors (Figure 7.1; Whiten et al., 1999, 2001). These behavioral variants are thus unlikely to be explicable by straightforward differences in ecology, a conclusion reinforced by the geographical closeness of some of the contrasts. In addition, genetic explanations for the variation were considered unlikely, partly because of these contrasts between populations of the same subspecies and their geographic proximity and partly because many of the behaviors, notably those involving tool use, are known to be susceptible to learning by these apes in captivity. Additional reasons to believe the variants identified as putative traditions are indeed socially learned include the intense visual attention that young chimpanzees devote to them, apparently avidly taking in social information, together with evidence from experiments in captivity that suggest the existence of several grades of social learning in apes from enhancement of target objects to the copying of action patterns (Whiten et al., 2004).

The majority of the variants identified involve tool use, but others include food processing without tools (such as pounding hard-shelled fruit on substrates), ways of dispatching ectoparasites removed in grooming (three variants involve either stabbing onto the forearm, squashing on a leaf, or inspecting on a leaf before eating or discarding); forms of grooming (clasping hands overhead, or not); and ways of obtaining attention in courtship efforts (including clipping leaves in the teeth, or stripping leaves off stems). Tool uses include the application of heavy sticks to break into termite mounds; fine probes for a variety of purposes; leaf masses chewed to make sponges to obtain water; wooden and stone hammers and anvils to crack nuts; leaves used as seats, for nests, or for wiping blood, semen, and other materials; leafy twigs used to fan away flies; and clubs.

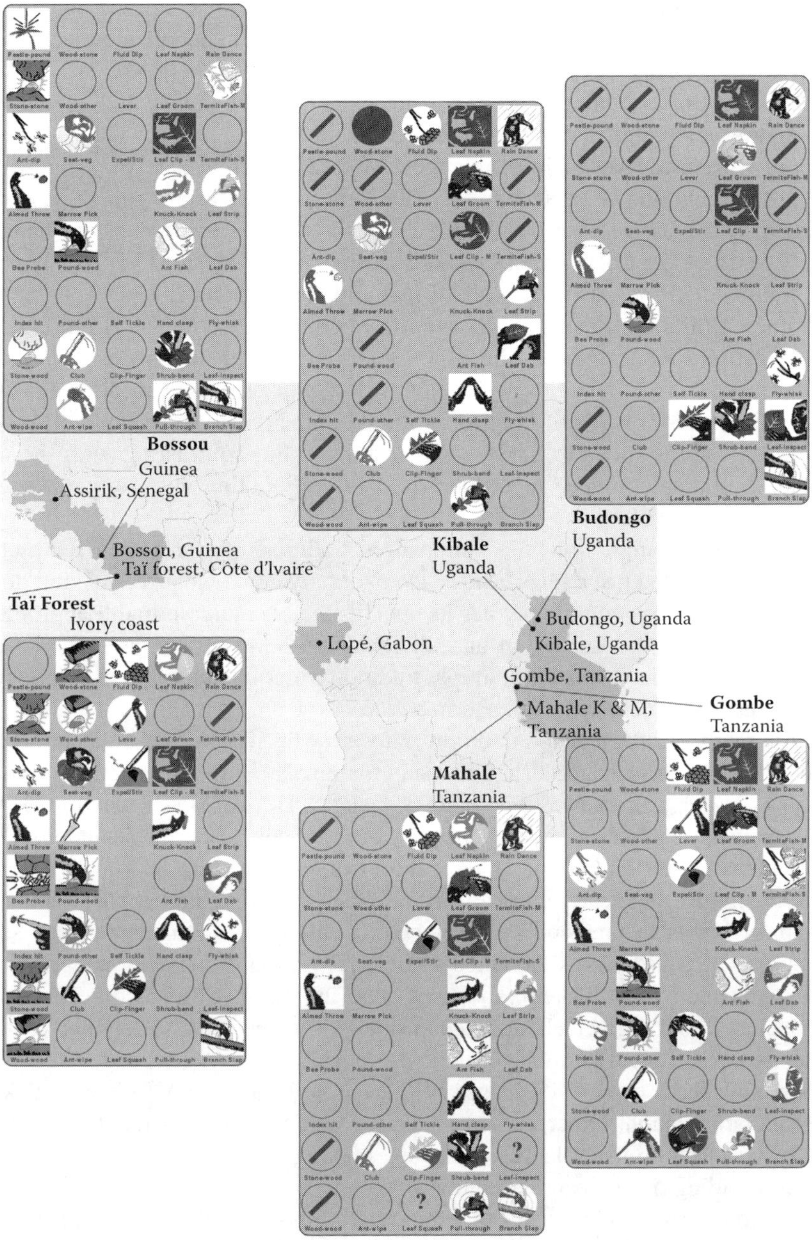


**FIGURE 7.1** Distribution of 39 chimpanzee behavior patterns classed as cultural variants at six long-term study sites. Cultural variants are those customary or habitual in at least one community but absent in at least one other without apparent environmental explanation. Square icons, Customary; circular icons, Habitual; bar, absent with environmental explanation; open circles, absent with no apparent environmental explanation. For further explanation and discussion, see text. After Whiten et al., 1999.

Some behavior patterns occur at only one site. An example is pestle pounding, in which the chimpanzee stands bipedaly and uses a large palm petiole to pound into the palm's growing point, permitting extraction of a rich source of nutrients. This has been found only at Bossou although similar palms occur at other sites including Taï, just a few hundred kilometers away, where the technique appears absent. Other behaviors are more widespread yet regional, the clearest case being nut cracking, occurring at both Bossou and Taï yet absent from other parts of the western subspecies' range separated by a massive river despite the existence of appropriate nuts and hammers there (Boesch et al., 1994). Other behavior patterns are widespread yet clearly absent from well-studied sites, as in the lack of leaf clipping at Gombe.

Some cultural contrasts are all-or-none, such as the existence of nut cracking in the west and its absence in the east. Others are better described as variations on a theme, where a behavior is done at two or more sites yet in different ways. One example is the ways of dealing with ectoparasites, outlined above. Another is the use of short sticks to harvest driver ants at Taï, contrasting with the more complex bimanual "swipe" technique used in conjunction with longer wands at Gombe.

In sum, the complex of wild chimpanzee traditions now revealed represents a richness unsuspected just a few decades ago. Not only do chimpanzees exhibit dozens of traditions, but each community has its own distinctive multiple-tradition profile, on the basis of which the locality of an individual can be identified, as is the case for human cultures. Indeed, it is this multiple-tradition patterning of different communities that is perhaps best deserving of the term *culture*, as so amply manifested in the human case. Although various isolated traditions have now been quite widely identified in a range of species that include fish, birds, and mammals (Heyes & Galef, 1996; Avital & Jablonka, 2000; Fragaszy & Perry, 2003; Galef & Heyes, 2004; Laland & Hoppitt, 2003), there is relatively little evidence to date that any other species approach the multiplicity of tradition that marks the cultures of chimpanzees.

Interestingly, the closest comparison has emerged through an application to wild orangutans of the exact method we developed for chimpanzees. In this way, van Schaik and a group of collaborators identified 19 culturally variable behavior patterns with a further five tentative candidates (van Schaik et al., 2003). As in the case of chimpanzees these included tool use, which, although a smaller set, showed a variety not known in species other than apes: examples include using leafy branches to scoop water, probes for extracting seeds from fruits, leaves as "gloves" to handle spiny fruit, and sticks for autoerotic stimulation. Other variants included communication (modifying a "kiss-squeak" vocalization using leaves or hands), making sun covers over nests, and displaying by riding on swaying vegetation in distinctive ways.

To further test the hypothesis that these variations are cultural, van Schaik et al. (2003) conducted three statistical analyses, the first two with both their orangutan and our chimpanzee data. First, they predicted that if the distribution of the habitual and customary patterns is due to cultural diffusion, cultural similarity (proportion of patterns in common) should correlate with geographical proximity. For both apes, it did. Second, they predicted that a measure of social tolerance (social proximity) would correlate with the number of variants a community supported. Again, for both species, it did. Finally, it was expected that ecological contrasts (floodplains versus forests for the orangutans) would not predict behavioral differences, which was also upheld. These tests provide further reassurance that the putative cultural differences are just that.

Comparably comprehensive databases remain to be published for gorillas and the still relatively lesser-known bonobo (Hohman & Fruth, 2003). Among other nonhuman primates, capuchins currently offer the most evidence of culture-like multiple traditions (Panger et al., 2002; Boinski et al., 2003; Perry et al., 2003). Multiple traditions have also been outlined for whales and dolphins, but the extent to which these are truly traditions is particularly difficult to judge given the observation conditions that apply (Rendell & Whitehead, 2001).

The complexity of cultural transmission in chimpanzees is thus not unique among nonhumans, although currently the evidence is that it shows the closest similarity to what we recognize as culture in humans. All the cultural characteristics outlined above that chimpanzees and orangutans share with humans suggest that this scale of social information transmission permeated the lives of apes well before the time of our common ancestor with chimpanzees, even though to this day it remains uncommon in the animal kingdom.

What then, was its likely functional significance? Again, we can turn to living apes for data that can help answer this question. What are the implications of the social transmission of information for such fundamental biological matters in chimpanzees' lives as their nutrition and health?

### 7.2.1 The Adaptive Significance of Chimpanzee Cultures

A majority of the chimpanzee cultural variants identified to date concern subsistence activities, which in many cases are accompanied by the use of tools, such as stones used to crack nuts. It might be thought that such cases are manifestly adaptive in permitting the animals access to resources otherwise unavailable. In many cases, this is indeed surely plausible. However, to simply acquiesce in this conclusion courts the danger of the kind of indiscriminate adaptationism that assumes that all behavior in wild animals is, by and large, adaptive. In this paper my aim is to go beyond such assumptions, to consider evidence that more directly assesses the biological functions of social information transmission. I do this by considering three case studies.

#### 7.2.1.1 Case Study 1: Coping with Seasonal Fruit Scarcity

A study by Gen Yamakoshi appears, perhaps rather surprisingly, to stand by itself as an attempt explicitly to assess the adaptive significance of putative cultural behaviors (Yamakoshi, 1998). In this case, two such behaviors are of interest, nut cracking and pestle pounding, the particular adaptive context being seasonal scarcity of the chimpanzees' primary food source.

Yamakoshi studied chimpanzees at Bossou, Guinea, through their annual cycle. As elsewhere, chimpanzees' primary and preferred food source was fruit pulp. Yamakoshi showed that through the whole year, fruit occupied 61% of feeding time, and of a total of 160 food types, 80 were fruit pulp. However, the availability of fruit fluctuated markedly through the year (Table 7.1). Fruit was abundant from the

**TABLE 7.1**
**Percentage of Time Spent Feeding on Fruit Pulp, Cracked Nuts, and Palm Pith Obtained Through Pestle Pounding in Periods 1, 2, and 3 at Bossou (after Yamakoshi)***

| | Period 1 | Period 2 | Period 3 |
|---|---|---|---|
| fruit pulp | 71.4 | 52.9 | 59.6 |
| nut cracking | 0.5 | 8.3 | 8.5 |
| pestle pounding | 0.1 | 10.4 | 3.9 |

*See text for further explanation.

late dry season to the early rainy season (January-April, Period 1 in the table), but this was followed by a second period of marked scarcity in May-August, followed by partial recovery in the remainder of the year. In the period of scarcity, the number of fruiting trees fell to less than half those fruiting in the peak month of March.

Yamakoshi found there was a high correlation between availability of fruit pulp and the percentage of time spent feeding on it, confirming that fruit is the preferred primary food source for Bossou chimpanzees, consistent with what we know of chimpanzee preferences across Africa.

The chimpanzees did not cope with fruit bottlenecks by broadening their diet to accept more diversity, as some optimal foraging theories would predict (MacArthur & Pianka, 1966). Instead, they devoted more time to harvesting two food types that are available through the year but that require the application of kinds of tool use that represent putative cultural variants in the study illustrated in Figure 7.1. One of these is nut cracking, which although done throughout the year is performed particularly during Periods 2 and 3 (Table 7.1). At Bossou, this involves gathering oil palm nuts (*Elaeis guineensis*) and cracking them by use of stone hammers and stone anvils. The other technology is pestle pounding, which is particularly prevalent in Period 2 (Table 7.1). In this technique, at present identified only at the Bossou site, the chimpanzee climbs to the top of an oil palm tree, and uses one of the large, stiff petioles of the tree as a ramrod to pound into the growing point. This creates a mushy mass likely to be highly nutritional, which the chimpanzee then scoops out and eats, before pounding further and repeating the process. The pounding behavior is dramatic, with the chimpanzee standing bipedaly, holding the pestle tool bimanually, and beating down forcefully and vertically into the point of contact. Nothing resembling this extraordinary motor pattern is seen in chimpanzees at other sites, and this novelty, combined with the implausibility of the behavior being invented anew by each chimpanzee and its absence at the other West African site of Taï Forest, leads field workers to infer this is a socially transmitted behavior.

Together, nut cracking and pestle pounding compensate for the lowered opportunities to feed on fruit in Periods 2 and 3 (Table 7.1). Coupled with the fact that these sources are available all year, Yamakoshi concludes they act as "keystone resources," permitting the chimpanzees to cope with the seasonal bottlenecks in their primary food, fruit. Obtaining these crucial foods, however, is highly dependent on the two aspects of material culture outlined above. Tool use accompanies 97% of feeding time on nuts and 47% of feeding time on oil palm growing-point pith. The latter figure reflects the fact that some growing material can be harvested directly with the hands, the pestle then being used to burrow ever deeper to obtain more pith. The two forms of tool use together occupy about 10% of feeding time through the year, but in the month with the greatest dearth of fruit, June, this figure rose to 32%. Interestingly, there was a stronger negative correlation between fruit availability and obtaining pith through pounding than for feeding on this source generally, underlining the fact that poor fruit availability is coped with by application of the pestle pounding technology.

The evidence from this case study, then, is consistent with the hypothesis that the two cultural patterns concerned are adaptive for these apes in permitting them to feed on alternative resources during seasonal fruiting bottlenecks. More studies of this kind are needed.

#### 7.2.1.2 Case Study 2: Sex Differences in Foraging Niche

Termite fishing at Gombe was one of the first reported cases of chimpanzee tool use, Goodall classing it as cultural on the basis of its apparent mode of acquisition, with youngsters intently peering at adult fishing and then beginning to perform the actions themselves (Goodall, 1964). However, there is a marked sex difference in the use of this technology, analyzed in depth by McGrew (1979, 1992) based on 7,500 hours of observation by field researchers at Gombe. The essential contrast is that termite fishing was performed three times more often by females than by males. This was apparent in their overt behavior and confirmed by fecal analysis. This same preferential investment by female chimpanzees in harvesting invertebrates through tool use is now known to extend to other activities such as ant dipping, and has also been documented at other field sites. According to McGrew, "the picture for insectivory, both specifically and generally, is clear: females in several widely separated populations of wild chimpanzees tend to specialize in a variety of prey species of insects. In no populations has the reverse been seen" (McGrew, 1979, p. 97).

By contrast, hunting for vertebrate prey is predominately a male activity (Table 7.2), particularly concerning the preferred prey of red colobus monkeys. The reason

**TABLE 7.2**
**Relative Numbers of Vertebrate Prey Recorded as Taken by Male and Female Chimpanzees at Gombe**

| | on colobus monkeys | on bushpigs | on bushbucks |
|---|---|---|---|
| by females | 24 | 14 | 10 |
| by males | 197 | 52 | 39 |

*Source:* McGrew, 1992, Table 5.5, page 102.

appears fairly clear: females are typically encumbered by one or more offspring that they are either carrying or closely associated with, making it difficult to indulge in the vigorous and often somewhat dangerous activity of pursuing prey, particularly when it is a group of monkeys fleeing through the tree tops. McGrew notes that one adult but sterile female, Gigi, has attended twice as many hunting and meat-eating sessions as any other female.

Putting these observations together, McGrew suggests that what we see here is the exploitation of two different but complementary foraging niches. Males predominantly obtain their animal protein and other nutrients from vertebrate prey, a niche to which females have limited access for the reasons outlined above. Although females occasionally catch prey and may gain some of males' meat through begging, their principal compensation lies in obtaining animal nutrients from invertebrates, through tool use. Nothing like this division of labor is known in other nonhuman primates. As McGrew notes, the parallel with the pronounced division of labor documented in foraging peoples, with males predominantly responsible for hunting and females for gathering, suggests that we have here another indication of patterns of behavior likely to have existed in our ancient common ancestors.

A most interesting and relevant twist to this story has recently arisen through the work of Lonsdorf and colleagues (2004, 2005). Lonsdorf observed and videotaped the activities of mothers and their developing offspring as they termite fished through a series of consecutive annual termiting seasons at Gombe. Young females were found to spend more time at an early age attending to termite fishing activity, and they began to fish successfully themselves much earlier (on average at 31 months) compared to their male peers (58 months; Figure 7.2). Moreover, adult females were observed to vary in the depth to which they typically inserted the fishing probes into the termite mounds, and the extent to which they did so predicted this same characteristic in their female offspring (Figure 7.2b). It seems difficult to imagine this can reflect a genetic correlation, suggesting this is evidence consistent with social information transmission; more particularly, that it involves actually copying this aspect of the mother's technique (Whiten et al., 2004). Young males not only pick up the behavior later, they show no sign of this correlation with their mother's preferred technique (Figure 7.2c).

These new findings are intriguing when put alongside the data showing that as adults, females depend much more than males on successfully exploiting the insect-harvesting niche. The latter suggests that for females this particular tradition has special adaptive significance; in addition, the new data from Lonsdorf et al. indicate the relevant social information transfer is accorded special attention by the young females for whom such accelerated acquisition is likely to be of special benefit.

#### 7.2.1.3 Case Study 3: Self-Medication

Over 20 years ago, Wrangham and Nishida (1983) drew attention to the peculiar way in which chimpanzees at both Gombe and Mahale ingested whole *Aspilia* leaves. The initial suspicion was that the leaves were being swallowed because of their pharmacological properties, but subsequent work has failed to support this. An extensive series of investigations by Huffman and his colleagues (1997, 2001) has

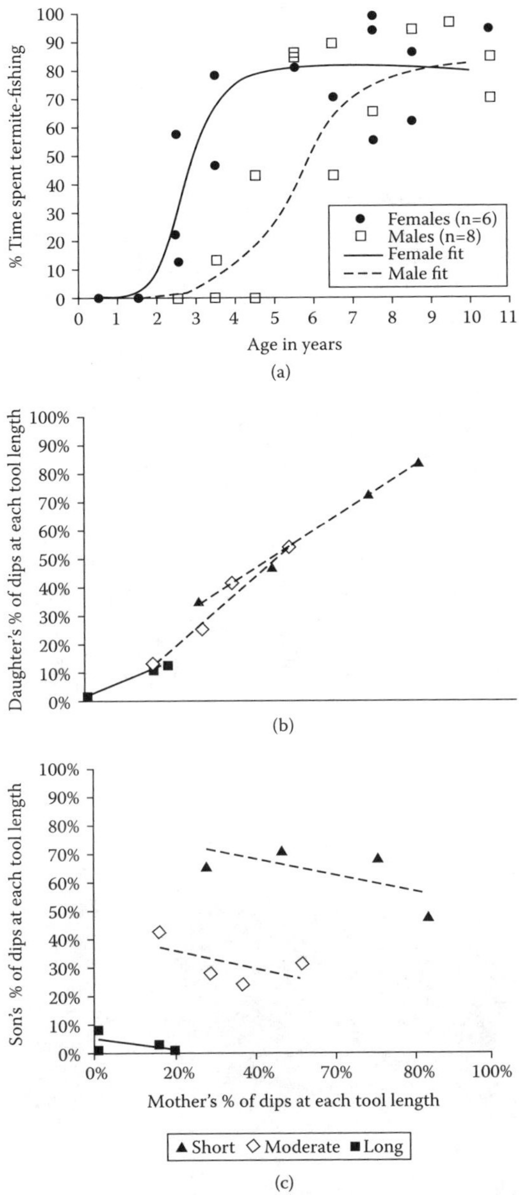


**FIGURE 7.2** (a) Percentage of time at termite mound spent in fishing-related activities for young male and female chimpanzees. Each point represents a given individual at a given age, with logistic growth curves fitted. (b) Regression of daughter's values on mother's for percentage of termite fishings using short, medium (3—5 fist lengths) or long insertions into the mound. There is a strong and significant relationship in each case. (c) As for (a) but showing that for sons, no relationship between maternal and offspring behavior is apparent. After Lonsdorf et al., 2004, reproduced with permission.

(a)

(b)

**FIGURE 7.3** Chimpanzees engaged in whole leaf swallowing (images courtesy of Michael Huffman).

clarified the functional significance of leaf swallowing as a form of self-medication, as well as another class of behavior, chewing bitter piths.

Leaf swallowing (Figure 7.3) has now been described in several communities of chimpanzees, as well as bonobos and gorillas at 13 study sites across Africa. As many as 34 different plant species have been implicated, including herbs, vines, shrubs, and trees, but all share a common property: the leaves are rough and hairy.

Under the microscope, the surfaces may evidence robust barbs. Chimpanzees ingest these leaves whole, carefully folding them with lips and tongue, and the leaves typically pass through the gut to emerge whole in the dung. From five to as many as 55 leaves may be swallowed in a sitting.

Such behavior occurs relatively rarely but rises markedly in frequency with the onset of the rainy season and is correlated with sickness, including diarrhea, lethargy, and signs of abdominal pain. Huffman's investigations at Mahale showed there is a strong association between such events and nematode infections, recorded through dung samples; moreover, dung evidencing such infections are highly and significantly more likely to contain whole leaves. On average, dung of infected animals contained 10 worms, along with 20 leaves. Reviewing all the evidence about the associations and sequelae of leaf swallowing, Huffman's conclusions are that the passage of these rough whole leaves through the gut likely controls infections by (i) causing adult worms to be dislodged and expelled, (ii) flushing out incoming larvae before they can attach, and (iii) reducing the pathology of nodular cysts by inducing the next larval stage to emerge from them and be flushed out. At Kibale, Wrangham (1995) found that tapeworm proglottids were also more common in dung containing whole leaves. Expelled nematodes tend to be alive and motile, further evidence that pharmacological agents are not critical to the function of leaf swallowing.

The chewing of plant parts with especially bitter piths appears to have a similar function and indeed may cooccur with periods of leaf swallowing and act in concert to reduce worm infections. Huffman notes that in the case of the primary source at Mahale, *Vernonia*, "chimpanzees meticulously remove the outer bark and leaves to chew on the exposed pith, from which they extract extremely bitter juice and residual amounts of fibre" (Huffman, 1997, p. 655). Chimpanzees engaging in such behavior typically exhibit diarrhea, malaise, weight loss, and nematode infections. However, two cases recorded in detail showed that recovery from these symptoms was evident 20–24 hours after chewing bitter pith. In one case, nematode infection as measured by eggs per gram of dung dropped from 130 to 15 over this period (whereas in seven individuals used as controls it rose).

Intriguingly, extracts from *Vernonia* and other plants with bitter products are used in human herbal medicines, providing yet another ape–human parallel indicating ancient evolutionary roots to a broad, functional class of behavior. Chemical analyses of *Vernonia* have confirmed the existence of pharmacologically active compounds that in turn have been shown to control nematode activity and reproduction (reviewed in Huffman, 2001). The bark and leaves contain exceptionally high levels of toxin, but these are avoided by chimpanzees in favor of the pith. As Huffman emphasizes, this suggests a sophisticated and critical knowledge among chimpanzees, which is unlikely to be acquired by trial and error. This leads us to our central concern in this matter: does self-medication represent socially transmitted knowledge, as this last point might suggest?

The principal argument against these behaviors being traditions is that they have been found to be common across Africa, not only in chimpanzees but in other apes, as noted above. Accordingly, Huffman suggests that in all these cases there is some preparedness either to seek the kind of experiences involved at times of sickness or to at least rapidly associate leaf swallowing with relief. However, he also emphasizes that the plants concerned vary locally, and, as noted above, the particular parts to be utilized are often not only very specific but must be differentiated from extremely poisonous alternatives. One must remember that wild chimpanzees are surrounded by literally thousands of plants species, each with numerous components, from which they select only a relatively small percentage (even though this may in absolute terms extend to hundred of types) for ingestion as food or for self-medication. The working hypothesis is thus that socially transmitted information is crucial in channeling attention to the critical environmental features ("localized stimulus enhancement," Caldwell & Whiten, 2004), creating local traditions of self-medication.

Huffman and Hirata have investigated this issue further by offering local leaves with the requisite properties to captive chimpanzees in Japan (Huffman & Hirata, 2004). Although none had parasite infections, 6 of the 11 chimpanzees were observed to swallow whole leaves, which passed out in their dung. Two chimpanzees spontaneously swallowed leaves, which is consistent with the notion outlined above that an innate tendency favoring such behavior may exist (although the suggestion from the wild data was that this is occasioned by sickness, so it is surprising these captive chimpanzees engaged in the behavior). However, the other four chimpanzees closely observed the first leaf swallower and only then did three of them begin to swallow leaves, having earlier either rejected them or only chewed on them. These responses thus indicate social learning (the fourth observer chimpanzee had not yet taken leaves but swallowed them when she later had her first encounter). In sum, the experiment provided data consistent with both parts of Huffman's hypothesis based on the evidence from wild chimpanzees, which suggests a basic preparedness for the behavior ready to be refined by the acquisition of socially transmitted information about what elements to ingest and perhaps how to do so.

## 7.3 CONCLUSION

Three different kinds of case study have been reviewed, each offering evidence that aspects of chimpanzee feeding or self-medication are made adaptive (i.e., fitness is likely to be enhanced) through the employment of traditional forms of behavior adopted through the acquisition of existing knowledge manifest in the community concerned — socially transmitted information. We have reviewed evidence that such traditions can allow chimpanzees to respond adaptively to three different kinds of conditions: (i) seasonal shortages of primary foodstuffs, when embedded alternative resources can be unlocked; (ii) constraints of an individual's condition, such as being encumbered by infants, when alternative embedded resources can be gained through skilled tool use; and (iii) sickness, when precise knowledge of appropriate antidotes is available. Each of these parallels human traditions, whereby people (i) use traditional techniques to cope with seasonal challenges; (ii) employ sexual division of labor that include adaptations to similar childcare constraints; and (iii) employ traditional medicines. The underlying patterns here are thus inferred to have ancient origins, at least as far back as our common ancestor with chimpanzees and probably older.

The convergent support for the functionality of traditions gained through examining these case studies makes more persuasive the general working hypothesis noted at the beginning of this chapter, that all the examples of traditional subsistence activities are in fact adaptive in the sense that those chimpanzees who perform them are likely to enjoy a higher biological fitness than if they did not have such traditions. At the outset I argued that such a hypothesis should not be accepted uncritically as an adaptationist assumption. These studies do not extend to directly measuring effects on long-term reproductive success, but the evidence they muster is consistent with this assumption. In this light, note that of the behavior patterns we earlier identified as putative cultural variants (Figure 7.1), half (18) represent foraging techniques that may represent adaptive traditions of this kind.

Does this imply that in those cases where we identified one or more communities where the variant is common, and at least one in which it is absent, the latter chimpanzees are at a disadvantage? Probably the answer to this should be a cautious yes (although our particular methods and criteria perhaps make this artificially likely, as discussed below). An example of such disadvantage, so far as we can see, is that chimpanzees in many areas could be eating the nuts that are all around, and the only reason they do not is that the relevant socially transmitted knowledge about how to crack nuts open has not reached them. The reason for caution is that as we saw in the case of the Bossou chimpanzees, the demonstrated functionality of the behavior is linked to a seasonal stress: might it perhaps be that chimpanzees in other areas do not have the same need? In fact that seems unlikely, when it has been demonstrated that on one side of a single large river, nut cracking is customary, and on the other it does not occur (Boesch et al., 1994).

What of the situation where a tradition may actually be adaptive in one region but not in another? That is surely far from implausible — and after all, much of the significance of human culture is that it permits such local adaptation, for example to life in the arctic. However, the particular criterion we adopted to sieve out chimpanzee cultural variants does militate against identifying such cases, insofar as we ruled out those where the absence of a behavior appears due to straightforward ecological conditions! This has served as a useful heuristic to identify the patterns in Figure 7.1, but there is clearly a sense in which it is overstrict and likely to omit true traditions that are locally adaptive. For example, we set aside "algae-scoop" (scooping algae off pools using a stick), despite the fact it is customary at Bossou, because its absence elsewhere seemed explicable either through the absence of algae or its accessibility without recourse to a tool. However, there is really no particular reason to think that at Bossou, algae scooping is any less of a tradition than pestle pounding, nut cracking, or ant dipping. If it *is* a local adaptation of this kind, then chimpanzees algae scooping at Bossou may or may not be at an advantage over chimpanzees elsewhere that do not know this technique, for those chimpanzees may simply not be in a situation where it could benefit them.

Another swathe of behavior patterns our rules set aside are those that occur at all sites, for our approach focused on only cultural *variants*. But again, our heuristic may actually discriminate against behaviors that are truly traditions. In this case, they just happen to have become universal ones (as in traditions of knife use by humans of all cultures, which is no less functional for being universal). A likely chimpanzee example is leaf sponging to obtain water from holes, which although occurring everywhere it can be of use, has other hallmarks of relying on socially acquired knowledge such as being closely observed by youngsters until it enters their repertoire.

In addition, there are cases in our database that are defined not so much by presence versus absence at different sites but rather distinguish between the same broad behavior done in different ways. In these cases, the alternative techniques may or may not confer the same degree of advantage. In the case of ant dipping, there are data that show such a difference exists: the more complex bimanual technique at Gombe has been shown to be about four times more efficient (in ants per minute) than that used at Taï (Boesch, 1996). By contrast, it is difficult to imagine that the different ways of dealing with ecto-parasites, in which they are either squashed on leaves or examined on leaves, are anything other than equally beneficial cultural alternatives. Doing either may confer more benefits than not doing them, but they may be equal to each other in worth. Similarly, all the various forms of gaining the attention of a sexually attractive individual may represent no more than different ways of achieving equivalent benefits. The same may be true of hand-clasp grooming versus the alternative seen elsewhere of grasping an overhead branch.

In sum, an enormously rich picture of ape culture has been gained in recent studies that could scarcely have been imagined half a century ago. At present, only a few studies can be called upon to explicitly address the question of the adaptive significance of the social information transfer implied for fundamental biological outcomes concerning nutrition and health, but these studies converge in supporting the view that anthropoid traditions offer important adaptive payoffs. Given this, we must acknowledge that many of the other putative cultural variants identified are no less consistent with a hypothesis of adaptiveness.

Given the manifest similarities to equivalent phenomena in human cultures highlighted at various points through this paper, ape cultures and their biological significance are likely to reflect forms of life that existed before the time of our common ancestors, laying the ground plan for many phenomena analyzed in the other papers of this symposium that focus more on our own species.

## ACKNOWLEDGMENTS

This paper was prepared while the author was supported by a Major Research Fellowship from the Leverhulme Foundation. For comments on an earlier version of this manuscript, I am grateful to Debbie Forde, Elizabeth Lonsdorf, Bill McGrew, Mike Huffman, Antoine Spiteri, and Gen Yamakoshi.

## REFERENCES

Avital, E., and Jablonka, E., *Animal traditions: behavioral inheritiance in evolution*, Cambridge University Press, Cambridge, 2000.

Boesch, C., The emergence of cultures among wild chimpanzees, In *Evolution of social behavior patterns in primates and man*, Runciman, W.G., Maynard-Smith, J., and Dunbar, R.I.M., Eds., Oxford University Press, Oxford, 251—268, 1996.

Boesch, C., et al., Is nut cracking in wild chimpanzees cultural behavior? *J. Hum. Evol.*, 4, 325–338, 1994.

Boinski, S., et al., Do brown capuchins socially learn foraging skills? In *The biology of traditions: models and evidence*, Fragaszy, D.M., and Perry, S., Eds., Cambridge University Press, Cambridge, 365–390, 2003.

Caldwell, C.A., and Whiten, A., Testing for social learning and imitation in common marmosets, *Callithrix jacchus*, using an artificial fruit, *Anim. Cogn.*, 7, 77–85, 2004.

Dawkins, R., *The ancestors tale: a pilgrimage to the dawn of life,* Weidenfeld and Nicolson, London, 2004.

de Waal, F.B.M., *The ape and the sushi master: cultural reflections of a primatologist*, Basic Books, New York, 2001.

Fragaszy, D.M., and Perry, S., Eds., *The biology of traditions: models and evidence*, Cambridge University Press, Cambridge, 2003.

Galef, B.G. Jr., The question of animal culture, *Hum. Nature,* 3, 157–178, 1992.

Galef, B.G. Jr., and Heyes, C.M., Eds., Social learning and imitation, Special issue, *Learning & Behavior*, 32, 1, 2004.

Goodall, J., Tool-using and aimed throwing in a community of free-living chimpanzees, *Nature*, 201, 1264–1266, 1964.

Goodall, J., *The chimpanzees of Gombe: patterns of behavior*, Harvard University Press, Cambridge, MA, 1986.

Heyes, C.M., and Galef, B.G. Jr., Eds., *Social learning in animals: the roots of culture*, Academic Press, London, 1996.

Hohman, G., and Fruth, B., Culture in bonobos? Between-species and within-species variation in behavior, *Curr. Anthropol.*, 44, 563–571, 2003.

Huffman, M.A., Current evidence of self-medication in primates: a multidisciplinary perspective, *Yrbk. Phys. Anthropol.*, 40, 171–200, 1997.

Huffman, M.A., Self-medicative behavior in the African great apes: an evolutionary perspective into the origins of human traditional medicine, *Bioscience*, 51, 651–661, 2001.

Huffman, M.A., and Hirata, S., An experimental study of leaf swallowing in captive chimpanzees: insights into the origin of a self-medicative behavior and the role of social learning, *Primates*, 45, 113–118, 2004.

Laland, K. N., and Hoppitt, W., Do animals have culture? *Evol. Anthropol.*, 12, 150–159, 2003.

Lonsdorf, E.V., Sex differences in the development of termite-fishing skills in the wild chimpanzees (*Pan troglodytes schweinfurthii*) of Gombe National Park, Tanzania, *Anim. Behav.*, 70, 673–683, 2005.

Lonsdorf, E.V., Pusey, E.A., and Eberly, L., Sex differences in learning in chimpanzees, *Nature*, 428, 715–716, 2004.

MacArthur, R.H., and Pianka, E.R., An optimal use of a patchy environment. *Am. Nat.* 102, 381–383, 1966.

McGrew, W.C., Evolutionary implications of sex-differences in chimpanzee predation and tool use, In *The great apes*, Hamburg, D.A., and McCown, E.R., Eds., Benjamin Cummins, Menlo Park, CA, 440–463, 1979.

McGrew, W.C., *Chimpanzee material culture: implications for human evolution*, Cambridge University Press, Cambridge, 1992.

Panger, M.A., et al., Cross-site differences in foraging behavior of white-faced capuchins (*Cebus capuchinus*), *Am. J. Phys. Anthropol.*, 119, 52–56, 2002.

Perry, S., et al., Traditions in white-faced capuchin monkeys, In *The biology of traditions: models and evidence*, Fragaszy, D. M., and Perry, S., Eds., Cambridge University Press, Cambridge, 391–425, 2003.

Rendell, L., and Whitehead, H., Cultures in whales and dolphins, *Behav. Brain Sci.*, 24, 309–382, 2001.

Slater, P.J.B., The cultural transmission of bird song, *Tr. Ecol. Evol.*, 1, 94–97, 1986.

van Schaik, C.P., et al., Orangutan cultures and the evolution of material culture, *Science*, 299, 102–105, 2003.

Waddell, P.J., and Penny, D., Evolutionary trees of apes and humans from DNA sequences, In *Handbook of symbolic evolution*, Lock, A., and Peters, C.R., Eds., Oxford University Press, Oxford, 53–73, 1996.

Whiten, A., Primate culture and social learning, *Cog. Sci.,* 24, 477–508, 2000.

Whiten, A., Horner, V., and Marshall-Pescini, S., Cultural Panthropology, *Evol. Anthropol.*, 12, 92–105, 2003.

Whiten, A., et al., Cultures in chimpanzees, *Nature*, 399, 682–685, 1999.

Whiten, A., et al., Charting cultural variation in chimpanzees, *Behavior,* 138, 1489–1525, 2001.

Whiten, A., et al., How do apes ape? *Learn. Behav.,* 32, 36–52, 2004.

Wrangham, R.W., Relationship of chimpanzee leaf swallowing to a tapeworm infection. *Am. J. Phys. Anthropol.*, 37, 297–303, 1995.

Wrangham, R.W., and Nishida, T., *Aspilia* spp. leaves: a puzzle in the feeding behavior of wild chimpanzees, *Primates*, 24, 276–282, 1983.

Yamakoshi, G., Dietary responses to fruit scarcity of wild chimpanzees at Bossou, Guinea: possible implications for ecological importance of tool use, *Am. J. Phys. Anthropol.*, 106, 283--295, 1998.

# 8 Language: Costs and Benefits of a Specialized System for Social Information Transmission

*Daniel Nettle*

## CONTENTS



## ABSTRACT

Language is often thought of as the crowning human adaptation, the one that allowed *Homo sapiens sapiens* to conquer the globe. The assumption underlying such ideas is that verbal transmission of information provides unalloyed benefits by reducing the costs of learning about the environment. However, this raises the question of why no other species has discovered such a good trick. I argue that verbal transmission is only likely to be adaptive in a restricted range of circumstances. Even then it cannot be exclusively relied on, and it causes problems of deceit and instances of maladaptation. We should expect natural selection to have made us discriminating evaluators of verbal information who ultimately trust the evidence of our senses.

Nonetheless, once language has become widespread, it can increase human adaptability by increasing the efficiency of individual learning.

## 8.1 THE GREAT LEAP FORWARD?

In language, human beings have a unique system for the cheap, reliable exchange of learned information. This sets them apart from all other species. A number of other sophisticated information exchange systems have been documented in nature, but each is limited to a particular specialized domain, such as predator types for vervet monkeys (Seyfarth et al., 1980), food sources for bees (Von Frisch, 1967), or clan membership in cetaceans (Rendell and Whitehead, 2003). Human language, on the other hand, through the arbitrary and learned nature of its vocabulary and the combinatorial properties of its syntax, allows an unlimited number of messages about an unbounded array of topics to be exchanged. Language is underlain by specialized neural structures and thus appears to be a human-specific adaptation for generalized information exchange (Pinker and Bloom, 1990). I will call such exchange *verbal transmission*, and henceforth assume verbal transmission to be the primary function of language. Thus, henceforth, I will refer to language as a set of capacities and verbal transmission as the activity that those capacities allow.

It is not difficult to identify putative benefits of verbal transmission. Indeed, accounts of the evolution of language have been more concerned with the origin of the necessary neural and morphological mechanisms than bothered by the question of what the adaptive value of the system might be — as if the latter were self-evident. Thus, for Pinker and Bloom (1990, p. 712): "[There is] an obvious advantage to being able to acquire … information second-hand: by tapping into the vast reservoir of knowledge accumulated by other individuals, one can avoid having to duplicate the possibly time-consuming and dangerous trial-and-error process that won that knowledge."

Many accounts of human evolution have assumed or implied that the origin of language allowed the explosion in human numbers and dispersal that began some time before 50,000 years ago and led to the colonization of every continent of the globe (Klein, 2000). It is certainly true that humans produce sophisticated and cumulative technological and social traditions and that in virtue of these, they are able to successfully inhabit a vast range of environments and gradually increase their exploitation of those environments. This is only possible because of their ability to learn culturally, in the broad sense of transmission of information between individuals (Boyd and Richerson, 2004; Henrich and McElreath, 2003). Verbal transmission is a key component of cultural abilities because it allows information to be encoded and transmitted with relatively high fidelity and low cost.

However, the advantages of verbal transmission immediately raise what is known as the "why not baboons?" problem (Henrich and McElreath, 2003, p. 126). If verbal transmission is so fantastically adaptive, why has only one of several highly social, relatively encephalized, group-living primates evolved it? Why, in short, is such a good trick so rare?[1] Answers to this question that rely exclusively on the contingent lack of availability of the relevant neural or articulatory machinery do not convince.

For one thing, highly complex adaptations such as they eye have originated several times independently in nature, despite their reliance on rare and specialized tissues. For another, such a view would imply a Whig view of history, as if evolution were always trying to get to language, but somehow most species were in limbo waiting for the necessary mutations to arise.

In this paper, instead, I argue that reliance on verbal transmission has costs as well as benefits, costs active both in the ancestral past and today. This means that it can only invade a nonverbal population under restricted circumstances. These restricted circumstances must have, at some point, obtained for a prolonged period during human evolution, and during this period the psychological adaptations for language were put in place. Once established, language clearly increased mean fitness and human adaptability, as the explosion of humans across the world attests. However, the question of exactly how adaptability was increased is not as simple as it may seem, as I shall show below.

In Section 8.2, I review the costs and benefits of verbally transmitting information rather than representing it genetically or producing it by individual learning. I use the considerations developed in Section 8.2 to develop a formal model of the evolution of verbal transmission in Section 8.3. In Section 8.4, I consider how verbal transmission increases human adaptability, and Section 8.5 presents the overall conclusions.

## 8.2 COSTS AND BENEFITS OF LANGUAGE

### 8.2.1 Benefit 1: Flexibility

Evolutionary theorists have addressed the question of when evolution should favor reliance on learning rather than provide behavioral strategies that are encoded genetically (Bergman and Feldman, 1995; Stephens, 1991). Fixed, innate strategies have the advantage of being available reliably whatever the contingencies of learning history happen to be, and are potentially available immediately at birth. Mutation notwithstanding, innate programs produce the optimal behavior more reliably and cheaply than learning, which involves more extra stages both inside and outside the organism. In general, these benefits will mean that behaviors will be encoded genetically unless there are counteracting advantages. Defensive reflexes, crying, and suckling are innate for these reasons.

Reliance on learned information can evolve where the rate of environmental change and, thus, change in the behavior that is locally optimal is too rapid for selection to assemble innate adaptations to produce the behavior directly (Stephens, 1991). The disadvantages in terms of learning time and the risk of mislearning are outweighed by the more rapid tracking of environmental variation in time or space. However, this leaves open the question of what kind of learning will be favored. In general, there are two possibilities. In individual learning, the organism bases its behavior on its own history of interaction with the environment. In social transmission, locally optimal strategies are passed from one individual to another either by imitation or by encoding them in language.[2] Verbal transmission is thus a subset of social information transmission.

### 8.2.2 Benefit 2: Reduced Learning Costs and Increased Fidelity

Let us assume, then, that reliance on some kind of learning will be favored when the environment changes too fast for selection to track the variation genetically. This raises the issue of what factors favor individual learning versus social, and particularly verbal, transmission. It has generally been assumed, as in the Pinker and Bloom quotation reproduced in Section 8.1, that the chief advantage of verbal transmission is that it reduces the cost, in time or energy, of acquiring information. It is much quicker and safer to be told which of two plant species is toxic than to have to discover this by experiment. A benefit of verbal transmission will thus be cost reduction.

A second plausible benefit of verbal forms of information transmission is that fidelity of knowledge about the environment is increased. For example, an environment may contain a species of which 30% of examples are toxic. A simple statistical power calculation shows that the number of individual experiments required for there to be a 95% chance of finding out that this plant can be toxic is nine. This may be too many, in terms of risk of exposing oneself to a fatal level of toxin. Nine trials could also accumulate by observation, but encoding the information verbally avoids the unnecessary repetition of eating trials by new individuals who have not made enough observations to be sure. The information can be handed down through time and out across space at minimal extra cost.

### 8.2.3 Cost 1: Reliance on Socially Transmitted Information

The discussion so far has focused on the clear benefits to individuals of receiving verbal information. However, there are potential costs too. Reliance on such information, as Emile Durkheim first pointed out nearly 100 years ago, binds human beings into an immense web of cooperation, stretching not only into space but through time as well (Durkheim, 1982, p. 248). Cooperation can bring adaptive benefits, but a huge literature in behavioral biology has concentrated on the difficulty of making cooperation evolutionarily viable (for some classic sources, recent developments, and reviews, see Axelrod and Hamilton, 1981; Boyd and Richerson, in press; Doebili, Hauert, and Killingback, 2004; Dugatkin, 1997; Hamilton, 1964; Panchanathan and Boyd, 2003; Trivers, 1971). The essential problem is that cooperation involves taking a personal cost (or at the very least, an opportunity cost) to reap a larger benefit. This leads to the evolution of strategies that take the benefit of cooperation without paying the costs. Such free-riding strategies soon invade a cooperating population, leading to the extinction of cooperation, because however badly a free-rider does, a cooperator in the same population will always do even worse.

In the evolution of signaling, this problem is particularly acute, because signals can be faked and thus are potentially misleading. Individuals maximize their own fitness, not that of the population, and where individuals have conflicting interests it is often adaptive for them to withhold information or deceive their rivals. As long as signals are cheap to produce, as language is, then under many conditions signals will become worthless as the system is invaded by misleading information (Dawkins and Krebs, 1978; Maynard-Smith, 1982; Silk, Kaldor, and Boyd, 2000).

Humans have somehow overcome this problem, at least often enough for language as a system to be viable. How they might have done so is discussed below. Certainly, humans show a general willingness to provide benefits to each other that is greater than that which seems optimal from an immediate fitness perspective (Boyd and Richerson, in press; Gintis et al., 2003). In general, to rely on information cheaply signaled by conspecifics is to expose oneself to the possibility that that information is false. To provide truthful information to conspecifics where there is any cost at all of doing so is a cooperative act. The development of language would thus be unlikely to be adaptive in a context where very extensive cooperation had not become prevalent, and such cooperation is rare in nature. Moreover, although the problem in general may have been overcome in humans, experience shows that there is still residual deception of individuals by others. Thus, the cost of relying on verbal information is still operative because it might have been provided maliciously or transmitted sincerely by someone who got it from someone who provided it maliciously, etc. Thus, in the model in Section 8.3, it is necessary to include a parameter that represents the probability of information being reliable.

There are several conditions that allow the problem of honest signaling to be overcome. Where individuals are closely related, they share sufficient genetic interests for a higher level of cooperation to be maintained than where they are unrelated (Hamilton, 1964). This principle accounts for the existence of cheap but honest signaling about the whereabouts of resources in social insects, where levels of relatedness are high. Individuals that encounter each other frequently can develop cooperation based on reciprocity, with one providing a benefit to the other in return for the reverse at a later point in turn (Trivers, 1971). Honest signaling can arise in this manner (Silk et al., 2000). However, such reciprocal altruism is easily disrupted by free-riders and unstable in sizable groups (Boyd and Richerson, 1988). For these reasons it is thought to be quite rare in nature and insufficient to account for the general human propensity to cooperate.

In sizeable groups, it is more plausible that cooperation is maintained by indirect reciprocity than by dyadic reciprocation. Indirect reciprocity refers to situations where noncooperators become publicly known as such and are excluded from future cooperation not just by the individuals they failed to cooperate with but by everyone in the group (Panchanathan and Boyd, 2003). The maintenance of indirect reciprocity is easiest when there is a mechanism of social information transmission about noncooperators, and it has been argued that one of the primary functions of language is to exchange social information (gossip) of this kind (Dunbar, 1996). However, this obviously poses a second-order problem: how to know that gossip is reliable and not malicious (Power, 1998). Cooperation is a *prerequisite* for language to be reliable, so if language is a prerequisite for indirect reciprocity and indirect reciprocity sustains human cooperation, we have a paradox. In small communities, indirect reciprocity could be sustained by observational learning, and then language and more extensive cooperation bootstrap off each other. Though indirect reciprocity is theoretically plausible, it cannot be assumed that it is universally stable, and thus the stability of cooperation cannot be taken for granted.

Two other factors have been proposed for the maintenance of cooperation. One is punishment; if individuals punish nonconformists then cooperation, or indeed any other behavior, can be stabilized within sizeable groups (Boyd and Richerson, 1992). This too raises a second-order problem. If individuals have to punish at a cost to themselves, then there will arise free-riders who benefit from being in a group with punishment but do not contribute to the punishment activity. Soon punishment goes extinct as individuals who skimp on the cost of punishing outcompete punishers. Panchanathan and Boyd (2003) have shown that the second-order problem can be solved where punishment takes the form of the withholding of benefits, rather than the inflicting of costs, because in this case the costs of punishment are negative and there is no temptation to skimp on them. In this scenario, punishment and indirect reciprocity are two sides of the same coin. However, even in this scenario, though cooperation is an evolutionarily stable strategy, so is a population of noncooperators. Cooperators can only invade a population composed of noncooperators if they are given an initial advantage by being allowed to preferentially assort with each other or their kin when they are rare. This is a common feature of many models of the evolution of cooperation (see, e.g., Nettle and Dunbar, 1997).

There is a final factor that has been argued to potentially account for the propensity of human beings to cooperate so widely — group selection. Group selection occurs when the total population is divided into interacting groups that persist or go extinct at different rates according to the properties of the individuals within them. Group selection can, in principle, favor the proliferation of traits that help the whole group to survive even at cost to the individual. However, the conditions for group selection to be important are extremely restrictive. For one thing, selection between individuals within each group will tend to reduce the proportion of group altruists, and because the extinction rate of individuals is much faster than that of groups, this will be the dominant force. Moreover, migration among groups reduces group-level variation, which is required for selection to operate. Thus, group selection is generally swamped by individual-level selection, and the result is the persistence of no more cooperation than would be predicted by assuming groups did not exist.

However, the argument has been made that group selection based on groups sharing cultural characteristics is plausible (Soltis, Boyd, and Richerson, 1995). Because human societies enforce conformity in cultural practices, by punishment or conformist adoption of norms, group traits can be maintained homogenously even in the face of a flow of migrants or differential reproductive success of individuals within the group. As long as cultural groups sometimes die out or disperse, which is true for example of New Guinea clans (Soltis, Boyd, and Richerson, 1995), and as long as the probability of this is related to the proportion of prosocial behaviors within the group, group selection will be operative. However, it is still likely to be a weak force compared to individual-level selection, and no one claims that humans are unconditional progroup altruists. Furthermore, the possibility of cultural group selection depends on the prior existence of cultural, perhaps including linguistic, transmission of norms.

The problem of human cooperation is far from solved, and it is beyond the scope of this paper to discuss it further. However, the brief review given so far allows a number of conclusions relevant to the current thesis that language is not unconditionally adaptive. These conclusions are:

1. The cooperativeness of unrelated individuals with one another should not be assumed. For cooperation to persist in sizeable groups, one or several of a number of specific and rare conditions must obtain. Widespread cooperation is rare in nature.
2. The benefits of verbal transmission depend on the stability of widespread cooperation; otherwise the information received verbally, which is cheap, fakeable, and often not immediately verifiable, would not be reliable.
3. Several of the factors that might explain human cooperativeness, such as indirect reciprocity and cultural group selection, are dependent on or at least facilitated by the existence of language, thus causing a chicken-and-egg problem. It may be that language began to be used in small groups characterized by high levels of genetic relatedness and limited opportunity to defect but that, once in place, language and culture potentiated the expansion of groups, which further enhanced the selective benefits of using verbally transmitted information, and so on in a feed-forward loop.
4. Even in fully modern populations, humans are not unconditional cooperators, and thus a potential drawback of using verbal information is that it may be unreliable. This should be considered in any model of the evolution of language.

### 8.2.4 Cost 2: Environmental Heterogeneity and Change

Under benefit 1, I argued that an advantage of using learned over innate information was that it allowed the tracking of environmental change too rapid for selection to follow. This is certainly an advantage of language relative to innate information. On the other hand, a cost of using linguistically transmitted learned information rather than individually learned information is that it allows slower response to environmental change. To see why this is the case, consider information about a foraging strategy that is successful in the local environment. If this information is acquired by the individual in the place where he currently is, there will be a costly learning process, but on the other hand, the information will be guaranteed to be locally appropriate and up to date. If the information is acquired verbally, then the source of the information will have either learned it individually himself in a previous time step or in turn acquired it verbally from someone else. If he acquired it verbally, then the person from whom he acquired it will in turn have either learned it individually or acquired it verbally, and so on *ad infinitum*.

The longer the chain of verbal links is before someone is reached who actually learned the information first hand, the greater the probability is that the environment in which the information was originally learned has changed or is different from the one at the current location. If environmental change is very fast or, equivalently, if the habitat is spatially very heterogeneous, then individual learning is favored over social transmission. This is because, despite the costs of individual learning, it is better to learn in the current place or time than be given hand-me-down information that is outdated or locally nonoptimal. This result can be shown in a wide variety of models (see Section 8.3), and leads to the general conclusion that social transmission is only adaptive where the environment changes too fast to be tracked by selection on genes but not so fast that constant individual learning is required (Boyd and Richerson, 1985; Henrich and McElreath, 2003). The late Pleistocene was characterized by dramatic climate fluctuations over the order of a few decades — fast in evolutionary terms, but long relative to human lifespan. This leads to the influential idea that human reliance on language and culture is an adaptation to Pleistocene climatic variability (Boyd and Richerson, 2004).

### 8.2.5 Cost 3: The Chinese Whispers Problem

As described above, the more verbal transmission there is in the population, the longer on average the chain of verbal information links is before someone is reached who actually acquired the information first hand. This leads to a puzzling result (Boyd and Richerson, 1995; Rogers, 1988). As the proportion of reliance on verbal information (versus individual learning) increases, the value of that information drops, and the fitness of verbal learners decreases. This is shown analytically in Section 8.3, but to see why it must be the case, consider the extremes. When verbal learning is extremely rare, then any information acquired verbally will almost certainly have been learned individually by the next person in the chain and thus is likely to be up to date and locally apt. On the other hand, consider the case where verbal learning has gone to fixation in a population. Now, everyone simply relies on verbal accounts of what has been done before, and no new learning brings better information into the system. This becomes like the old parlor game of Chinese Whispers; a sentence is passed round and round a circle of people, gradually becoming more distorted. In this case, it becomes distorted not because of the fidelity of language, but because as the environment changes, the information becomes less and less useful.

Two things follow from the Chinese Whispers problem. One is that 100% verbal transmission of information is not adaptive (Rogers, 1988). There will instead always be a mixed equilibrium of some verbal transmission and some individual learning (see Section 8.3). Someone somewhere has to actually be testing things out, otherwise unreliable information is just recycled. Second, the mean fitness of a population at equilibrium can never be increased by the adoption of verbal transmission alone. This is because reliance on verbal information increases to the point where the payoff of the verbal strategy is identical to that of individual learning (see Section 8.3). This is a very important point because those who argue that it was the adoption of language that allowed humans to dramatically expand their numbers across the globe are assuming that using language enhanced mean fitness. It may well be that it did, but for that to be the case it must have done more than just eliminate learning costs. I consider the question of what else it does for humans in Section 8.4.

### 8.2.6 Summary of Costs and Benefits

To summarize Section 8.2, verbal transmission has a number of benefits and costs. It is more flexible than reliance on innate information, though not as flexible as individual learning. We should thus expect it to be favored in situations where the environment changes rapidly, but not too rapidly. It saves the costs of individual learning. However, on the cost side, it makes organisms reliant on information that is fakeable and potentially incorrect. Thus, it is only likely to be adaptive either where individuals are closely related or where widespread cooperation has somehow become established. Even then, there is a residual risk of deception. Moreover, because of the Chinese Whispers problem, the whole population can never be exclusively reliant on verbally transmitted information. There must always be a component of individual learning updating information within the system. Finally, by reducing learning costs alone, verbal transmission can never increase the mean fitness of a population. Any absolute increment in fitness and adaptability due to language must come from another source.

The inventory of costs and benefits laid out in this section leads naturally to the statement of a more formal model of the evolution of verbal transmission. This model, which draws heavily on those of Rogers (1988) and Boyd and Richerson (various papers, especially 1995), allows more precise predictions to be made about the general conditions for the evolution of language.

## 8.3 EVOLUTION OF VERBAL TRANSMISSION: A FORMAL MODEL

Let us assume that there is a temporally variable environment with an infinite number of states. The organisms must forage in that environment as best they can. There is a skill available that increases foraging efficiency, so that unskilled individuals have fitness $w$ but skilled individuals have fitness $w + b$. Each generation there is a probability $u$ that the environment switches to another state in which the old skill is no longer useful though there exists a new skill for the new environment that gives fitness $w + b$.

We will consider two genotypes with different learning rules. *Individual learners* acquire the skill individually, with probability $d$ and cost $c$. The parameter $d$ is the efficiency of learning; where it is low, even prolonged trial and error does not guarantee that the optimal skill will be found. The parameter $c$ is the cost, in time or energy, of learning something for oneself. *Verbal learners* obtain their information verbally from another in the community, chosen at random. There is no learning cost for them. (We can make the model more complex by making verbal learning have a cost, but because we generally assume that the costs of verbal learning are less than those of individual learning this case can be ignored for now.) More complex genotypes that use a combination of individual and verbal learning would be more realistic, but the use of the two simple genotypes allows the model to be simple and does not affect the general conclusions. Similarly, a more complex model would also allow possibilities intermediate between individual and verbal learning, such as social enhancement of individual learning, but for simplicity, the only comparison here is between the two extreme alternatives.

Consider a population with a frequency $p$ of verbal learners and $(1 - p)$ of individual learners. The fitness of the individual learners is:

$$w_I = w + db - c \tag{8.1}$$

Verbal learners acquire behaviors that were originally individually learned at some earlier time (through a chain of intermediaries). The probability that the behavior was originally learned one generation ago would be $(1 - p)$. The probability that it was learned a generation before that would be $p(1 - p)$, and the generation before that, $p^2(1 - p)$. Thus, in general, the probability that the behavior was originally learned $r$ generations ago[3] is $p^{(r-1)}(1 - p)$.

The fitness of a verbal learner will be $w + b$ if the skill they learn about verbally is the optimal one for the current environment, and $w$ if it is not. There is some chance $e$ that they will be fed misleading or malicious information. The probability of the information being sincere is thus $(1 - e)$. The information will also only be optimal if there has been no environmental perturbation since the behavior was originally learned. The probability that there has been no perturbation in $r$ generations is $(1 - u)^r$. Therefore, the fitness of a verbal learner will be given by Equation 8.2:

$$w_v = w + (1-e)b\sum_{r=1}^{\infty} p^{r-1}(1-p)\,(1-u)^r \quad (8.2)$$

Equation 8.2 can be rearranged to give Equation 8.3:

$$w_v = w + b(1-e)(1-p)(1-u)\sum_{r=1}^{\infty} p^{r-1}(1-u)^{r-1} \quad (8.3)$$

Using the formula for the sum of an infinite geometric series, Equation 8.3 gives Equation 8.4:

$$w_v = w + b\cdot\frac{(1-e)(1-p)(1-u)}{1-p(1-u)} \quad (8.4)$$

First, let us derive the Chinese Whispers problem by a simple combination of parameters where verbal information is perfectly honest, the environment changes moderately fast, and individual learning is moderately costly (see Figure 8.1). The result shown in Figure 8.1 is that, given we have assumed that verbal information is generally reliable and environmental change is quite slow, verbal learners have a large fitness advantage when rare. However, this advantage diminishes as their numbers increase. At fixation, their fitness is less than that of individual learners, and, as a consequence, the evolutionary equilibrium is always a mixture of strategies.[4] There is another important consequence; at equilibrium the mean fitness of an individual in the population, regardless of which strategy they follow, is exactly the same as if everyone was an individual learner. Verbal learning has become widespread but not increased the average fitness in the population.

Now consider the conditions for a rare verbal learner to become numerous in a population of individual learners. Thus we set $p$ close to zero. Verbal learners will increase in numbers where $w_v > w_I$, that is where Equation 8.5 is satisfied.

$$w + b(1-u)(1-e) > w + db - c \quad (8.5)$$

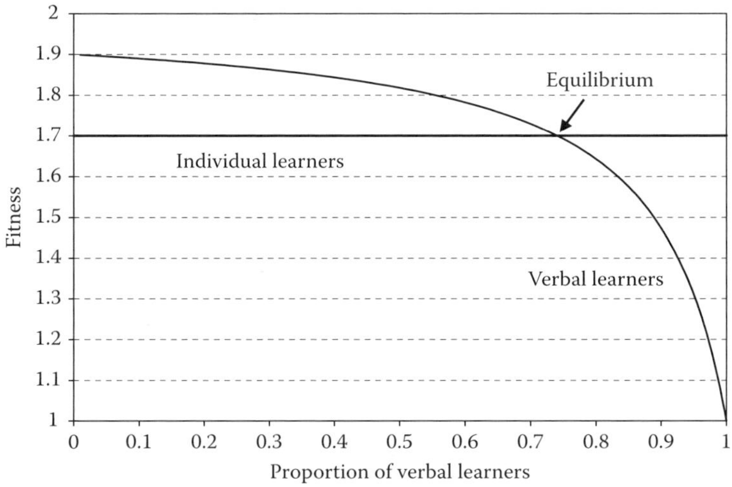


**FIGURE 8.1** The Chinese Whispers problem. Using the model outlined in the text with, for simplicity, $e$ at 0, $w$ and $b$ at 1, $u$ at 0.1, $d$ at 0.9, and $c$ at 0.2. $p$ is allowed to vary from 0 (no verbal learners) to 1 (all verbal learners). Whereas the fitness of individual learners is constant, fitness of verbal learners declines as their frequency increases, such that at fixation their fitness would be at background level $w$. As a result, the evolutionary equilibrium is always a mixture of strategies with proportions determined by the point of crossing of the two lines (in this instance, about 75% verbal learning). Making verbal information less reliable or the speed of environmental change faster shifts the equilibrium point to the left. Making individual learning more costly or less reliable shifts it to the right.

Equation 8.5 can be arranged to give Equation 8.6:

$$b(1-u)(1-e) > db - c \tag{8.6}$$

Equation 8.6 defines an area of the parameter space where verbal transmission can become more common when rare. Some illustrative results are given in Figure 8.2. The model shows that there are combinations of parameters where verbal transmission can spread, and many others where it cannot. In general, other things remaining constant:

- As environmental change gets faster, or malicious information becomes more common, it gets more difficult for verbal transmission to spread.
- As the efficiency of individual learning diminishes, verbal transmission becomes more likely to spread.
- As the costs of individual learning increase, the fitness advantage of verbal transmission increases.

The various parameters are traded off against each other so that, for example, where the cost of individual learning is low, it is hard for verbal transmission to

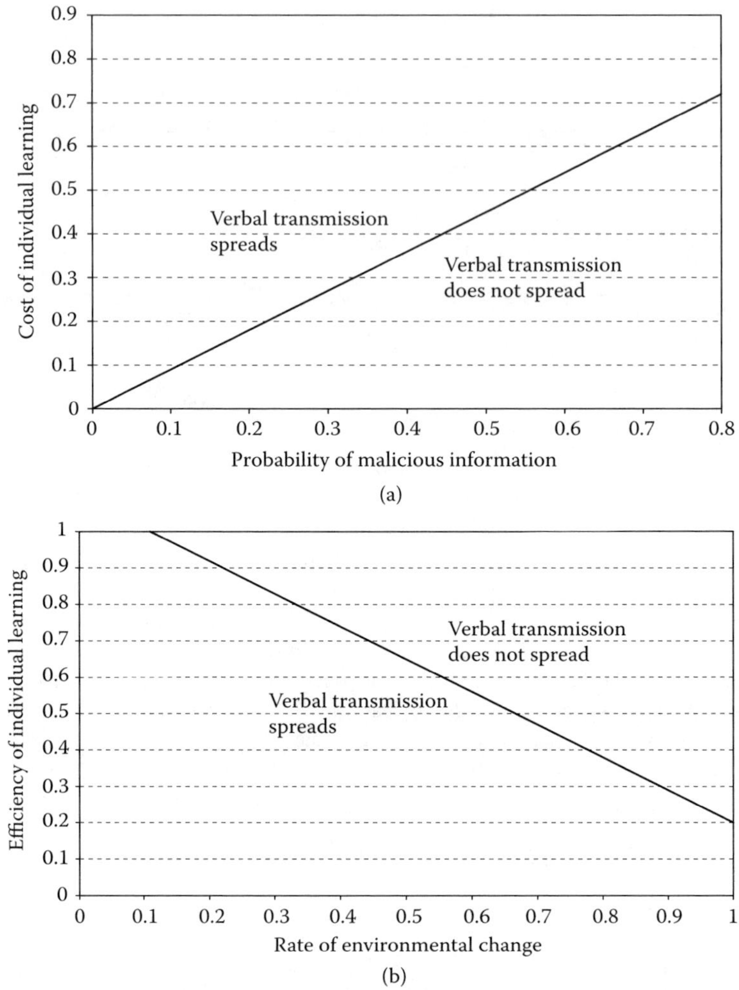


**FIGURE 8.2** Regions of parameter space where verbal transmission can and cannot spread when it is rare. The parameters $w$ and $b$ are set at 1. (a) $u = 0.1$ and $d = 0.9$. Verbal transmission can spread if the cost of individual learning is high and/or the probability of malicious information is low. (b) $e = 0.1$ and $c = 0.2$. Verbal transmission cannot spread if environmental change is very fast and/or individual learning very efficient.

evolve even if verbal information is very reliable, but as the cost of individual learning increases, the reliability threshold required for verbal transmission gets lower (Figure 8.2a). Similarly, if environmental change is very fast, then verbal transmission cannot spread even if the efficiency of individual learning is very low (Figure 8.2b).

## 8.4 LANGUAGE AND COMPLEX CULTURAL ADAPTATIONS

We have seen that linguistic transmission can proliferate under a certain set of conditions and equilibrate at the point where the mean fitness of individuals is exactly the same whether they use language or individual learning. However, the argument usually made is that language *increased* average human fitness, allowing the population to explode and prosper in new habitats all over the world. Boyd and Richerson (1995) have argued, correctly, that this must mean that language does more than just reduce the costs of learning that individuals could achieve on their own. For language or culture to increase human adaptability, they must actually allow better adaptations to be acquired than individual learning could ever achieve.

Language surely does this, not by replacing individual learning, but by canalizing it. Language provides an enormous compression of information and, because it is categorical, a huge reduction of ambiguity. Verbal information is available as background for individual experiment. In skilled domains such as foraging, medicine, or construction of technology, verbal information interacts dynamically with individual learning, preguiding it to fruitful regions and prewarning it of possible pitfalls. This allows more reliable discovery of complex local solutions in a multidimensional fitness landscape where individual trial and error might get stuck in a valley. Individuals add their learning to the existing structure and pass it back into the mix, so that cognition becomes distributed across many individuals (Hutchins, 1995). Just as adding neurons to a neural network increases its computational power, so adding individuals to a community increases its cultural learning potential. Note again, though, that this relies on verbal transmission never being a pure strategy but individuals retaining discriminating ultimate reliance on their individual experience.[5]

Language may also have benefits internal to the individual. There has been much interest within linguistics in the notion that language is a system of internal representation as much as interindividual communication (Hauser, Chomsky, and Fitch, 2002). Because linguistic representations are discrete and systematically combinable, they can be used internally for processes of logical reasoning and off-line behavior planning. This does not mean that the language of mental representations is narrowly tied to any particular natural language, but rather that the evolution of a discrete combinatorial communication system made additional representational possibilities available to internal as well as communicative processes, which in turn increased the efficiency and scope of individual learning.

Language as a system of representation and distributed cognition, means that cultures can acquire adaptations that no nonlinguistic individual could ever have come up with. Put together they allow cumulative and increasingly sophisticated knowledge-based local adaptations. They unite powerfully when cultures develop external means of information storage, in tallies, painting, writing, and ultimately computers. These developments increase the potential for discovery of good local adaptations by a further order of magnitude (Donald, 1991).

## 8.5 CONCLUSION: WHY LANGUAGE IS RARE

We have seen that there are great rewards in store for any species that gets an efficient system of social information transmission running widely. As so often argued, this may be the key to the human demographic and cultural explosion. Language would allow this not by replacing individual learning but by allowing it to become more efficient and better guided. But if the fitness benefits are so overwhelming, why is language so rare? In short, why not baboons?

We have seen that the rarity of language is not the simple contingency of the right mental machinery not having been widely available. Selection is a powerful mechanism for producing adaptations, and where there is a fitness payoff, it has had no trouble producing the discrete alarm calls of vervet monkeys or the waggle dance of bees, to cite but two examples. Instead, we have seen that verbal transmission would only be adaptive in a restricted set of circumstances: where environmental change is fast enough but not too fast; where individual learning is not too cheap or efficient; and above all, where cooperation is widespread. It is probably rare for these circumstances to prevail, but once they do, the existence of language reinforces its benefits, for example by allowing cooperation to be policed and group norms to be agreed upon (Boehm, 1996). Thus language is a good illustration of the path-dependent nature of evolution.

Moreover, the initial transition to widespread language would not immediately lead to the human explosion because, where language simply reduces learning costs, we have seen that at equilibrium it does not increase average fitness. Rather, the transition to language to save learning costs made possible later cognitive adaptations whose benefit was to make individual learning more sophisticated. Distributed cognition leads to cumulative increases in the sophistication of learned adaptations, which gives human cultures their initially slow, but undeniably directional, increase in overall complexity.

However, even with all these benefits in place, language is a costly business. Partly because of language, people can be persuaded to behave in ways detrimental to their fitness and can end up with beliefs and values far from optimal in their environments. Even where cultural transmission is generally adaptive, it can lead to many instances of groups behaving in a collectively maladaptive way, especially if conformism is strong (Boyd and Richerson, 1985). In the United States in the first half of the 20th century, there was a craze for radioactive drinks, cures, and spas. People spent considerable amounts of money on irradiating themselves, at least in part because of verbal transmission. A Dr. C. G. Davis had written in the *American Journal of Clinical Medicine* that "Radioactivity prevents insanity, rouses noble emotions, retards old age, and creates a splendid youthful joyous life."

This maladaptative cultural practice, which did not die out until the late 1950s, could never have become prevalent by trial and error learning because it had no positive effects whatever. Deceit and credulity must have been recurring problems in human evolution, and continue to be so today. As a result, we have sophisticated intuitions about social cheaters and deception and are not cultural dupes (as some social constructionists seem to imply) but generally skeptical evaluators of verbal information who ultimately trust most in the evidence of our senses. Language is the crowning human adaptation, but like any other adaptation, it had costs as well as benefits and is only adaptive in relation to a particular set of circumstances.

## NOTES

1. Cultural variation of a basic sort is relatively common in nature (Boyd and Richerson, 1996; Heyes and Galef, 1996), with numerous examples of socially transmitted local foraging practices. However, such variation has two features that differ from verbal transmission in humans. First, it is brought about by performance of the behavior itself, not by conversion of information about the behavior into another format (words) and back again. Second, the behavioral traditions are not cumulative in the sense of leading to progressively more complex behaviors that no individual could have come up with alone.
2. There are also intermediate cases between individual and social learning, such as social enhancement of individual learning (see Boyd and Richerson, 1996; Heyes and Galef, 1996). This occurs where the activity of older animals increases the probability that naïve ones will manage to learn a particular behavior, for example, by causing them to be in particular locations. Many animal cultural traditions belong to this weaker class than fully symbolic, cumulative human social information transmission. For clarity of argumentation in this chapter, I contrast pure verbal transmission with pure individual learning though recognizing that there are in fact intermediate possibilities.
3. It is assumed here that natural selection is weak, that is, cultural change is much faster than genetic change, so that its effect on p can be ignored in this calculation.
4. In this version of the model, a mixture of genotypes but a single genotype that relied on a mixture of verbal and individual learning would also be an ESS.
5. Indeed, if conformism is very strong within a group, then adding more individuals is not guaranteed to lead to the discovery of better adaptations. The viability of cultural group selection demands high conformism, but the efficiency of distributed cognition demands skepticism and independent mindedness. These conflicting influences can be seen in many cultural traditions, including of course science itself.

## REFERENCES


Axelrod, R. and Hamilton, W.D., The evolution of cooperation, *Science*, 211, 1390–1396, 1981.

Bergman, A. and Feldman, M.W., On the evolution of learning: representation of a stochastic environment, *Theor. Pop. Biol.*, 48, 251–276, 1995.

Boehm, C., Emergency decisions, cultural selection mechanics, and group selection, *Curr. Anthropol.*, 37, 763–793, 1996.

Boyd, R. and Richerson, P.J., *Culture and the evolutionary process,* University of Chicago Press, Chicago, 1985.

Boyd, R. and Richerson, P.J., The evolution of reciprocity in sizable groups, *J. Theor. Biol.*, 132, 337–356, 1988.

Boyd, R. and Richerson, P.J., Punishment allows the evolution of cooperation (or anything else) in sizable groups, *Ethol. Sociobiol.*, 13, 171–195, 1992.

Boyd, R. and Richerson, P.J., Why does culture increase human adaptability? *Ethol. Sociobiol.*, 16, 125–143, 1995.

Boyd, R. and Richerson, P.J., Why culture is common but cultural evolution is rare. *Proc. Br. Acad.*, 88, 73–93, 1996.

Boyd, R. and Richerson, P.J., *Not by genes alone*, University of Chicago Press, Chicago, 2004.

Boyd, R. and Richerson, P.J., Solving the puzzle of human cooperation, In *Evolution and culture*, Levinson, S., Ed., MIT Press, Cambridge, MA, Forthcoming.

Dawkins, R. and Krebs, J.R., Animal signals: information or manipulation? In *Behavioral ecology: an evolutionary approach,* Krebs, J. R., Davies, N.B., Eds., Blackwell, Oxford, 282–309, 1978.

Doebili, M., Hauert, C., and Killingback, T., The evolutionary origin of cooperators and defectors, *Science*, 306, 859-862, 2004.

Donald, M., *The origins of the modern mind*, Harvard University Press, Cambridge, MA, 1991.

Dugatkin, L. A., *Cooperation among animals: an evolutionary perspective,* Oxford University Press, New York, 1997.

Dunbar, R. I. M., *Grooming, gossip and the evolution of language,* Faber, London, 1996.

Durkheim, E., *The rules of the sociological method*, Macmillan, London, 1982.

Gintis, H., Bowles, S., Boyd, R., and Fehr, E., Explaining altruistic behavior in humans. *Evol. Hum. Behav.,* 24, 153–172, 2003.

Hamilton, W.D., The genetical evolution of social behavior, Parts I and II. *J. Theor. Biol.*, 7, 1–52, 1964.

Hauser, M., Chomsky, N.A., and Fitch, W.T., The faculty of language: what is it, who has it, and how did it evolve? *Science*, 298, 1569–1579, 2002.

Henrich, J. and McElreath, R., The evolution of cultural evolution. *Evol. Anthropol.,* 12, 123–135, 2003.

Heyes, C. and Galef, B.G., *Social learning in animals: the roots of culture,* Academic Press, San Diego, CA, 1996.

Hutchins, E., *Cognition in the wild*, MIT Press, Cambridge, MA, 1995.

Klein, R.G., Archeology and the evolution of human behavior, *Evol. Anthropol.*, 9, 17–36, 2000.

Maynard-Smith, J., Do animals convey information about their intentions? *Journal of Theor. Biol.,* 97, 1–5, 1982.

Nettle, D., and Dunbar, R.I.M., Social markers and the evolution of reciprocal exchange, *Curr. Anthropol.,* 38, 93–99, 1997.

Panchanathan, K. and Boyd, R., A tale of two defectors: the importance of standing for the evolution of indirect reciprocity, *J. Theor. Biol.*, 224, 115–126, 2003.

Pinker, S., and Bloom, P., Natural language and natural selection. *Behav. Brain Sci.,* 13, 707–726, 1990.

Power, C., Old wives' tales: the gossip hypothesis and the reliability of cheap signals, In *Approaches to the evolution of language,* Hurford, J.R., Studdert-Kennedy, M., and Knight, C., Eds., Cambridge University Press, Cambridge, 111–129, 1998.

Rendell, L.E. and Whitehead, H., Vocal clans in sperm whales (Physeter macrocephalus), *Proc. Roy. Soc. Lond. Series B*, 270, 225–231, 2003.

Rogers, A. R., Does biology constrain culture? *Am. Anthropol.,* 90, 819–831, 1988.

Seyfarth, R.M., Chenery, D.L., and Marler, P., Vervet monkey alarm calls: semantic communication in a free-ranging primate, *Anim. Behav.*, 28, 1070–1094, 1980.

Silk, J. B., Kaldor, E., and Boyd, R., Cheap talk when interests conflict, *Anim. Behav.,* 59, 423–432, 2000.

Soltis, J., Boyd, R., and Richerson, P.J., Can group-functional behaviors evolve by cultural-group selection — an empirical test, *Curr. Anthropol.,* 36, 473–494, 1995.

Stephens, D.W., Change, regularity, and value in the evolution of animal learning. *Behav. Ecol.,* 2, 77–89, 1991.

Trivers, R.L., The evolution of reciprocal altruism, *Q. Rev. Biol.,* 46, 35–57, 1971.

Von Frisch, K., *The dance language and orientation of bees*, Harvard University Press, Cambridge, MA, 1967.

# 9 The Evolution of Social Information Transmission in *Homo*

*Steven Mithen*

## CONTENTS



## ABSTRACT

The social transmission of information provides the basis for the development of cultural traditions, as found not only in human communities but those of a variety of other species, notably chimpanzees. The cultural traditions of modern humans are far more complex and diverse than those of other species, which reflects the significance of socially transmitted information within and between generations. Modern humans use a vast array of methods for transmitting such information, only some of which operate with awareness and/or intentionality. Three keys modes of such information transmission are body language, spoken language, and material culture. This paper considers aspects of human evolutionary history that are of relevance for how such modes of information transmission evolved.

## 9.1 THREE MODES OF INFORMATION TRANSMISSION

Modern humans transmit more information by nongenetic means than any other species. Three key modes of such information transmission are body language, vocal utterances, and material culture. The boundaries between these can be somewhat fuzzy: body painting, tattooing, and piercing, for instance, are means of information transmission that involve both body language and material culture. Indeed, the transformation of the body from a purely physiological to a cultural entity will be one the most important issues to address in this article. Its aim is to consider selected aspects of human evolution that are of significance for how these modes of information transmission evolved within the *Homo* lineage.

Body language is taken here to include the physical appearance of the static unadorned body, which is replete with information regarding health, nutrition, life history, and so forth. As soon as bodies move, a great deal more information becomes available, whether from deliberate gesture, unconscious posture, or the style of movement. Although body language is important in many animal species, ranging from threat displays to subtle gestures, the body language of humans is particularly diverse, complex, and information laden (Beattie, 2003). This derives from the muscular control we have over our bodies, the potential for body language that arises from being a large, bipedal primate, and how our bodies have become perceived, exploited, and manipulated as cultural entities.

Sounds emanating from the body form a second mode of information transmission. These include deliberate sounds such as spoken language, uncontrollable sounds such as spontaneous screams, laughing, and farting, and sounds created by the contact of the body with other substances, such as footsteps or drumming. Of these, spoken language is by far the most important and, hence, addressing its evolution is a key task for any study for how human information transmission evolved.

Both of the above modes are intimately related to the body and are shared, to a greater or lesser extent, with other species. But humans are unique in their use of material culture for transmitting information, as epitomized by cave paintings and written texts that can literally transmit information across vast expanses of time and space. Some forms of material culture are deliberately used for transmitting information, others arise unintentionally but become highly valued — such as those footprints of past humans that are meticulously excavated and recorded by archaeologists.

The three modes of information transmission I have briefly summarized exclude the role of human odors, which, as with appearances, can be manipulated, for example by the use of perfumes. Smell is perhaps the least refined means of information transmission that our species utilizes, one far less developed than that found in other species. We should also note that the most potent forms of information transmission often arise when these three modes are combined, such as when highly decorated, dancing bodies are engaged in musical performance.

Only a small fraction of the information we transmit is done so intentionally. Body language is the classic example of when we may be entirely unaware of the "messages" being sent. Some messages can become under our control following awareness of the information being transmitted, such as when we deliberately change our posture during a conversation to pretend we are not bored. Other messages remain involuntary such as those from blushing or pupil dilation. Similarly, we are only aware of a small proportion of the information that we are unconsciously processing and using when making decisions about how to behave, especially regarding social interaction. When we have a conversation with another person, for instance, we may learn as much from their gestures and posture as we do from the words they are saying, although it may only be the latter that we are consciously attending to. More generally, we grow up acquiring the culture of our society and family with hardly any awareness of doing so.

We sometimes transmit information to deceive others: spoken untruths are an obvious example but so too are physical "untruths," such as various forms of cosmetic surgery. Sometimes we misinterpret the information that someone is trying to transmit and the same information can have a multitude of different implications and meanings to different people. In fact, all information requires interpretation of some kind — which is a simple reason, and just one of many, why the analogy of memes with genes is in my view entirely fatally flawed (see Lake, 1998, for a discussion of this problem).

## 9.2 THE FOSSIL AND ARCHAEOLOGICAL RECORD

To trace when, how, and why these three modes of information transmission evolved within the *Homo* lineage, we must examine the fossil and archaeological record for human evolution. The significant time period for evolution of unique and/or distinctive human traits is that after the date of the common ancestor we share with the chimpanzee at between six and eight million years ago. Although the fossil record becomes richer year by year and occasionally throws up a surprise such as the discovery of *Homo floresienses* in 2003 (Brown et al., 2004), the basic outline of human evolution is unlikely to be substantially altered by new discoveries, especially as it is increasingly supported by evolutionary genetics (Jobling et al., 2003).

In essence, there were numerous species of bipedal primates in Africa between six and two million years ago (Johansen and Edgar, 1996; Lewin and Foley, 2003). These display considerable morphological variation, which most likely relates to the exploitation of specific niches within the African landscape. These hominins have been placed into three genera, *Ardipithecus*, *Australaopithecus*, and *Homo*, with the latter constituted by two purported species, *H. habilis* and *H. rudolfensis*. Flaked stone tools are known from at least 2.5 million years ago, but in light of the repertoire of tools used by chimpanzees (Whiten, 2005) it seems likely that hammer stones, sticks, leaves, and other minimally modified materials were used long before flaked stone artifacts appeared.

The two species of early *Homo* are characterized by relatively larger brains than the other hominins, up to 750 c.c. rather than the 450 c.c. that is also characteristic of chimpanzees today, together with smaller teeth and a flatter face. These features may be within the range of variation for this grade of hominin without necessarily indicating any evolved cognitive or linguistic abilities. The key problem we face with assessing the significance of brain size is the rarity of the postcranial skeletons for earliest *Homo*, which might indicate that the relatively large brains are simply a product of large body size rather than encephalization. Indeed, a strong case can be made for reclassifying *H. habilis* and *H. rudolfensis* as australopithecines and identifying *H. ergaster*, appearing by 1.8 million years ago, as the first member of the *Homo* genus (Wood and Collard, 1999).

*H. ergaster* may mark an evolutionary transition to a type of hominin for which behavioral analogies with living nonhuman primates are of limited value. With a near fully modern stature and bipedal gait, this species is most likely the first to disperse out of Africa (Straus and Bar-Yosef, 2001). Brain size reached up to 900 c.c., although some specimens show the maintenance of relatively small brain capacities — some specimens from the site of Dmanisi in Georgia are no more than 650 c.c. (Gabunia et al., 2000). The Asian lineage of this species evolved into *Homo erectus* and appears to have made at least one water crossing on rafts to reach Flores Island by 850,000 years ago (Morwood et al., 1998).

In Europe, the middle Pleistocene is marked by a succession of hominins that are claimed to constitute at least three species: *H. antecessor*, *H. heidelbergensis*, and *H. neanderthalenis*. These appear to show a gradual increase of brain size until a capacity equivalent to, and in some cases exceeding, that of *Homo sapiens* is attained in late Pleistocene specimens. Stone artifacts appear to increase in technological complexity from Oldowan-like flakes associated with *H. antecessor*, Acheulian handaxes with *H. heidelbegensis*, and levallois technology with *H. neanderthalensis*. Examples of bone and wooden tools are exceedingly rare, but the discovery of the Schöningen spears (Thieme, 1997) indicates that this is most likely a consequence of preservation and discovery. Traces of structures and nonutilitarian artifacts are effectively absent (Gamble, 1999), with the few ambiguous examples claimed by some merely emphasizing their extreme rarity and unsophisticated nature.

The hominin species of Europe may form a single evolving lineage or multiple dispersals into that continent of species that evolved from *H. eragster* in Africa. Within Africa there is evolutionary continuity from *H. ergaster* to *H. sapiens* (McBrearty and Brooks, 2000), with the earliest specimens of the latter dating to c. 200,000 years ago (McDougall et al., 2005), a date that effectively coincides with an estimate for this species appearance from the study of modern day genetic diversity (Ingman et al., 2000; Jobling et al., 2003). Although *H. sapiens* specimens in Israel indicate initial dispersal out of Africa around 100,000 years ago (Lahr and Foley, 1994), the genetic evidence indicates that it was after 60,000 years ago that major dispersals into Asia and Europe occurred and that it was these that gave rise to the extant populations today (Ingman et al., 2000).

Immediately before such dispersals, we find evidence in South Africa for new types of material culture, often assumed to reflect the appearance of symbolic thought and language. Most notable are the incised ochre nodules and shell beads from Blombos Cave dating to 70,000 years ago (Henshilwood et al., 2002, 2004), while red ochre is prevalent in Middle Stone Age deposits at South African sites reaching back to 100,000 years ago (Knight et al., 1995).

The dispersals of *H. sapiens* out of Africa resulted in the colonization of Australia by at least 30,000 years ago, and most probably between 40–45,000 years ago (O'Connell and Allen, 2004), and that of Europe by 40,000 years ago (Gamble, 1999; Mellars, 2004). The latter is associated with major technological innovations characterized by the Aurignacian culture. The Neanderthals may have attempted to imitate the culture of the incoming *H. sapiens* although d'Errico et al. (1998) argue that they independently developed the Upper Palaeolithic-type elements that are present in the Chatelperronian culture, but they were either outcompeted for resources or were unable to survive the major climatic fluctuations of the late Pleistocene (Stringer et al., 2000; D'Errico and Sanchez Goni, 2003). By 30,000 years ago, the *H. sapiens* in Europe were engaging in cave painting, carving intricate bone figurines, and making elaborately decorated burials. Similarly, the *H. sapiens* in Australia and Africa were likely to have been engaged in both abstract and figurative rock art at this date.

The Neanderthals became extinct soon after 30,000 years ago. Precisely when *H. erectus* in Asia became extinct remains unclear although a similar date to that of the Neanderthals is likely, although *H. floresiensis* may have survived on Flores island as late as 14,000 years ago (Morwood et al., 2004). The climatic deterioration of the last glacial maximum of 20,000 years ago caused *H. sapiens* to abandon northern landscapes that became polar desert and those areas in low latitudes that became extremely arid. As global warming began, *H. sapiens* recolonized those landscapes and proceeded to colonize the rest of the world either in the late Pleistocene or early Holocene — the far north, the Americas, and the islands of the Pacific (Mithen, 2003).

It was only in the Holocene, beginning a mere 10,000 years ago, that agricultural economies developed, initially in the Near East at c. 9000 years ago and then quite independently at several locations elsewhere in the world — including rice farming in China at 7000 years ago, maize and squash in Central America at 8000 BC, and domesticated camelids (llamas and alpacas) in the Peruvian Andes by 7000 BC (Smith, 1995; Mithen, 2003). Farming provided the economic foundation for the development of towns, and within a few thousand years the first "civilizations," within which writing was independently invented. Since these developments, cultural complexity, economic efficiency, and population size appear to have grown exponentially along with artistic achievement and scientific knowledge.

It is within this span of human evolution that I have so briefly summarized that the nature of information transmission has changed from that used by the common ancestor of six million years ago, most likely similar to that of chimpanzees today, to the extraordinarily complex and diverse character it takes among modern humans. Many aspects of our evolutionary history could potentially be selected for particular consideration. I have chosen three, each relating to one of the three key modes of information transmission I discussed above: bipedalism because of its significance for both body and spoken language, spoken language itself because of its preeminent role in information transmission, and material culture because of how this transmits information across extended periods of time and space.

## 9.3 BIPEDALISM AND BODY LANGUAGE

Bipedalism — walking on two legs — is critical for the evolution of information transmission in a myriad of ways. It was critical in freeing the hands to enable more effective use of gesture and for a far more sophisticated form of body language than can be found in a quadruped. In essence, the new type of muscular control that was required to walk and run on two legs would have also enabled a range of expressive postures and movements to have been adopted that may have even involved dance at an early stage of human evolution. The anatomical adaptations for bipedalism also had consequences for the vocal tract and enabled an increased diversity of sounds to be uttered, while the freed hands could be used for manufacturing and carrying items of material culture. In the longer term, the loss of body hair ultimately required the development of clothing, perhaps the major source of information transmission involving material culture. Indeed, there is little if anything in the course of human evolution that cannot find a source in the anatomical and cognitive developments associated with bipedalism.

The origin of bipedalism has been debated for many years, with theories being gradually revised or dramatically falsified as new evidence has appeared (Lewin, 1999). It was once argued that bipedalism arose to free the hands so that stone artifacts could be manufactured. This had to be rejected when the evidence for australopithecine bipedalism was pushed back far beyond 2.5 million years ago, the date at which the first stone tools were made. Similarly, the idea that bipedalism originated when hominins began living in savannah environments, perhaps to allow them to stand and watch for predators or prey over the top of long grass, could not cope with the evidence that partially bipedal ardipithecus and australopithecines inhabited wooded landscapes. The notion that bipedalism arose because male hominins needed to carry foodstuffs to the females with whom they had pair bonded had to be similarly rejected when anthropologists recognized that the extent of body size differences between males and females — the sexual dimorphism — indicated that such pair bonding, and hence provisioning, is unlikely (McHenry, 1996; Mithen, 2005).

The most persuasive current argument for the origin of bipedalism is that it involved two distinct stages each with their own selective pressure, the first leading to the partial bipedalism of the australopithecines and the second to the fully modern bipedalism of *H. ergaster.*

With regard to the first, Hunt (1994) argued that fruit consumption may have played a crucial role. He made a detailed study of the situations in which chimpanzees stand and occasionally walk on two legs and found that they do so principally to harvest fruit from small trees using their hands for balance and slowly shuffling from tree to tree. By standing on two legs both hands can be used, while the bipedal shuffling avoids the need to keep raising and then lowering their body weight, which delivers a major saving in energy expenditure. Hunt suspects that the australopithecines were behaving in a similar fashion arguing that the anatomy of *Australopithecus afarensis*, as represented by "Lucy," was more suited to standing than walking. The broad pelvis, he suggests, made a stable platform, while the curved fingers and powerful arms should be seen as adaptations not for climbing trees but for hanging by one hand while standing on two legs and feeding with the other hand.

The stage two shift to a fully modern bipedalism is most likely related to the spread of savanna-like environments in East Africa soon after two million years ago owing to major climatic change (DeMenocal, 1995). But rather than being driven by the need to see across tall grass or for carrying, full bipedalism probably relates to the need to reduce heat stress owing to increased exposure to the sun in landscapes that had markedly fewer trees (Wheeler, 1984, 1991, 1994).

Heat stress is a problem to all animals living in savanna environments, largely because brains begin to malfunction if raised 2°C above their normal temperature. By standing erect, a hominin would have absorbed sunlight only on the top of its head and shoulders; by knuckle walking its whole back would have been exposed. Moreover, the air is cooler away from the soil itself, and wind velocities are significantly higher, which would have improved the cooling effectiveness of bodily evaporation.

By using a range of computer and engineering models, Wheeler demonstrated that by walking bipedaly, savanna-living hominins would have significantly reduced both heat stress and water consumption. This may have played a major role in "opening up" the new dietary niche of meat eating as such reductions allowed hominins to travel greater distances across savanna landscapes in search of carcasses, possibly at times of the day when competing carnivores would be resting in the sun.

Aiello (1996) built on Wheeler's ideas to consider the relationship between bipedalism, brain, and language. She argued that brain expansion could be explained by the new demands on sensory–motor control that bipedalism both required and enabled. Standing or walking on two legs requires that the center of gravity is constantly monitored and small groups of muscles frequently recruited and changed to correct its position; the movement of the legs has to be integrated with that of the arms, hands, and trunk to maintain a dynamic balance. And once arms and hands are freed from a locomotory role, they can be used independently from the legs, such as for carrying, signaling, throwing, and making tools. Bipedalism requires, therefore, a larger brain and more complex nervous system just to attain the more complex degree of sensory–motor control.

The evolution of complex vocalization would have also been facilitated by the impact of a bipedal anatomy on the vocal tract. The human larynx is found much lower in the throat than that of the chimpanzee, and this enables a much wider array of sounds to be produced. Anthropologists had traditionally assumed that strong selective pressures for spoken language had "pushed" the larynx into the throat, which then had the ill effect of creating a risk of choking on food. But Aiello (1996) argued that the low larynx was a consequence of the anatomical adaptations necessary for bipedalism rather than having been selected itself for enhanced vocalization. Because the spinal cord now had to enter the brain case from below rather than behind (as shown by the position of the foramen magnum), there was less space between the spinal cord and the mouth for the larynx. This space had been further reduced by the changes in the hominin face and dentition that arose with greater degrees of meat eating. Consequently, the larynx had to become positioned lower in the throat, which had the incidental effect of lengthening the vocal tract and increasing the diversity of possible sounds.

Aiello (1996) also suggested that not only the position but also the makeup of the larynx itself may have been changed by the new bipedal anatomy, resulting in the production of a less harsh, more melodious sound than in the australopithecines. The larynx of modern apes and humans is essentially a valve — the vocal cords — that can be closed off to allow the air to build up and then opened to create a sound.

This "valvular larynx" also has a quite different function, one to do with movement. Air pressure behind a closed larynx stabilizes the chest and provides a fixed basis for arm muscles — this is why we hold our breath when about to use our arms vigorously, such as when throwing a ball. Chest stabilization is really important for primates who use their arms for knuckle walking or climbing. As a consequence they have relatively thick, cartilaginous vocal cords that produce rather harsh sounding vocalizations. The vocal cords of modern humans, however, are more membranous.

Bipedalism may have caused the change by relaxing the selective pressure on the locomotor function of the valvular larynx. If so, then by becoming less rigid, *H. ergaster's* vocal cords may have further enhanced the diversity of sounds now being made by this bipedal hominin, even though there had been no selective pressures for enhanced vocalization itself.

## 9.4 INFORMATION TRANSMISSION BY VOCALIZATION: WHAT WAS THE NATURE OF PROTOLANGUAGE?

Although bipedalism would have enabled a significant increase in the transmission of information by gesture and body language, the significance of these were overshadowed by the development of vocal communication and, ultimately, language. Although the latter is most likely restricted to *Homo sapiens*, the cultural accomplishments of early humans were of a nature to suggest they possessed a sophisticated form of protolanguage. These accomplishments included the dispersal out of Africa into both northern latitudes and southeast Asia, the production of complex stone artifacts such as handaxes and levallois points, big game hunting, and survival during the dramatic environmental changes of the Middle Pleistocene. To have attained such accomplishments, some form of vocal communication more complex than that found in the great apes today appears essential.

Language is a particularly complex system of communication. It has to have evolved gradually in a succession of ever more complex communication systems used by the early humans and their early *Homo* and australopithecine ancestors (Pinker and Bloom, 1995). Academics refer to these communication systems by the catchall term *protolanguage*, and identifying the correct nature of protolanguage is the most important task facing anyone attempting to understand how language evolved. Current theories for protolanguage fall into two camps — those that believe that protolanguage was compositional in character, and those who believe it was holistic.

The essence of compositional theories is that protolanguage consisted of words with limited, if any, grammar. The champion of this view is Bickerton (e.g., 1990, 1998, 2000), who argues that human ancestors and relatives such as the Neanderthals may have had a relatively large lexicon of words, each of which related to a mental concept such as meat, fire, hunt, and so forth. They were able to string such words together but could only do so in a near arbitrary fashion. Bickerton recognizes that this could result in some ambiguity. For instance, would *man killed bear* have meant that a man has killed a bear, or that a bear has killed a man? Jackendoff (1999, 2000) suggests that simple rules such as agent first (i.e., the man killed the bear) might have reduced potential ambiguity. Nevertheless, the number and complexity of potential utterances would have been severely limited. The transformation of such protolanguage into language required the evolution of grammar — rules that define the order in which a finite number of words can be strung together to create an infinite number of utterances, each with a specific meaning.

Compositional theories of protolanguage have dominated studies of language evolution for the past decade; they have been highly influential but have now been challenged by holistic theories, as proposed by Wray (1998, 2000, 2002) and Arbib (2002, 2003). By using the term *holistic*, Wray means that the precursor to language was a communication system composed of "messages" rather than words; each hominin utterance was uniquely associated with an arbitrary meaning, as are the words of language today and indeed those of a Bickertonian-like protolanguage. But in Wray's protolanguage, hominin multisyllable utterances were not composed out of smaller units of meaning (i.e., words) that could be combined together in either an arbitrary fashion or by using rules to produce emergent meanings. In this view, modern language evolved only when such holistic utterances were segmented to produce words that could then be composed together to create statements with novel meanings. Arbib (2002) uses the term *fractionation* in a similar manner. Hence whereas Bickerton believes that words were present in the early stages of language evolution, Wray and Arbib believe that they appeared only in its latter stages.

Whether early human protolanguage was compositional, holistic, or some combination of both in character, the anatomical changes evident from the fossil record suggest that strong selective pressures had existed for enhanced vocal communication. The Kebara burial provided anthropologists with the first discovery of a Neanderthal hyoid bone — in fact the first from any premodern human (Arensburg et al., 1989, 1990). The hyoid bone is attached to the cartilage in the larynx and anchors the muscles necessary for speech. The Kebara specimen is morphologically similar to that of a modern human and has been used to argue that the Neanderthal vocal tract was also essentially the same as that which you or I possess (see discussion between Arensburg et al. 1990; Lieberman, 1990, 1992). It was certainly not identical, and contrasts in facial morphology probably led to nasal-like vocalizations. The claim by Lieberman and Crelin (1971) that the Neanderthal larynx had been positioned high in the throat — equivalent to that of a modern human baby or a chimpanzee — have now been thoroughly dismissed. So although the Neanderthal vocal tract may have been unable to produce exactly the same range of sounds as that of a modern human, they would certainly have been sufficiently diverse to enable speech if the neural circuitry for language had been present within the Neanderthal brain.

Two further aspects of anatomical evidence support the claim for vocal capabilities equivalent to those of modern humans. First, is the dimension of the hypoglossal canal, which carries the nerves from the brain to the tongue. The width of this is larger in modern humans than in chimpanzees and gorillas; Kay et al. (1998) argue that this reflects the larger supply of nerves required for the complex motor control necessary for speech (see Degusta et al., 1999, for an opposing view). The canal is partially preserved at the base of the cranium; Kay et al. (1998) found that those from *Australopithecus africanus* and *Homo habilis* specimens fell into the chimpanzee and gorilla size range, whereas those from Neanderthals were equivalent to the canal size of modern humans.

A second nerve-carrying canal measurable from fossilized remains is that of the thoracic vertebrae through which the nerves for controlling the diaphragm, and hence breathing, pass. MacLarnon and Hewitt (1999) found that this has had a similar evolutionary history as the hypoglossal canal: the vertebral canals of australopithecines and *Homo ergaster* are similar in size to those of African apes, whereas those from Neanderthals are larger and match the size of those from *Homo sapiens*. MacLarnon and Hewitt considered and rejected explanations for its increase in size relating to bipedalism, breathing for running, or the avoidance of choking on food. Their conclusion was that the increased number of nerves passing through the thoracic vertebrae arose for an increased control of breathing for speech.

The evidence from the hypoglossal and vertebral canals indicates that Neanderthals had a similar motor control of their tongues and breathing to that of modern humans. This also appears to be the case for the perception of sound owing to evidence from the inner ear. Whereas australopithecines appear to have been more sensitive to higher frequency sounds than are modern humans (Moggi-Cecchi and Collard, 2002), the sound perception of *H. heidelbergensis* appears to have been equivalent to that of modern humans (Martínez et al., 2004). As *H. heidelbergensis* was most likely a direct ancestor of the Neanderthals, we can conclude that they too had an auditory capacity similar to that of modern humans.

Unlike earlier anatomical developments relating to vocal communication, the enhanced motor control of the tongue and breathing and the modern-like auditory capacity possessed by the Neanderthals are most likely explained as having been specifically selected by evolution for vocal communication. The selective pressures for this are likely to have been numerous including the need for enhanced communication with infants (Dissanayake, 2000, and see below), hunting and gathering (Bickerton, 2003), and the demands of living in larger and more complex social groups (Dunbar, 1993).

Mithen (2005) followed Wray by proposing that protolanguage was holistic in nature and argued that early human vocalization was highly musical in character, making extensive use of variations in melody and rhythm to express emotion and induce emotional states in others. He suggested that early human vocal and gestural communication should be thought of as *protomusic* as much as protolanguage. While messages would have consisted of arbitrary strings of syllables, early human utterances are also likely to have made extensive use of vocal imitation of natural sounds — especially animal calls, onomatopoeias — and to have had the phenomenon of sound synesthesia, as described by Berlin (2005). The latter is a correspondence between a physical quality of an animal, such as its size or type of movement, and the sound made to refer to it: a broad generalization is that large animals tend to be referred to by words or expressions that make use of the vowel sounds 'o', 'u' and 'a', whereas 'i' and 'ee' are used for small animals, these arising because the aural cavity appears to be mimicking the size of the animal. And as Donald (1991) argued, mimesis is likely to have played a key role in hominin communication.

However complex early human vocal communication may have been, it seems unlikely to have possessed the key characteristics of modern day language, namely, a set of grammatical rules that allow an infinite set of phrases to be generated from a finite set of words. Modern language generates creative thought and enables rapid cultural transmission and cultural change through time (see below). These are precisely the features that appear absent from the archaeological record of early humans. Their whole period of existence is characterized by immense cultural stability — a limited number of technological themes are simply used in different ways in different times and places. Other than a few highly ambiguous examples, there are no traces of symbolic behavior. Even with the big-brained Neanderthals, which have a fully evolved vocal tract, there are no traces of language-mediated thought and behavior (Mithen, 2005). It seems most likely that their communication system was based on a suite of relatively fixed holistic phrases appropriate for frequently recurring situations (such as communal hunting, food sharing, tool making, and social interactions), the emotional impact of which could have been changed by variations in pitch, rhythm, and melody (Mithen, 2005).

## 9.5 ORIGINS OF THE MODERN MIND AND ITS IMPLICATIONS FOR INFORMATION TRANSMISSION

The evolutionary transition from holistic protolanguage to compositional language of the type we use today is most likely associated with the origin of *Homo sapiens* in Africa at c. 200,000 years ago and the emergence of a modern mentality.

The date of 200,000 years ago is one on which both the fossil and genetic evidence increasingly converge (McBrearty and Brooks, 2000; White et al., 2003; McDougall et al., 2005; Ingman et al., 2000; Jobling et al., 2003). It is not until 70,000 years ago, however, that we find unambiguous traces of symbolic activity (McBrearty and Brooks, 2000; Henshilwood et al., 2002, 2004). This lapse in time might be a consequence of our limited knowledge of the African archaeological record in the critical intervening period. Alternatively, 70,000 years ago may mark a date at which human populations pass a demographic threshold that enabled enhanced cultural activity (Shennan, 2000) or when a further genetic mutation happens that changed the nature of human mentality by enabling cognitive fluidity. In the context of this article, there are two key issues to address regarding the transmission of information: the evolution of language and the "extension of mind."

Wray (1998, 2000) uses the term *segmentation* to describe the process whereby hominins began to break up holistic phrases of protolanguage into separate units, each of which had their own referential meaning and could then be recombined with units from other utterances to create any infinite array of new utterances. This is the emergence of compositionality, the feature that makes language so much more powerful at information transmission than any other communication system.

Wray suggests that segmentation may have arisen from the recognition of chance associations between the phonetic segments of the holistic utterance and objects or events to which they related. Once recognized, these associations might then have been used in a referential fashion to create the new, compositional phrases. Bickerton (2003) questioned the feasibility of segmentation as the process by which words were invented/discovered. He had several pointed criticisms: the likelihood of such chance associations arising are remote; if holistic utterances had existed as Wary describes, then they are unlikely to have been long enough to contain multiple phonetic segments that could have been partitioned in this fashion; Wray's theory of segmentation has to assume the prior existence of discrete segments, which then invalidates her conception of holistic utterances.

Such criticisms do not take into account the likely presence of onomatopoeia, vocal imitation, and sound synesthesia in early human holistic utterances. These would have created nonarbitrary associations between phonetic segments of holistic utterances and certain entities in the world, notably species of animals with distinctive calls, environmental features with distinctive sounds, and bodily responses (Mithen, 2005). These nonarbitrary associations would have significantly increased the likelihood that the particular phonetic segments would eventually come to refer to the relevant entities and hence exist as words. Once some words had emerged, others would have followed with greater ease by the segmentation process that Wray describes.

Further confidence in the process of segmentation comes from the use of computer models to simulate the evolution of language. Kirby (2000, 2002) is one of several linguists who have begun to explore the evolution of language by the use of computer simulation models (e.g., see also Batali 2002; Nowak and Komarova, 2001; Komarova and Nowak, 2003). Kirby creates populations of agents — simulated people — who communicate with each other by using strings of symbols for language. Each simulated population is multigenerational, with each new "learning agent" acquiring its language from the already existing "speaking agents." In general, Kirby's simulations indicate that grammatical structure can be an emergent phenomenon of the cultural transmission of language between generations.

Although the arguments of Wray and Kirby help us to understand how compositional language evolved from holistic phrases, we must ask why this appears to have happened in Africa only after 200,000 years ago. There are two possibilities, one relating to social life and one to human biology. With regard to the first, we should initially note that Kirby found holistic languages remained stable in those simulations in which learning agents hear so much of the speaking agents utterances that they learn every single association between symbol string and meaning. In other words, there is no learning bottleneck for language to pass through and hence no need for generalization. It may be the case, therefore, that the social arrangements of *Homo* populations other than *H. sapiens* were such that infants had intense and continuous exposure to a limited number of holistic speakers, resulting in the acquisition of a whole suite of utterances intact with no need for generalization.

This would indeed have been quite likely for early humans as they most likely lived in socially intimate groups with limited, if any, need for the type of novel utterances that could be produced only by compositional language. Moreover, there would have been little need and few opportunities to communicate with people from outside one's own group — although some contacts involving the movement of people would have been essential to maintain demographic and genetic viability. But little need have been said on such occasions.

It may only have been within the earliest *H. sapiens* communities in Africa that people began to adopt specialized economic roles and social positions that trade and exchange with other communities, and that "talking with strangers" became an important and pervasive aspect of social life. Such developments would have created pressures to exchange far greater amounts of information than was previously necessary in the socially intimate, undifferentiated groups of early humans. Only then would there have been the need for generalization, in the manner that Kirby describes within his simulation, and to continually generate novel utterances at a rate and of a type beyond the capability of holistic protolanguage.

The archaeological record of Africa relating to the appearance of *H. sapiens* certainly suggests that such social developments occurred (McBrearty and Brooks, 2000). The dilemma, of course, is whether we are dealing with cause or effect: one might argue that the development of economic specialization and exchange relations between groups were a consequence of compositional language that enabled the necessary communication to be undertaken. My guess is that we are dealing with strong feedback between the two — they "bootstrapped" each other to create rapid changes in both society and communication.

The kick start to such developments may have been a chance genetic mutation. This may have provided the ability to identify phonetic segments in a holistic utterance, an ability that had previously been absent. Some aspects of language are dependent on the possession of the specific gene FOXP2, the modern human version of which seems to have appeared in Africa sometime 200,000 years ago (Enard et al., 2002). Perhaps the process of segmentation was dependent upon this gene in some manner that has yet to be discovered. Indeed, it may be significant that those members of the KE family that were afflicted by a faulty version of the FOXP2 gene not only had difficulties with grammar but also with understanding complex sentences and judging whether a sequence such as "blonterstaping" is a real word (Bishop, 2002). These difficulties seem to reflect a problem with the segmentation of what would have sounded to them as holistic utterances. So perhaps it was only with the chance mutation of the FOXP2 gene to create the modern human version that segmentation became possible. Alternatively, there may have been other genetic mutations at a similar date that enabled the transition from holistic phrases to compositional language, perhaps by the appearance of a general purpose statistical learning ability.

Once the process of segmentation had begun, we should expect a rapid evolution of grammatical rules. Such rules would have evolved via the process of cultural transmission in the manner that Kirby describes and perhaps through natural selection leading to the appearance of genetically based neural networks that enable more complex grammatical constructions. Although the evolution of compositional language in this manner would have enhanced communication, it may have also led to a more fundamental change in the nature of human thought — the transition from a domain-specific to a cognitively fluid mentality.

Mithen (1996) argued that cognitive fluidity was a consequence of language: spoken and imaginary utterances acted as conduits for ideas and information to flow from one intelligence/module/cognitive domain to another, a view supported by Carruthers (2002). He drew on latest research in neuroscience and psychology to argue that the "imagined sentences" we create in our minds allow the outputs from one intelligence/module/cognitive domain to be combined with those from one or more others, and thereby create new types of conscious thoughts. Carruthers placed considerable emphasis on syntax — an essential part of compositional language. Syntax allows for the multiple embedding of adjectives and phrases, the phenomenon of recursiveness. According to Carruthers, syntax allows one imaginary sentence generated by one type of cognitive module/intelligence to be embedded into that of another imaginary sentence coming from a different module/intelligence. By so doing, a single imaginary sentence will be created that generates an "intermodular" or cognitively fluid thought, one that could not have existed without compositional language.

Any cognitive impact that language has on an individual is magnified a multitude of times by how language connects people's minds into a single network allowing ideas to "migrate" between minds. Indeed, language creates a communal mind/brain. The philosopher Andy Clark (1996) explains this remarkable power of language:

> Migrations [of ideas by language] may allow the communal construction of extremely delicate and difficult intellectual trajectories and progressions. An idea that only Joe's prior experience could make available, but that can flourish only in the intellectual niche currently provided by the brain of Mary, can now realize its full potential by journeying between Joe and Mary as and when required. The path to a good idea can now crisscross individual learning histories so that one agent's local minimum becomes another's potent building block … culturally scaffolded reason is able to incrementally explore spaces which path dependent reason could never hope to penetrate. (p. 206)

## 9.6 MATERIAL CULTURE AND THE LONG-TERM TRANSMISSION OF INFORMATION

The evolution of compositional language provided *Homo sapiens* with enhanced communicative abilities and changed the capacity for thought by enabling cognitive fluidity. One consequence of this was a change in the relationship with material culture, one that had profound influence for the development of human history. The use of material culture had, of course, been a fundamental aspect of hominin lifestyles for millions of years, probably long before the first stone artifacts were manufactured in light of the use of tools by African apes. But until 70,000 years ago — on the basis of current evidence — we have limited, if any, evidence that material culture was used in a manner other than to extend the physical capabilities of the human body. There are no unambiguous traces of artifacts that were used for information transmission, that may have carried symbolic meanings, and that may have enhanced the computational powers of the human mind. Today, of course, such artifacts are pervasive in all of our activities ranging from the Post-its that extend our memories, calculators that extend our computational powers, written texts that transmit detailed information across generations, and symbolic images that extend our abilities at metaphorical thought. Our reliance on technology is now so great that Clark (2003) recently described *Homo sapiens* as "natural born cyborgs."

The emergence of cognitive fluidity lies at its root, as this both created the need for material culture to play this role and enabled the necessary technological creativity for it to do so. The development of religious thought, involving a belief in supernatural beings, provides a particularly good illustration.

Ideas about supernatural beings are a natural consequence of the cognitive fluidity that language delivered to the human mind (Mithen, 1996). By combining, say, what one knows about people (from social intelligence) with what one knows about animals (from natural history intelligence), one can imagine an entity that is part human and part animal — such as the lion/man figurine from Hohlenstein Stadel in Germany dating to 33,000 years ago. Similarly, by combining what one knows about people with that about physical objects, one can create ideas of beings that can live forever (like rocks), walk on water (like a floating branch), or are invisible (like the air we breath).

Ideas about supernatural beings are the essence of religious thought (Boyer, 2001), but they are unnatural in the sense that they conflict with our deeply evolved domain-specific understanding of the world. As a consequence they are difficult to hold within our minds and transmit to others — try, for instance, explaining to someone the concept of the Holy Trinity or the Aboriginal Dreamtime or try understanding these ideas as someone explains them to you. As Day (2004, p. 116) has recently written, "one of the bedeviling problems about dealing with gods is that … they are never really *there*" and hence we have difficulty in knowing not only how to communicate with them, but also how to think about them.

Modern humans compensate for this by the use of material symbols that provide cognitive anchors for such difficult ideas, whether about supernatural beings or scientific theories (Mithen, 1998a). Whether supernatural beings are made tangible in a representative manner, as we suppose the lion/man from Hohlensetin-Stadel is doing, or in abstract form as in the Christian Cross, such material symbols function to help conceptualize and share the religious entities and ideas that one believes. In this regard, such objects constitute an extension of the human mind and a critical means of information transmission for conceptually difficult ideas. As Day (2004, p. 116) explains, "the broad spectrum of rituals, music, relics, scriptures, statues and buildings typically associated with religious traditions are no longer seen as mere ethnographic icing on the cognitive cake. Rather than thin cultural 'wrap arounds' that dress up the real cognitive processes going on underneath, they begin to look like central components of the relevant machinery of religious thought."

The use of material culture and music to extend the capabilities of the human mind by providing anchors for ideas about supernatural beings that would be otherwise difficult to maintain, comprehend, and transmit is just one example of a much more general phenomenon. We can, for instance, consider how the use of shell beads, such as those dating to 70,000 years ago at Blombos Cave (Henshilwood et al., 2004), and other forms of body ornamentation function to mediate social relationships by indicating social identity and/or status. Rock art enables information and ideas to be recalled and communicated across time and space far in excess of what a human body and brain is able to do alone, indeed paintings from more than 30,000 years ago are still communicating ideas to us today. The invention of pictograms and writing provide a means by which narratives having a level of detail beyond that which can be depicted in art alone can be recorded, communicated, and analyzed in a manner that is not possible with oral recitation alone. The invention and development of mathematics and science are entirely dependent upon offloading information and ideas into material culture, which can then be examined further and developed by either the same mind or a different mind at a different time and place.

In this regard, the quotation from Clark given above concerning language can also apply to material culture; in fact, it is even more appropriate for material culture (Mithen, 1998b). Although language is effective at allowing ideas to migrate between minds, it has a major constraint — utterances are not durable. If there is no one around to hear what one says, the idea within the utterance is lost, or at least remains within the mind of the speaker. But once ideas are encoded into durable media they become part of the world and can be communicated across vast time spans. As such, material culture is the prime means by which minds are extended out of the body and connected between individuals who may never meet each other or even know of each other's existence.

I have so far stressed how cognitive fluidity enabled material culture to extend the capabilities of the human body and mind. The converse is also true and of equivalent importance: the human body became perceived, exploited, and manipulated as a cultural artifact. This is the case whether the body was still a living entity, such as by tattooing, piercing, scarification, and deformation, or a dead entity, such as by bones becoming religious relics.

## 9.7 SUMMARY

Although this article has addressed the evolution of the three key modes of information transmission in modern humans — body language, spoken language, and material culture — there are other modes that also require consideration to gain a full understanding, notably smell. Moreover, each of the three I have considered require examination at greater length and there are key cognitive processes that underlie information transfer that have not been addressed, notably theory of mind. This is the ability to appreciate that other individuals have beliefs and desires different from one's own (Carruthers and Smith, 1996), and hence is a cognitive prerequisite for the evolution of intentional information exchange (Dunbar, 1998; Mithen, 1999).

The evolution of dance as a form of body language is one of the most under-researched topics, as are the deliberate alterations we make to our bodies by piercing, tattooing, and enlarging/reducing areas of muscle, fat, and bone by what is now termed cosmetic surgery. The transmission of information by sound has only been addressed within this article with regard to language, with no attention paid to the important involuntary utterances — ranging from screaming to laughing — that play a key role in information transmission or to music. Material culture as a medium for information transmission is perhaps the subject most in need of further consideration, whether we are dealing with Oldowan stone tools or the latest uses of the internet.

Nevertheless, while recognizing the need for a far longer and more in depth treatment of this subject, this article has attempted to review the key developments in human evolution that enable us today to use body language, spoken language, and material culture as such effective means of information transmission.

## ACKNOWLEDGMENTS

I am grateful to Jonathan Wells for inviting me to the conference from which this volume has arisen, and to Mark Collard to comments on an earlier version of this article.

## REFERENCES


Aiello, L.C., Terrestriality, bipedalism and the origin of language, In *Evolution of social behavior patterns in primates and man*, Runciman, W.G., Maynard-Smith, J., and Dunabr, R.I.M., Eds., Oxford University Press, Oxford, 269–290, 1996.

Arbib, M.A., The mirror system, imitation and the evolution of language, In *Imitation in animals and artifacts*, Nehaniv, C., and Dautenhahn, K., Eds., MIT Press, Cambridge, MA, 229–280, 2002.

Arbib, M.A., The evolving mirror system: a neural basis for language readiness, In *Language evolution*, Christiansen, M.H., and Kirby, S., Eds., Oxford University Press, Oxford, 182–200, 2003.

Arensburg, B., Schepartz, L.A., Tillier, A.M., Vandermeersch, B., and Rak, Y., A reappraisal of the anatomical basis for speech in Middle Palaeolithic hominids, *Am. J. Phys. Anthropol.*, 83, 137–156, 1989.

Arensburg, B., Tillier, A.M., Duday, H., Schepartz, L.A., and Rak, Y., A Middle Palaeolithic human hyoid bone, *Nature*, 338, 758–760, 1990.

Batali, J., The negotiation and acquisition of recursive grammars as a result of competition among exemplars, In *Linguistic evolution through language acquisition: formal and computational models*, Briscoe, E., Eds., Cambridge University Press, Cambridge, 111–172, 2002.

Beattie, G., *Visible thought: the new language of body language*. Routledge, London, 2003.

Berlin, B., "Just another fish story?" Possibile size-symbolic properties of fish names, In *Animal names*, Mineli, A., Ortalli, G., and Sanga, G., Eds., Instituto Veneto di Scienze, Lettere et Arti, Venice, 9–21, 2005.

Bickerton, D., *Language and species*, Chicago University Press, Chicago, 1990.

Bickerton, D., Catastrophic evolution: the case for a single step from protolanguage to full human language, In *Approaches to the evolution of language: social and cognitive biases*, Hurford, J., Studdert-Kennedy, M., and Knight, C., Eds., Cambridge University Press, Cambridge, 341–358, 1998.

Bickerton, D., How protolanguage became language, In *The evolutionary emergence of language*, Knight, C., Studdert-Kennedy, M., and Hurford, J., Eds., Cambridge University Press, Cambridge, 264–284, 2000.

Bickerton, D., Symbol and structure: a comprehensive framework for language evolution, In *Language evolution*, Christiansen, M.H., and Kirby, S., Eds., Oxford University Press, Oxford, 77–93, 2003.

Bishop, D.V.M., Putting language genes in perspective, *Trends in genetics*, 18, 57–59, 2002.

Boyer, P., *Religion explained*, Basic Books, New York, 2001.

Brown, P., Sutikna, T., Morwood, M.J., Soejono, R.P., Jatmiko, Wayhu Saptomo, E., and Rokus Awe Due, A new small-bodies hominin from the Late Pleistocene of Flores, Indonesia, *Nature*, 431, 1055–61, 2004.

Carruthers, P., The cognitive functions of language, *Brain Behav. Sci.*, 25, 657–726, 2002.

Carruthers, P., and Smith, P., Eds., *Theories of theories of minds*, Cambridge University Press, Cambridge, 1996.

Clark, A., *Being there: putting brain, body and world together again*, MIT Press, Cambridge, MA, 1996.

Clark, A., *Natural born cyborgs: minds, technologies and the future of human intelligence*, Oxford University Press, Oxford, 2003.

Day, M., Religion, off-line cognition and the extended mind, *J. Cogn. Cult.*, 4, 101–121, 2004.

DeGusta, D., Gilbert, W.H., and Turner, S.P. Hypologossal canal size and hominid speech, *Proc. Nat. Acad. Sci.*, 96, 1800–1804, 1999.

DeMenocal, P.B., Plio-Pleistocene African climate, *Science*, 270, 53–59, 1995.

D'errico, F., and Sanchez Goni, M.F., Neanderthal extinction and the millennial scale climatic variability of OIS 3, *Quat. Sci. Rev.*, 22, 769–788, 2003.

D'errico, F., Zilhão, J., Julian, M., Baffier, D., and Pelegrin, J., Neanderthal acculturation in Western Europe, *Curr. Anthropol.*, 39, S1-S44, 1998.

Dissanayke, E., Antecedents of the temporal arts in early mother–infant interaction, In *The origins of music*, Wallin, N.L., Merker, B., and Brown, S., Eds., MIT Press, Cambridge, MA, 389–410, 2000.

Donald, M., *Origins of the modern mind*, Harvard University Press, Cambridge, MA, 1991.

Dunbar, R.I.M., Coevolution of neocortical size on group size in primates, *J. Hum. Evol.*, 20, 469–493, 1993.

Dunbar, R.I.M., Theory of mind and the evolution of language, In *Approaches to the evolution of language*, Hurford, J.R., Studdert-Kennedy, M., and Knight, C., Eds., Cambridge University Press, Cambridge, 92–110, 1998.

Enard, W., Przeworski, M., Fisher, S.E., Lai, C.S., Wiebe, V., Kitano, T., Monaco, A.P., and Paabo, S., Molecular evolution of FOXP2, a gene involved in speech and language, *Nature*, 418, 869–872, 2002.

Gabunia, L., Vekua, A., Lordkipanidze, D., Swisher III, C.C., Ferring, R., Justus, A., Nioradze, M., Tvalchrelidze, M., Antón, S.C., Bosinski, G., Jöris, O., de Lumley, M.-A., Majsuardze, G., and Mouskhelishvlli, A., Earliest Pleistocene hominid cranial remains from Dmanisi, Republic of Georgia: taxonomy, geological setting and age, *Science*, 288, 1019–1025, 2000.

Gamble, C., *The Paleolithic societies of Europe*, Cambridge University Press, Cambridge, 1999.

Henshilwood, C.S., d'Errico, F., Vanhaeren, M., van Niekerk, K., Jacobs, Z., Middle stone age shell beads from South Africa, *Science*, 304, 404, 2004.

Henshilwood, C.S., d'Errico, F., Yates, R., Jacobs, Z., Tribolo, C., Duller, G.A.T., Mercier, N., Sealy, J.C., Valladas, H., Watts, I., and Wintle, A.G., Emergence of modern human behavior: Middle Stone Age engravings from South Africa, *Science*, 295, 1278–1280, 2002.

Hunt, K.D., The evolution of human bipedality: ecology and functional morphology, *J. Hum. Evol.*, 26, 183–202, 1994.

Ingman, M., Kaessmann, H., Paabo, S., and Gyllensten, U., Mitochondrial genome variation and the origin of modern humans, *Nature*, 408, 708–713, 2000.

Jackendoff, R., Possible stages in the evolution of the language faculty, *Trend. Cogn. Sci.*, 3, 272–279, 1999.

Jackendoff, R., *Foundations of language: brain, meaning, grammar, evolution*, Oxford University Press, Oxford, 2000.

Jobling, M.A., Hurles, M.E., and Tyler-Smith, C., *Human evolutionary genetics*, Garland Publishing, New York, 2003.

Johansen, D., and Edgar, B., *From Lucy to language*, Weidenfeld & Nicolson, London, 1996.

Kay, R.F., Cartmill, M., and Balow, M., The hypoglossal canal and the origin of human vocal behavior, *Proc. Nat. Acad. Sci.*, 95, 5417–5419, 1998.

Kirby, S., Syntax without natural selection: how compositionality emerges from vocabulary in a population of learners, In *The evolutionary emergence of language: social function and the origins of linguistic form*, Knight, C., Studdert-Kennedy, M., and Hurford, J.R., Eds., Cambridge University Press, Cambridge, 303–323, 2000.

Kirby, S., Learning, bottlenecks and the evolution of recursive syntax, In *Linguistic evolution through language acquisition: formal and computational models*, Briscoe, E., Ed., Cambridge University Press, Cambridge, 173–204, 2002.

Knight, C., Powers, C., and Watts, I., The human symbolic revolution: a Darwinian account, *Cam. Arch. J.*, 5, 75–114, 1995.

Komarova, N.L., and Nowak, M., Language learning and evolution, In *Language evolution*, Christiansen, M.H., and Kirby, S., Eds., Oxford University Press, Oxford, 317–337, 2003.

Lahr, M.M., and Foley, R., Multiple dispersals and modern human origins, *Evol. Anthropol.*, 3, 48–60, 1994.

Lake, M., Digging for memes: the role of material objects in human evolution, In *Cognition and material culture: the archaeology of symbolic storage*, Renfrew, C., and Scarre, C., Eds., 1998.

Lewin, R., *Human evolution*, Blackwell Science Inc., London, 1999.

Lewin, R., and Foley, R., *Principles of human evolution*, Blackwell Scientific Inc., Oxford, 2003.

Lieberman, P., Hyoid bone position and speech: reply to Dr. Arensberg, *Am. J. Phys. Anthropol.*, 94, 275–278, 1990.

Lieberman, P., On Neanderthal speech and Neanderthal extinction, *Curr. Anthropol.*, 33, 409–410, 1992.

Lieberman, P., and Crelin, E., On the speech of Neanderthal man, *Linguistic Inquiry*, 2, 203–222, 1971.

MacLarnon, A., and Hewitt, G.P., The evolution of human speech: the role of enhanced breathing control, *Am. J. Phys. Anthropol.*, 109, 341–343, 1999.

Martínez, I., Rosa, M., Arsuaga, J.-L., Jarabo, P., Quam, R., Lorenzo, C., Gracia, A., Carretero, J.-M., Bermúdez de Castro, J.-M., and Carbonell, E., Auditory capacities in Middle Pleistocene humans from the Sierra de Atapuerca in Spain, *Proc. Nat. Acad. Sci.*, 101, 9976–9981, 2004.

McBrearty, S., and Brooks, A., The revolution that wasn't: a new interpretation of the origin of modern human behavior, *J. Hum. Evol.*, 38, 453–563, 2000.

McDougall, I., Brown, F.H., and Fleagle, J.G., Stratigraphic placement and age of modern humans from Kibish, Ethiopia, *Nature*, 433, 733–736, 2005.

McHenry, H. M., Sexual dimorphism in fossil hominids and its sociological implications, In *The Archaeology of human ancestry*, Steele, J., and Shennan, S., Eds., Routledge, London, 91–109, 1996.

Mellars, P., Neanderthals and the modern human colonization of Europe, *Nature*, 432, 461–465, 2004.

Mithen, S.J., *The prehistory of the mind: a search for the origin of art, science and religion*, Thames and Hudson, London, 1996.

Mithen, S.J., The supernatural beings of prehistory and the external storage of religious ideas, In *Cognition and material culture: the archaeology of symbolic storage*, Renfrew, C., and Scarre, C., Eds., McDonald Institute of Archaeological Research, Cambridge, 97–106, 1998a.

Mithen, S.J., A creative explosion? Theory of mind, language and the disembodied mind of the Upper Palaeolithic, In *Creativity in human evolution and prehistory*, Mithen, S., Ed., Routledge, London, 165–192, 1998b.

Mithen, S.J., Palaeoanthropological perspectives on the theory of mind, In *Understanding other minds: perspectives from autism and cognitive neuroscience*, Baron-Cohen, S., Flusberg, H.T., and Cohen, D., Eds., Oxford University Press, Oxford, 494–508, 1999.

Mithen, S.J., *After the ice: a global human history, 20,000–15,000 BC*, Weidenfeld and Nicolson, London, 2003.

Mithen, S.J., *The singing Neanderthals: the origin of music, language, mind and body*, Weidenfeld and Nicolson, London, 2005.

Moggi-Cecchi, J., and Collard, M., A fossil stapes from Stekfontein, South Africa, and the hearing capabilities of early hominids, *J. Hum. Evol.*, 42, 259–265, 2002.

Morwood, M.J., O'Sullivan, P.B., Aziz, F., and Raza, A., Fission-track ages of stone tools and fossils on the east Indonesian island of Flores, *Nature*, 392, 173–176, 1998.

Morwood, M.J., Soejono, R.P., Roberts, R.G., Sutikana, T., Turney, C.S.M., Westaway, K.E., Rink, W.J., Zhao, J.-X., van den Bergh, G.D., Rokus Awe Due, Hobbs, D.R., Moore, M.W., Bird, M.I., and Fifield, L.K., Archaeology and the age of a new hominin from Flores in eastern Indonesia, *Nature*, 431, 1087–1091, 2004.

Nowak, M.A., and Komarova, N.L., Evolution of universal grammar, *Science*, 291, 114–118, 2001.

O'Connell, J., and Allen, J., Dating the colonization of Saul: a review of recent research. *J. Arch. Sci.*, 31, 835–853, 2004.

Pinker, S., and Bloom, P., Natural language and natural selection, *Behav. Brain Sci.*, 13, 707–784, 1995.

Shennan, S.J., Population, culture history and the dynamics of culture change, *Curr. Anthropol.*, 41, 811–835, 2000.

Smith, B.D., *The emergence of agriculture*, Scientific American Library, New York, 1995.

Straus, L., and Bar-Yosef, O., Eds., Out of Africa in the Pleistocene, *Quat. Int.*, 75, 2001.

Stringer, C.B., Barton, R.N.E., and Finlayson, C.B., Eds., *Neanderthals at the Edge*, Oxbow Books, Oxford, 2000.

Thieme, H., Lower Palaeolithic hunting spears from Germany, *Nature*, 385, 807–810, 1997.

Wheeler, P., The evolution of bipedality and the loss of functional body hair in hominids, *J. Hum. Evol.*, 13, 91–98, 1984.

Wheeler, P., The influence of bipedalism on the energy and water budgets of early hominids, *J. Hum. Evol.*, 21, 107–136, 1991.

Wheeler, P., The thermoregulatory advantages of heat storage and shade seeking behavior to hominids foraging in equatorial savannah environments, *J. Hum. Evol.*, 21, 107–136, 1994.

White, T.D., Asfaw, B., DeGusta, D., Gilbert, H., Richards, G.D., Suwa, G., and Clark Howell, F., Pleistocene *Homo sapiens* from Middle Awash, Ethiopia, *Nature*, 423, 742–747, 2003.

Whiten, A., Chimpanzee cultures, In *The world atlas of great apes and their conservation*, Cladecott, J., and Miles, L., Eds., University of California Press, Berkeley and Los Angeles, 2005.

Wood, B., and Collard, M., The human genus, *Science*, 284, 65–71, 1999.

Wray, A., Protolanguage as a holistic system for social interaction, *Lang. Comm.*, 18, 47–67, 1998.

Wray, A., Holistic utterances in protolanguage: the link from primates to humans, In *The evolutionary emergence of language: social function and the origins of linguistic form*, Knight, C., Studdert-Kennedy, M., and Hurford, J.R., Eds., Cambridge University Press, Cambridge, 285–302, 2000.

Wray, A., Ed., *The transition to language*, Oxford University Press, Oxford, 2002.

# 10 From Cultural History to Cultural Evolution: An Archaeological Perspective on Social Information Transmission

*Stephen Shennan*

## CONTENTS



## 10.1 INTRODUCTION

When prehistoric archaeology began as a discipline in Europe in the late 19th century, it was in a climate of nationalism. The nation states that were emerging, and agitating to emerge, from the decaying empires and fragmented principalities and dukedoms of Central Europe were looking for a charter to justify their existence and their romantic intellectuals had sought this in the past. It became apparent to them that the increasing quantities of what was now recognized as archaeological material that were being dug up as a result of expanding industrialization could potentially be used to trace a history of peoples into the remote past, if a relationship could be established between records of such peoples in ancient sources and sets of material remains whose origins could then be traced still further back into the preliterate past. It was soon clear that this involved two main descriptive tasks: identifying patterns of similarity and difference in space and placing them in time. This was archaeology as cultural history: identifying prehistoric cultures and putting them in order. It implied tracing histories of social information transmission and identifying breaks in transmission, spatial and chronological. The factors that were believed to produce the continuities and discontinuities in traditions were migration and diffusion. A new set of cultural material appearing in an area would be the result of a new migrating people. The spatial boundaries of the material would represent the territory occupied by that people, and when that type of archaeological material was in its turn replaced by something new this indicated the arrival of new people, with different traditions. Less radical changes, especially if they were apparently useful, such as the introduction of bronze metallurgy, could be explained by diffusion, the transmission of new information across existing populations.

This concern with cultural history was more or less universal in the first half of the 20th century and continues to be significant today. However, in the 1960s the main emphasis in Anglo-American archaeology shifted from a concern with the tracing of traditions, migration, and diffusion to a new interest in adaptation: explaining changes in material culture as the result of new social and economic patterns, from the emergence of social hierarchies to the adoption of new subsistence systems such as agriculture. The motor for change came from the environment: new conditions produced appropriate adaptive reactions. This new approach had little time for history. The task of this *New Archaeology* was to come up with "laws of cultural process."

The adaptationist perspective did not so much solve the problems that the cultural historians were interested in; by and large it simply ignored them. The emerging recognition in the 1980s that this was the case, together with a growing awareness of the sociology of knowledge and the impact of political interests on interpretation, led many archaeologists to a rejection of scientifically based empirical methods for making inferences about the prehistoric past and to the adoption of an "anything goes" approach to the production of stories about the past. This position should preferably be socially progressive, in the sense of countering oppression — whether based on gender, ethnicity, colonialism, or whatever — in some way. This was archaeology's own version of postmodernism.

A different perspective began to be taken by a much smaller group of archaeologists and anthropologists, who took the view that a scientific approach to the past remained possible and that evolutionary biology provided a model for how studies of adaptation and the history of cultural traditions could actually be integrated with one another in a way that did justice to both. Further than that, they argued, because the study of human culture from an evolutionary perspective is so new and underdeveloped, anthropological and archaeological case studies can contribute to the development of the necessary new cultural evolutionary theory. There are now several different versions of evolutionary anthropology and archaeology, with varying degrees of mutual hostility (see, e.g.,

Smith, 2000). In what follows I will present a perspective based largely, but not exclusively, on Boyd and Richerson's (1985) dual inheritance theory and developments of it.

## 10.2 EVOLUTIONARY PROCESSES AND CULTURAL TRADITIONS

Obviously, the basis for taking evolutionary theory as a starting point in this way is the establishment of a connection between the processes producing stability and change in human culture and relevant processes and entities in evolutionary theory. There must be an inheritance mechanism, a means of generating novelty, and processes that affect the frequency with which things are represented in future generations. A required starting point is an appropriate definition of culture and the one adopted here is as follows: "Culture is information capable of affecting individuals' phenotypes which they acquire from other conspecifics by teaching or imitation" (Boyd and Richerson, 1985, p. 33).

In the case of the cultural practices of interest to archaeologists, the inheritance mechanism in social learning and novelty is produced by innovation. The latter can be mere copying error but it can also arise from trial-and-error experimentation that leads individuals to give up what they were originally doing and do something else in the belief that it produces better results in some sense. This does not mean that problems always call into being their solutions: necessity may be one of the mothers of invention but it is not necessarily the mother of successful invention. Success depends on a variety of processes, one of which is natural selection.

The operation of natural selection needs some unpacking in the human case. In the first place it assumes that humans, like other animals, have evolved to have a propensity to maximize their reproductive success and to be able to recognize and respond to variations in their environment relevant to achieving it. This leads to the assumption that at the level of the day-to-day business of making a living the principles of optimal foraging should hold. Subject to these, if changed conditions of some kind alter the trade-offs involved in achieving reproductive success, people will respond to them, increasing their fertility in good times and decreasing it when times are hard.

But natural selection can also act on people via their cultural traditions. In other words, particular inherited cultural beliefs and practices may lead to some people having greater survival rates and reproductive success than others. If they also transmit these successful beliefs/practices to their children, then their prevalence will increase and will in turn again increase the reproductive success of the children that practice them for as long as those practices are advantageous.

But this is not the end of the story. A particular set of beliefs or practices may lead an individual not to greater reproductive success and associated transmission of the successful beliefs to children, but to greater success as a model to be copied in various respects. If the practices/beliefs that made the individual a successful model to be copied are adopted by the copiers, they, too, will be more likely to be models for the beliefs and practices of subsequent generations. Whereas in the first case there is competition between cultural attributes that affects who will be the biological parents of the next generation, in this case the competition is about who will be its "cultural parents." It is possible for the competition to become a cultural parent to override that to become a biological one (Boyd and Richerson, 1985, pp. 182–186).

However, the frequencies of beliefs and practices through time are also affected by so-called decision-making biases. These include the preferential copying of practices because they appear to give better results in some sense (directly biased transmission) and the preferential copying of practices simply because they are common (conformist transmission) or because the individual being copied is prestigious in some way (prestige-biased transmission). There may be complex feedbacks between all these different processes. A conformist bias, for example, may lead not simply to greater replicative success for one particular belief/practice as opposed to another, it might have deleterious effects on the reproductive success of members of a community if it means that they respond more slowly than would be beneficial to changes, for example, in the natural environment. On the other hand, if, in general, those who conform tend to be more successful in obtaining a partner and bringing up children, then a process of gene-culture coevolution may develop if relevant aspects of human psychology are genetically influenced.

A special position in analyses of natural selection and culture is occupied by evolutionary game theory (e.g., Skyrms, 1996), focused on the competition between strategies of social interaction in terms of the payoffs they give to the interactors in different social situations. Here, what is changing through time as a result of differential payoffs is the frequency of social strategies in a population, but obviously this too may have survival/reproductive success consequences for the practitioners of the strategies and also for the replicative success of the norms associated with the favoring of particular strategies.

Most recently, attention has been drawn to the evolutionary significance of the process of niche construction, or ecological inheritance (e.g., Laland et al., 2000) — the fact that it is not only socially transmitted traditions that are passed on through time, but novel humanly created social, cultural, and "natural" environments, which provide the selective conditions for cultural practices, social (including reproductive) strategies, and subsistence activities of future generations.

Last but not least, in terms of the evolutionary processes that affect the frequency of transmitted cultural practices over time is drift: the effect of chance variation in the frequency with which practices are copied, unrelated to their properties. The idea that drift is relevant to understanding cultural change has a long history in anthropology and archaeology (e.g., Rivers, 1926), but its significance was most explicitly argued by Dunnell (1978) in his distinction between style and function in artifacts, the latter referring to all variation that is under selection and the former to that which is not, i.e., is neutral. For Dunnell this was an absolute distinction — attributes were either under selection or they were not — but, in fact, even if particular attributes are under selection, their representation in subsequent generations may be almost entirely the result of chance factors if population sizes are small.

Clearly, if we think of archaeology as a field for developing cultural evolutionary theory or testing evolutionary models, as opposed to merely interpreting archaeological data in the light of evolutionary theory, then it has serious disadvantages: there is no possibility of experimentally manipulating relevant variables nor is it possible to observe mechanisms and processes directly. On the other hand, archaeology provides unique quantitative information on population-level distributions of cultural attributes over long periods of time. This information concerns not only socially transmitted cultural traditions but also the ongoing process of niche construction. In many cases we actually know far more about the day-to-day material life of communities living thousands of years ago whose settlements have been excavated than we do about more or less modern ones. Moreover, the archaeological data capture variation, essential to evolutionary analysis, as opposed to the generalized, normative accounts of most sociologists and ethnographers. The value of this information is now being greatly enhanced by the increasing quantities of data that are becoming available on potential external selective factors operating over long periods of time, for example, climate change.

## 10.3 IDENTIFYING SOCIAL INFORMATION TRANSMISSION IN THE ARCHAEOLOGICAL RECORD

If we are to take the social transmission of information as a foundation for adopting an evolutionary approach to archaeology, then the problem of identifying it arises immediately. Ethnographers can observe children learning from others, and psychologists can analyze learning processes experimentally. Archaeologically, the identification of continuities through time in cultural practices does not automatically imply that they are generated by social learning. They may arise, for example, from environmental continuities. Figure 10.1 shows through-time patterning in the proportional and absolute frequencies of wild and domestic animals at a series of sites located very close together on lake shores in eastern Switzerland and well dated by dendrochronology. The patterns of fluctuating high and low relative and absolute frequencies relate to climatic fluctuations and also to the increase in forest clearance over time, both independently evidenced, rather than to socially transmitted traditions of what proportion of the diet should come from eating hunted as opposed to domestic animals. Moreover, the patterns crosscut those identified on the basis of pottery-making traditions.

If we look at the patterning shown by a pottery making tradition (Figure 10.2), we see something different. This is an activity that ethnographic studies tell us is always socially transmitted. In these circumstances we do not expect to see radical changes from one short time step to the next. If we put the pottery assemblages in chronological order, we expect chronologically adjacent ones to be similar to one another and thus to show a relatively smooth pattern of changes through time (O'Brien and Lyman, 2000, Chapter 6). This is what Figure 10.2 shows. If adjacent

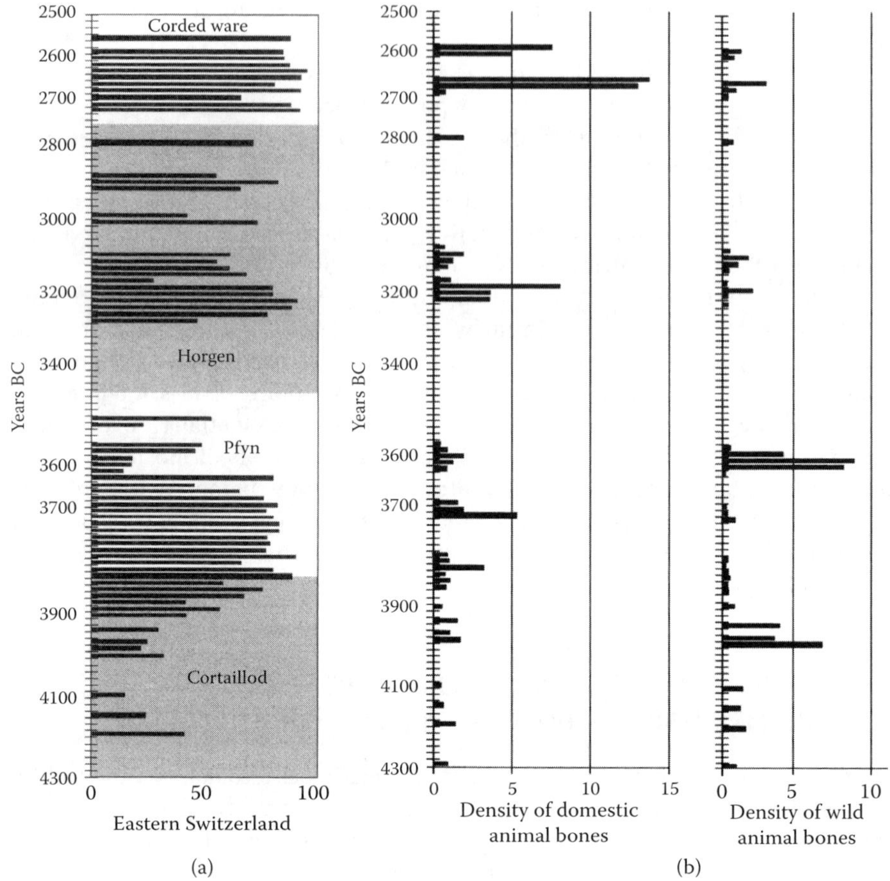


**FIGURE 10.1** (a) Percentages of domestic (as opposed to wild) animal bones from a series of Neolithic sites and layers within individual sites in eastern Switzerland. The order of the bars and the distances between them correspond to the date of the site or layer concerned. Cultural phases are also shown. (b) The absolute densities (i.e., numbers of bones per standardized excavation unit) of the domestic and wild animal bones found at Neolithic sites around Lake Zurich, showing rapid fluctuations over time together with a general trend for densities of domestic animal bones to increase and wild animal bones to decrease. From Schibler et al. 1997, used with permission.

assemblages do not tend to be similar, then we have to doubt whether we have evidence of a socially learned tradition. Equally, if we have a sudden very marked alteration in what appears to be such a tradition, then that is an indication that the tradition has been broken.

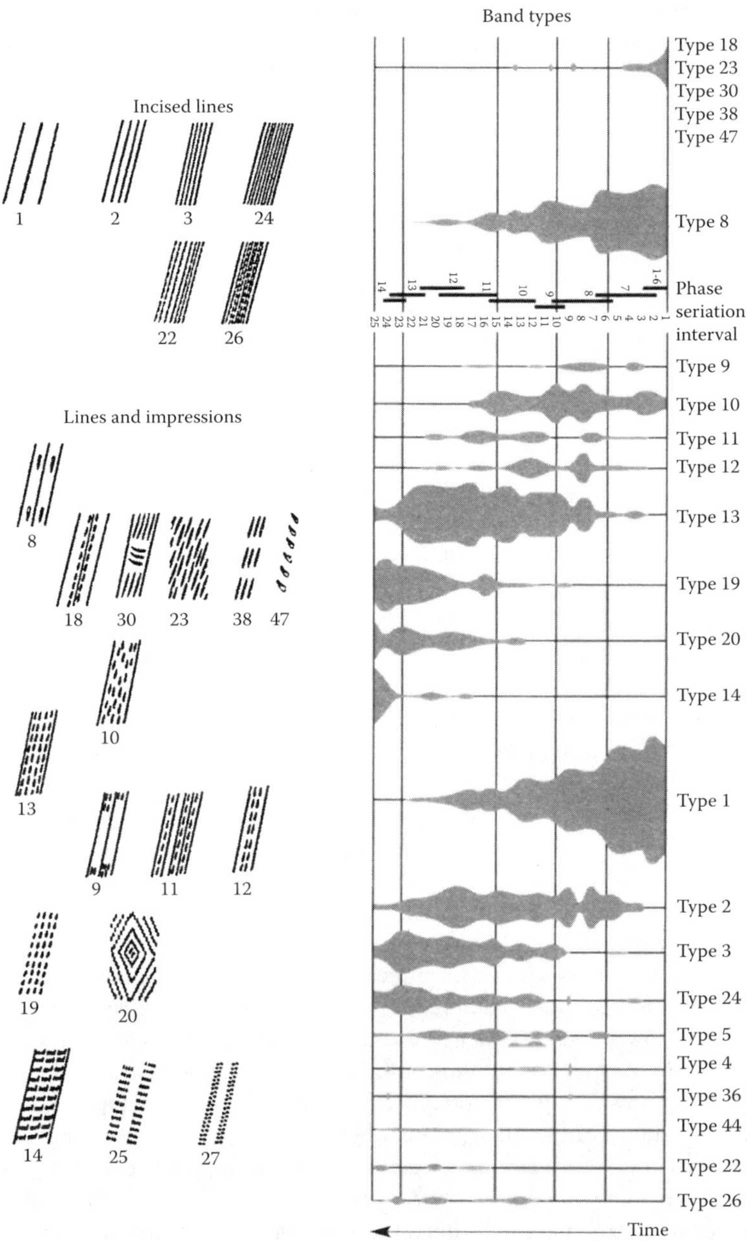


**FIGURE 10.2** The results of a seriation of pottery assemblages from a group of early farming sites in Germany, dating to c. 5300–4850 BC in terms of the frequency of vessels decorated with different types of impressed band motifs, which are also shown. The seriation was carried out by means of correspondence analysis and its chronological validity checked against independent evidence. The classic "battleship curve" pattern of the frequencies of the band motifs through time is readily apparent. From Frirdich (1994), used with permission.

## 10.4 ARCHAEOLOGICAL CASE STUDIES

In the remainder of this paper I will present three related archaeological case studies that adopt an evolutionary approach to accounting for historical patterns. They address a variety of processes affecting patterns of social information transmission in prehistory, starting with an example of the operation of natural selection and how culture affects survival and reproductive success then going on to illustrate the processes of niche construction and cultural drift.

### 10.4.1 The Spread of Cereal Agriculture into Europe, c.11,000–7000 BP

Farming in western Eurasia began in the so-called Fertile Crescent of the Near East c. 12,000 years ago. This is now clear both from archaeological evidence and from recent genetic studies that show that sheep, goats, cattle, and the key cereals were domesticated only once, or in some cases perhaps twice, in this region (Zohary, 1999; Bollongino et al., 2005). By 11,000 years ago early farming groups had colonized the island of Cyprus (Peltenburg et al., 2001), and this was only the beginning of an expansion that within less than 4000 years had reached the Atlantic coast of Portugal and the Low Countries of northwest Europe, not to mention areas far to the east of the Fertile Crescent (for an overview, see Harris, 1996). How this new adaptation emerged and the factors affecting its emergence remain unclear (see, e.g., Richerson et al., 2001; Colledge et al., 2004). However, once it did emerge it had the potential to sustain higher population densities than previous adaptations, meaning that for a time people locally had the chance to achieve greater reproductive success than previously, until a new ceiling was reached. New adaptations will be especially successful if dispersal opportunities are available to the populations practicing them (Voland, 1998), so that when a local ceiling has been reached expansion can continue elsewhere. The spread of farming into Europe can be regarded as a classic example of a dispersal opportunity.

In large parts of Europe, away from coastal and riverine areas with rich aquatic resources, Mesolithic hunter–gatherer population densities were very low. However, the areas with low population densities included zones that were very suitable for growing cereal crops and could thus sustain much higher densities of farmers than hunter–gatherers. Moreover, the combination of annual cereals and domestic animals, in addition to supporting higher population densities and, therefore, greater reproductive success before a new higher ceiling was reached, was extremely portable, far more so than many other agricultural systems. The result was a process of demic diffusion, which would have subsumed the small hunter–gatherer populations existing in the areas initially occupied by early farmers. In other words, this is a classic example of natural selection acting on people through an inherited cultural tradition, which gave a selective advantage to those who adopted it and passed it on to their children. In fact, the process involved not simply the inheritance of a tradition but also the transmission of a new niche, because the actual descendants of the cereal crops and animals that had originally been domesticated were being carried along as part of the dispersal and social strategies were affected by these new conditions (see below). In the course of this dispersal we can see the changes that the agricultural practices themselves underwent in a classic example of descent with modification affected by processes of transmission and selection, as Figure 10.3 shows (see also Colledge et al., 2004).

The visible evidence of the success of the agriculturally based dispersal process in natural selection terms for the human population is seen not only in the greatly increased density of human settlements in those areas where farming arrived, but also in the demographic profiles of prehistoric cemeteries in Europe and western Asia, as Bocquet-Appel (2002) has shown. Before the arrival of farming, population growth rates inferred from cemetery populations were at or just below replacement level; afterwards they were growing, albeit at varying rates. On the basis of studies of the number, size, and density of settlements of the first farmers in Central Europe, Petrasch (2001, 2005) has calculated population growth rates of between 0.9% and 2.7% for the first farming societies in this area.

In detail, the expansion process seems to have involved the colonization of patches favorable to the new economy rather than uniform expansion (van Andel and Runnels, 1995), hence the rapidity with which it occurred in many areas.

### 10.4.2 Patch Colonization and Its Consequences: Niche Construction

The principle of the "ideal despotic distribution" from population ecology (Sutherland, 1996) provides the basis for understanding the consequences of the patch colonization process. To understand its implications it is necessary to start with the more basic concept of the "ideal free distribution." This proposes that individuals occupy the resource patch that gives them the best returns. As more individuals occupy the patch the returns to each individual decline, to the point that the returns to an individual from the best patch are no better than those from the next best patch, which has no occupants. Now the returns from both patches are equal and they will be occupied indiscriminately until such time as the population grows to the point at which there is an equal benefit to be gained by occupying a still worse patch, and the process is repeated. When there is territoriality, however, the situation is different. Here the ideal despotic distribution applies. The first individual occupying the area is able to select the best territory in the best patch. Subsequent individuals settling there do not affect the first arrival but have to take the next best territory, and so on, until there comes a point where the next settler will do just as well by taking the best territory in the next best patch. Subsequent individuals will then take territories in either patch where the territories are equally suitable. In contrast to the ideal free distribution, where new settlers decrease the mean return for everybody, including those who arrived first, in the case of the ideal despotic distribution the returns depend on the order of settlement, so that the initial settlers of the best territory in the patch will do best, as long, that is, as they can defend the territory against anyone who might seek to take it from them.

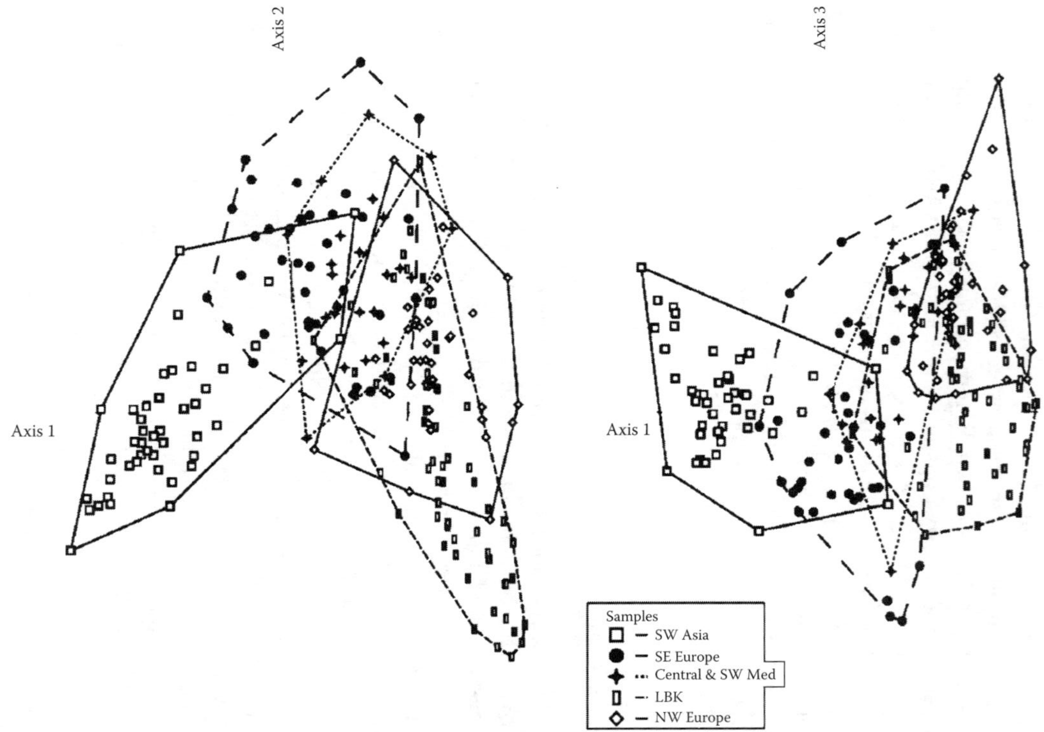


**FIGURE 10.3** The results of a correspondence analysis of data on the presence/absence of species from samples of archaeobotanical remains at early farming sites from southwest Asia to northwest Europe. The different regional composition of the assemblages is indicative of the way that farming practices changed as agriculture spread across Europe, adapting to new conditions. (a) Samples plotted against axes 1 and 2 of the analysis. (b) Samples plotted against axes 1 and 3. From Colledge et al., forthcoming.

It is apparent in those areas where detailed archaeological work has been carried out, particularly in Central Europe, that over time these local early farming societies became more unequal. The evidence for this comes from both their settlements and their cemeteries. In the settlements some houses were larger than others and had granaries (van der Velde, 1990). In some cases these houses also had more stone artifacts made of exotic raw materials that would have had to be obtained by exchange; in others there is evidence of higher proportions of domestic animal bones being associated with large houses and more remains of hunted animals with smaller ones (Hachem, 2000). It also seems that within settlements particular households and groups of households continued through time, with continuing inheritance of status witnessed by the rebuilding of houses of the same type in the same places. Over time, the proportional frequency of small houses as opposed to large ones increased, suggesting growing inequality. As far as burials are concerned, it is clear that the earliest cemeteries present a picture of relatively egalitarian societies, with indications of achieved status for older men, whereas the later ones tend to have a small group of graves, including child burials, clearly distinguished from the rest by the presence of markedly richer grave goods and probable symbols of power (Jeunesse, 1997).

These developments are best accounted for in terms of the logic of the ideal despotic distribution. The first comers to favorable settlement patches were able to settle in the best locations; indeed, the founding settlements invariably remained the most important ones. Despite the high population growth rate there was initially no competition between different groups because, as new households were formed, they were able to move to favorable locations elsewhere — indeed, this is what made the population growth possible. Relatively rapidly though individual microregions began to fill up. This did not affect the suitability of the area for those who had already established themselves there, because success depended on having a territory and they controlled the best ones. Cemeteries would have come into existence to represent an ancestral claim to territory (van der Velde, 2000, cited in Zimmermann, 2002) in the face of increasing competition as local carrying capacities began to be reached.

Strontium and oxygen isotope analyses of human skeletons make it possible to infer whether people spent the early part of their lives in the same region as they were buried. Analyses of the skeletons of the first farmers in Central Europe (Bentley et al., 2002) suggest that when farmers first arrived there was fairly general mobility but that in the later phases it was mainly women that moved. This would point to the emergence of patrilineal corporate groups. One can postulate that over time the senior line of the lineage in a given patch would have maintained control of the prime location and its territory and is represented archaeologically by the larger houses in the founding settlements. The junior branches, on the other hand, would be in increasingly inferior positions and would have relatively little option to go elsewhere because the same process was going on everywhere around them, hence the increasing number of smaller houses in the settlements.

One can further speculate that once local patches became full, contest competition between lineages would have become increasingly important and members of the senior line would increasingly have had to assert their position to maintain it. The deposition of rich grave goods in burials as a form of costly signaling (Bliege Bird, and Smith, 2005) would probably have had a role here. In this case, the number of rich burials would not simply be a reflection of the size of the senior lineage but of the competitive pressure it was under in particular places and times.

At this local level, then, the requirements of the agricultural economy via the ideal despotic distribution created a new type of niche with a new set of selection pressures that redefined the form and nature of social competition and by implication would also have had natural selection consequences leading to differential reproductive success among the descendants of particular lineage founders. Unfortunately, the data and methodologies currently available do not enable us to test this latter idea.

### 10.4.3 Cultural Drift

As already pointed out earlier, if we have a cultural tradition being passed on and modified through time, then, in addition to the various selection and biased transmission processes that may affect it, it is also subject to the process of cultural drift, random variation in the replicative success of particular cultural practices dependent on the sampling effects of copying in finite populations. Even if no other processes are operating, some variants will be copied more frequently and some less frequently than others. In the absence of innovation all the variants except one will eventually disappear, the time taken for this to happen depending on the size of the population (Neiman, 1995). If innovation is occurring, the success of any particular variant will be a result of chance processes that depend on the rate of innovation and the population size. These will also be the only factors that affect the diversity of a given population, i.e., how many variants are present in a particular population and how frequent they are proportionally. Neiman (1995) showed how the neutral theory of molecular evolution from which these principles and their mathematical formulation were derived could be used as a basis for generating quantitative predictions about the neutrality or otherwise of the processes affecting distributions of cultural variants through time. In effect, this becomes a kind of null hypothesis. If a particular distribution fails to depart from neutrality, there is no reason to postulate anything other than drift as the process producing it (Bentley et al., 2004). If it does then something further needs to be invoked to account for it (see Shennan and Wilkinson, 2001).

In the case to be described here a study was carried out of the changing distribution through time of pottery decoration motifs used by the first European farmers who have been the basis of the two previous case studies.

If neutrality holds, then the diversity of an assemblage, will be a function of the innovation rate and the effective population size. Because for any given assemblage we will know the number of different decorative variants present and the assemblage size, we can insert the relevant numerical values into an equation produced by Ewens (1972) to obtain the value of θ, predicted diversity if drift is the only process operating (Neiman, 1995):

$$E(k) = \sum_{i=0}^{n-1} \frac{\theta}{\theta + i}$$

where $E(k)$ is the expected number of different variants, and $n$ corresponds to the assemblage size.

This may be compared with the actual diversity value obtained from the data by squaring the proportional frequencies of each variant and summing the resulting values:

$$\theta = \frac{1}{\sum_i p_i^2} - 1$$

where $p_i$ represents the proportional frequency of the $i$th decorative variant.

In this particular analysis the predicted and observed values were obtained for the decorative motifs from a series of pottery assemblages from early farming sites in southwest Germany, where each assemblage came from a single phase at a single site (Shennan and Bentley, forthcoming). Some phases were represented at several sites, others at only one. Figure 10.4 shows the relationship between the logs of the actual and predicted diversity values: $r = 0.85$.

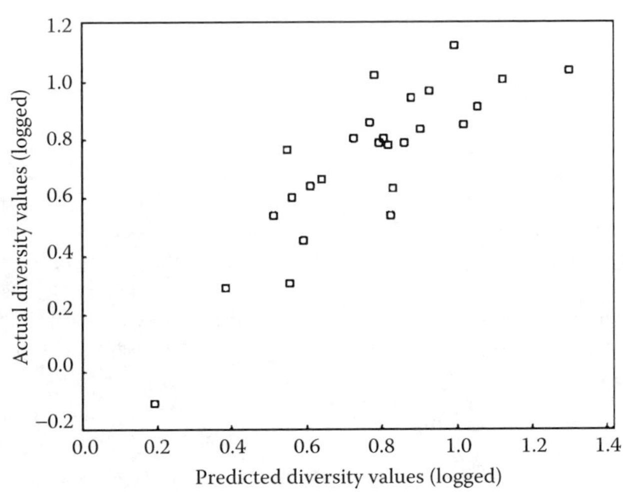


**FIGURE 10.4** Plot of the actual diversity values of the pottery assemblages from early farming sites in southwest Germany, based on the relative frequencies of different decorative motifs against the diversity values corresponding to the neutral model.

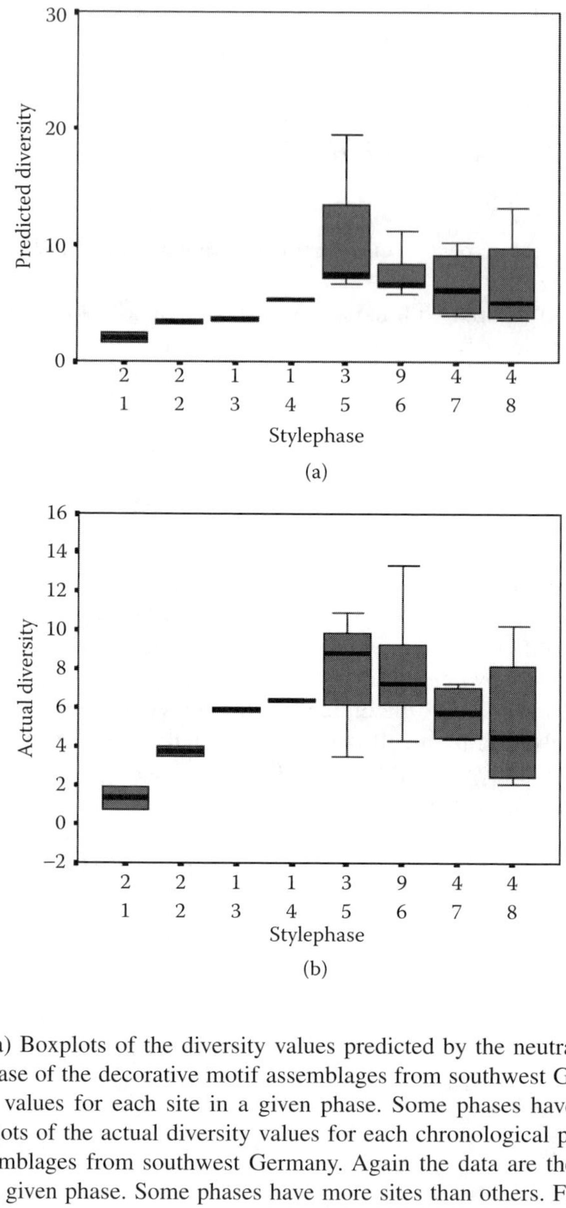


**FIGURE 10.5** (a) Boxplots of the diversity values predicted by the neutral model for each chronological phase of the decorative motif assemblages from southwest Germany. The data are the diversity values for each site in a given phase. Some phases have more sites than others. (b) Boxplots of the actual diversity values for each chronological phase of the decorative motif assemblages from southwest Germany. Again the data are the diversity values for each site in a given phase. Some phases have more sites than others. From Shennan and Bentley, forthcoming.

Figure 10.5 shows boxplots of the predicted and actual diversity values by phase. It can be seen that both show the same pattern of increasing diversity up to phase 5, gradually decreasing thereafter.

An alternative way of testing for neutrality is based on the actual frequencies of the different variants in the assemblages, rather than the assemblage diversity.

It has been shown by Bentley et al. (2004) that if a random copying process is in operation then the frequency distribution of the variants will correspond to a power law in which a very small number of variants occur at very high frequency but most of them have frequencies that are very low. For all eight phases the $r^2$ value of the fit of the distributions to the power law predictions was > 0.65, in several cases 0.9 or greater (Shennan and Bentley, forthcoming).

It appears that in this case the patterns in the frequency and diversity distributions of the decorative motifs from each site phase assemblage are neutral and therefore solely a function of the rate of innovation of new decorative motifs and of the effective population size. Looked at from the perspective of any single site, both the innovation rate and the effective population size depend to some extent on the amount of intersite interaction. If there is more interaction between pottery makers at different sites, then the effective population size will be greater. In addition though, under neutrality greater interaction will also mean that motifs currently not existing locally will be copied from other places and will thus represent local innovations. It appears that in the case of the pottery of the early farmers from southwest Germany we have evidence for a trend of gradually increasing then decreasing interaction, which may be at least partly related to a corresponding trend in absolute population size.

It is important to emphasize, however, that such analyses do not always fail to reject the null hypothesis of neutrality. Kohler et al.'s (2004) study of diversity in pottery styles in prehistoric New Mexico showed that the assemblages were less diverse than predicted by the model, despite increased local population size. The authors concluded that conformist transmission was operating as one element in a process of enhancing cooperation within larger groups. Shennan and Wilkinson's (2001) study, in contrast, showed an emerging pattern of excess diversity, pointing to a bias in favor of novelty.

## 10.5 CONCLUSION

Prehistoric archaeology started as a discipline concerned with describing cultural history, variation in space and time in the material recovered from the archaeological record. Unfortunately, the way its practitioners conceptualized that variation and the mechanisms they used to account for the patterns they described were very naïve. Among other things they failed to take into account the importance of adaptive processes of various kinds. The adaptationists, on the other hand, by assuming that necessity was the mother of invention, did not think history was important at all and ignored the process of social information transmission. By bringing the two together, dual inheritance theory and its niche construction extension do not simply effect some kind of conceptually satisfying synthesis but offer a set of theoretical tools developed from evolutionary biology that enable us to make predictions testable against the archaeological record. Social and ecological information transmission provide a basis for the operation of natural selection on human populations. Culturally inherited traditions are themselves modified by transmission biases, but they are not started afresh by each generation from first principles. Some of the patterning observed in those traditions can be accounted for in terms of drift, a process that can only exist in the context of an evolutionary framework.

It would be idle to pretend that disentangling the histories of cultural traditions and of human populations and their constructed niches is straightforward. Nevertheless, once we recognize the existence of these different histories and the evolutionary processes affecting them, we can switch our focus from one to another in illuminating ways. Archaeology does have some unique advantages as a domain of cultural evolutionary study through the access it gives us to the long term. Furthermore, to the extent that archaeologists succeed in offering satisfactory explanations for the patterns they observe, they are contributing to the justification and development of cultural evolutionary theory.

## REFERENCES

Bentley, R.A, Hahn, M.W., and Shennan, S.J., Random drift and culture change. *Proc. Roy. Soc. Biol. Sci.*, 271, 1443–1450, 2004.

Bentley, R.A., Price, T.D., Lüning, J., Gronenborn, D., Wahl, J., and Fullagar, P.D., Human migration in early Neolithic Europe, *Curr. Anthropol.*, 43, 799–804, 2002.

Bliege Bird, R., and Smith, E.A., Signaling theory, strategic interaction, and symbolic capital, *Curr. Anthropol.*, 46, 221–248, 2005.

Bocquet-Appel, J.P., Paleoanthropological Traces of a Neolithic Demographic Transition, *Curr. Anthropol.,* 43, 637–647, 2002.

Bollongino, R., Edwards, C.J., Bradley, D., Alt, K.W., and Burger, J., 2005. Phylogenie und Populationsgenetik neolithischer Haus- und Wildrinder, Paper presented at the conference *Die Neolithisierung Mitteleuropas*, Mainz, Germany, June 2005.

Boyd, R., and Richerson, P.J., *Culture and the Evolutionary Process*, University of Chicago Press, Chicago, 1985.

Colledge, S., Conolly, J., and Shennan, S., Archaeobotanical evidence for the spread of farming in the East Mediterranean, *Curr. Anthropol.*, 45, S4, S35-S58, 2004.

Colledge, S., Conolly, J., and Shennan, S., The evolution of early Neolithic farming from SW Asian origins to NW European limits, forthcoming.

Dunnell, R., Style and function: a fundamental dichotomy, *Am. Antiquity*, 43, 192–202, 1978.

Ewens, W.J., The sampling theory of selectively neutral alleles, *Theor. Popul. Biol.,* 3, 87–112, 1972.

Frirdich, C., Kulturgeschichtliche Betrachtungen zur Bandkeramik im Merzbachtal, In *Die Bandkeramik im Merzbachtal auf der Aldenhovener Platte*, Lüning, J., and Stehli, P., Eds., Habelt, Bonn, Germany, 207–394, 1994.

Hachem, L., New observations on the Bandkeramik house and social organisation, *Antiquity*, 74, 308–312, 2000.

Harris, D., Ed., *The origins and spread of agriculture and pastoralism in Eurasia*, UCL Press, London, 1996.

Jeunesse, C., *Pratiques funéraires au Néolithique ancien: sépultures et nécropoles danubiennes 5500–4900 av. J.-C.*, Errance, Paris, 1997.

Kohler, T.A., VanBuskirk, S., and Ruscavage-Barz, S., Vessels and villages: evidence for conformist transmission in early village aggregations on the Pajarito Plateau, New Mexico, *J. Anthropol. Archaeol.*, 23, 100–118, 2004.

Laland, K.N., Odling-Smee, J., and Feldman, M.W., Niche construction, biological evolution, and cultural change, *Behav. Brain Sci.*, 23, 131–145, 2000.

Neiman, F.D., Stylistic variation in evolutionary perspective: inferences from decorative diversity and inter-assemblage distance in Illinois woodland ceramic assemblages, *Am. Antiquity*, 60, 7–36, 1995.

O'Brien, M.J., and Lyman, R.L., *Applying evolutionary archaeology,* Plenum, New York, 2000.

Peltenburg, E., Colledge, S., Croft, P., Jackson, A., McCartney, C., and Murray, M.A., Neolithic dispersals from the Levantine corridor: a Mediterranean perspective, *Levant*, 33, 35–64, 2001.

Petrasch, J., Seid fruchtbar und mehret euch und fullet die Erde und machet sie euch untertan, *Archaeol. Korrespond.*, 31, 13–25, 2001.

Petrasch, J., Demografischer Wandel am Beginn der Bandkeramik, Paper presented at the conference *Die Neolithisierung Mitteleuropas*, Mainz, Germany, June 24–26, 2005.

Richerson, P.J., Boyd, R., and Bettinger, R.L., Was agriculture impossible during the Pleistocene but mandatory during the Holocene? A climate change hypothesis, *Am. Antiquity*, 66, 387–412, 2001.

Rivers, W.H.R., *Psychology and ethnology,* Kegan Paul, Trench, Trubner, London, 1926.

Schibler, J., Jacomet, S., Hüster-Plogmann, H., and Brombacher, C., Economic crash in the 37th and 36th centuries cal. BC in Neolithic lake shore sites in Switzerland, *Anthropozool.*, 25–26, 553–570, 1997.

Shennan, S.J., and Bentley, R.A., Style, interaction and demography among the earliest farmers of Central Europe, forthcoming.

Shennan, S.J., and Wilkinson, J.R., Ceramic style change and neutral evolution: a case study from Neolithic Europe, *Am. Antiquity*, 66, 577–593, 2001.

Skyrms, B., Evolution of the social contract, Cambridge University Press, Cambridge, 1996.

Smith, E.A., Three styles in the evolutionary analysis of human behaviour, In Adaptation and human behaviour, Cronk, L., Chagnon, N., and Irons, W., Eds., Aldine de Gruyter, New York, 27–46, 2000.

Sutherland, W.J., *From individual behaviour to population ecology*, Oxford University Press, Oxford, 1996.

van Andel, T.H., and Runnels, C.N., The earliest farmers in Europe, *Antiquity*, 69, 481–500, 1995.

van der Velde, P., Banderamik social inequality — a case study, Germania, 68, 19–38, 1990.

van der Velde, P., Dust and ashes: the two Neolithic cemeteries of Elsloo and Niedermerz compared, *Analecta Praehistorica Leidensia*, 25, 173–188, 2000.

Voland, E., Evolutionary ecology of human reproduction, *Annu. Rev. Anthropol.*, 27, 347–374, 1998.

Zimmermann, A., Landschaftsarchäologie I: Die Bandkeramik auf der Aldenhovener Platte, *Bericht der Römisch-Germanischen Kommission*, 83, 17–38, 2002.

Zohary, D., Monophyletic vs. polyphyletic origin of the crops on which agriculture was founded in the Near East, *Genet. Resource. Crop Evol.*, 46, 133–142, 1999.

# 11 The Uptake of Modern Contraception in a Gambian Community: The Diffusion of an Innovation over 25 Years

*Ruth Mace, Nadine Allal, Rebecca Sear, and Andrew Prentice*

## CONTENTS



## ABSTRACT

In this study we examine first, use of modern contraception in four Gambian villages over 25 years. This is the first such study showing microlevel change over time from the first availability of this new technological innovation. In 1975, a medical center was opened in one village providing contraceptive services free of charge to those who wished to use it. We examined determinants of women's age at first use of modern contraceptives from 1975 or from age 15 if younger than that in 1975. The ideal of large family size remains strong, and those at low parity are significantly less likely to start using contraception than those at high parity for their age. Wealth was also significantly related to the probability of contraceptive use, but negatively, with the wealthiest ranked women being the least likely to adopt the innovation. But we find that the largest effects on the probability of uptake were village and calendar year. Over the last 25 years, there is a doubling time of about 10 years in the risk of progressing to first use of contraception. Villages with strong social ties proceed at a similar rate, whereas one village that had fewer social ties with the others proceeded at a much faster rate. These patterns of uptake suggest that cultural transmission has an important effect on the spread of this technological innovation. We also compare the reproductive success (i.e., completed fertility) of users and nonusers and find that women using contraception actually have higher reproductive success than those that do not. The dynamics of uptake are discussed in the light of both evolutionary and social network models of cultural diffusion.

## 11.1 INTRODUCTION

The diffusion of innovations through communities has been a topic of interest for both sociologists (Rogers, 1995) and demographers and, in a parallel literature, for those interested in cultural evolution (Boyd and Richerson, 1985). Demographers, trying to understand the demographic transition to low fertility, have found the study of social networks a useful tool for understanding the diffusion of ideas about reducing fertility (Kohler, 1997) and about using modern contraceptives to achieve this. But although demographers are now reasonably happy to explain demographic transitions through proximate mechanisms, such as the spread of modern ideas about low fertility through populations, human evolutionary ecologists interested in trends in fertility are still unsatisfied with such explanations because they do not resolve a deeper problem of why anyone should voluntarily limit their fertility when this would ostensibly reduce their reproductive success and thus their Darwinian fitness. Economic demographers and evolutionary ecologists have both emphasized the quantity/quality trade-off inherent in raising children as a possible explanation (Borgerhoff Mulder, 1998; Kaplan, 1996). The higher the levels of parental investment required for children to be successful in marrying and establishing viable households of their own, the smaller the optimal family size that will maximize long-term reproductive success and wealth (Mace, 1998). However, although there is evidence that having a smaller family may promote economic success, the evidence that limiting fertility (other than by very small amounts) enhances long-term reproductive success is not evident (Kaplan et al., 1995). Gene-culture coevolutionary theorists have argued that innate biases to copy those in prestigious positions may have led to the trend to reduce fertility: if children hamper our ability to achieve wealth and status in the modern world, then copying the wealthy could lead to the spread of notions of low fertility (Boyd and Richerson, 1985). Thus, they argue an adaptive, evolved bias leads to a maladaptive outcome in a novel environment. Demographers have also stressed the importance of influential members of the community to help promote the spread of ideas about contraception. In this study, we follow the adoption of contraception in a rural Gambian community over 25 years. We examine determinants and patterns of uptake in the light of both sociological and evolutionary theories.

Whereas most developing regions of the world followed fairly rapid fertility decline during the 20th century, sub-Saharan African countries were the last to start on what is now a worldwide phenomenon and appear to be proceeding rather slowly. Typically, the onset of fertility decline is associated with mortality decline, but beyond that very broad generalization, the precise timing of the onset of fertility decline has proved hard to associate with any particular feature of the demographic or economic profile of a region (Cleland and Wilson, 1987). After decades of research on this topic, favor was found with the hypothesis that cultural diffusion of an idea was a major determinant of the variation in the timing of the onset of fertility decline (Bongaarts and Watkins, 1996).

Top down political programs urging small family size are less influential than one might expect, at least in the absence of coercion. It has been hypothesized that women in social groups who come into contact with other women who are using modern contraception successfully are far more likely to start using modern contraception themselves. There is evidence that ideas or information about contraception may diffuse horizontally between proximal regions (Bocquet-Appel and Jakobi, 1997) through social networks (Paz Soldan, 2004; Valente et al., 1997) and across regions speaking the same language (Amin, Basu, and Stephenson, 2002; Lesthaeghe, 1977). It is also possible that women who are in a position to observe the benefits to children and mothers of smaller family size are encouraged to change their preferences with respect to family size.

The adoption of technological innovations that arise from such cultural diffusion usually follow an S-shaped curve. Sociologists have typically interpreted this S-shape as a reflection of different levels of risk aversion in populations, with innovators acting first, followed by the less adventurous individuals, with conservatives and laggards taking much longer to adopt new technologies (Rogers, 1995). However, cultural evolutionary modelers have pointed out that even in a population of identical individuals, if all individuals are influenced by the frequency of their contacts with users of the new technology, then uptake rates will accelerate as the density of potential "cultural parents" that use the innovation increases, followed by a decline only when the number of nonusers becomes sparse (Boyd and Richerson, 1985; Henrich, 2001). Alternatively, if women are not responding to each other, but to advertising or other forms of public information, then uptake curves might be expected to be linear (if advertising rates are constant) or r-shaped, or even highest at the point of the advertisement, followed by decline. Thus, the shape of the uptake curves over time can give us information about how ideas about modern contraception are being propagated.

In the Gambia, fertility decline is only very recently observed (Cohen, 1998), and contraceptive use is low: the most recent data suggests only 8.9% of Gambian women are using modern methods of contraception (UN, 2003). The ideal of large family size persists, especially in rural areas, as it does in most of rural Africa although modern contraception nonetheless provides a useful alternative to sexual abstinence for those wanting to space births, preserve maternal health, and prevent extremely large family sizes (Bledsoe et al., 1994). One barrier to contraceptive uptake is fear of the new technology, often associated with the fear that fertility might not return after contraceptive use has stopped. Another is opposition from husbands: the women reported to us that their husbands were less concerned or aware of the high health costs to mothers and children of high fertility. However, some men reported that problems with school fees and clothing costs resulted in large families being problematic. Some religious leaders also oppose birth control.

Our study takes place in four neighboring villages in rural Gambia, three of which have unusually good access to health care due to the presence of a medical center. This was set up by the UK Medical Research Council (MRC) as part of a long-term medical research project. The majority of villagers are Mandinka farmers, who live and work in an environment characterized by strong seasonality and high disease burdens and are Muslim. Medical research started at the site in 1949, which, for the next 25 years, included the recording of dates of all births and deaths in the four villages but did not involve the year-round presence of medical staff in the villages. Prior to 1974, the villages experienced high fertility and high mortality, due to high levels of general parasitism and specifically malaria (Billewicz and McGregor, 1981). In 1975 the medical center was opened in one of the villages, providing free medical care, which was available throughout the year; mortality rates began to decline immediately (Lamb et al., 1984; Rayco-Solon et al., 2004) although fertility remained high. It was at that time that contraceptive services first became available. Before 1975, one village (D) had withdrawn from the research project, so only three villages had access to the contraceptive and other services provided by the medical center. Much of the research being conducted in Villages A, B, and C was on mothers and babies, so mothers had frequent contact with midwives, from whom contraceptive advice and the contraceptives themselves could be obtained. Our main interest here is in how contraceptive uptake spread through these villages over time.

Evolutionary ecologists (and hence evolutionary demographers) make a distinction between the proximate and ultimate causes of a behavior (Krebs and Davies, 1993). By ultimate causes, we mean the factors influencing the fitness of individuals following that behavior and, thus, the evolutionary forces that cause it to be maintained in the population over time. By proximate explanations, evolutionary ecologists mean the mechanisms that might drive the behavior, which include both the biological influences that influence fecundity and the mechanism by which a behavior is learned or enforced. We are interested in the ultimate explanations for fertility decline (i.e., its fitness implications) as well as the proximate mechanisms by which this decline is brought about (such as the diffusion of ideas); so we have also investigated whether contraceptive use appears to be maladaptive (reducing biological fitness) in this particular setting.

## 11.2 METHODS

A total of 760 married women between the ages of 15 and 92, currently living in the four villages, were interviewed in a single-round survey in the year 2001. This included all women living in the four villages who were present at the time of the survey (women who moved to the villages solely to work at the health clinic and research station were not included in the survey). The characteristics of the villages are given in Table 11.1. The survey included questions on whether they were currently using modern contraceptives and when, if ever, they had first used modern contraceptives. For the latter question, they could answer before their first birth or after which birth they had first used contraception if ever. Only those women whose year of birth was on record and whose children's years of birth were known were included in our analysis. Further, women who were 50 or over in 1975 were excluded as they would not have had need of contraception by the time the medical center was opened. This reduced the sample size to 707.

**TABLE 11.1**
**Characteristics of Study Villages**

| Village | Population in 2001* | No. of households in 2001 | No. of women included in sample | Contraceptive prevalence in under 50s in sample in 2001 |
|---|---|---|---|---|
| A | 1676 | 174 | 288 | 10.4 |
| B | 425 | 53 | 87 | 15.3 |
| C | 550 | 80 | 142 | 29.6 |
| D | 1044 | 117 | 190 | 0.9 |
| Totals | 3695 | 424 | 707 | 12.3(mean) |

*Population sizes are estimated to have increased by between 50%–90% since 1975.

It is assumed that women started using contraception in the year of the opening of the birth interval in which they stated they first used contraception. Event history analysis (a logistic regression of yearly probability of first use) was used to investigate which covariates influenced age at first use (using the proc logistic procedure on year-based data in SAS). Methods of modern contraceptive use were either injectables and pills. Our response variable was a simple dichotomy of use/nonuse of modern contraception, and we did not distinguish between contraceptive methods in our analysis. Records were right censored in 2001 or when the woman reached age 50; and records were left censored in 1975 or at age 15. Sample sizes of those included, by village, are shown in Table 11.1.

The households in which women lived were wealth ranked into three categories, according to the judgment of three independent villagers — the wealth rank used was the median rank. Covariates examined included age, age squared, parity, and date (which were entered as time-varying covariates) and village of residence, wealth rank of head of household in 2001, and comprehension of the English language (which were entered as fixed).

In order to see whether the shape of the uptake curves over time were following a different pattern in different villages, we repeated the model including only Villages A and B (together) and another model including only Village C.

We analyzed the reproductive success (controlling for maternal age) of users compared with nonusers. The general linear model (GLM) procedure in SPSS, Inc. was used to determine whether women who had ever used contraception had fewer living children (a measure of reproductive success) than women who had never used contraception.

## 11.3 RESULTS

There were 152 events of contraceptive uptake among the 707 women, representing a minority of the women interviewed (21.5%). Current use of contraception in these villages is 7.9% (see Table 11.1). However these figures mask a great deal of variation over space and time. Mean level of current contraceptive use is only marginally

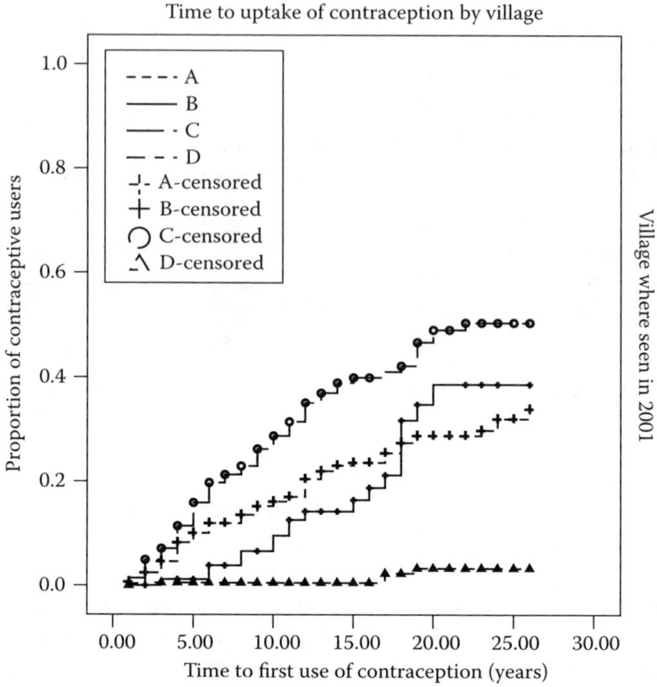


**FIGURE 11.1** The cumulative uptake rate of first use of contraceptives by age by birth cohort.

lower than that observed among married women in the Gambia as a whole, but it varied from 0.5% in Village D to 18.3% in Village C. All women stated they had not used contraception prior to their first birth. Furthermore, the first use of contraception was only rarely associated with an end to reproduction. Most women went on to have further births after they had become contraceptive users, consistent with the notion that women were mostly using contraception to space births or avoid postpartum sexual abstinence.

The cumulative probability that women will have become users at a certain age, over the decades, is shown in Figure 11.1. The probability of starting to use modern contraception at a given age is roughly doubling each decade since 1975.

Figure 11.2 shows the cumulative adoption rate by age and by village. Village D did not have access to the medical center in Village A and women living there had to travel to government health services 15 miles away in order to obtain contraceptives. The very low rates of uptake in this village thus illustrate the well-known finding that easy access to contraceptive services is an extremely important determinant of contraceptive prevalence (Tsui and Ochoa, 1992). Village D represents a control for the effect of access to the medical center. The very low rates of uptake indicate that there are indeed still areas of rural Gambia where effective access to contraceptive services is almost nonexistent. These women would not have had the regular contact with midwives, who were an important way of accessing knowledge

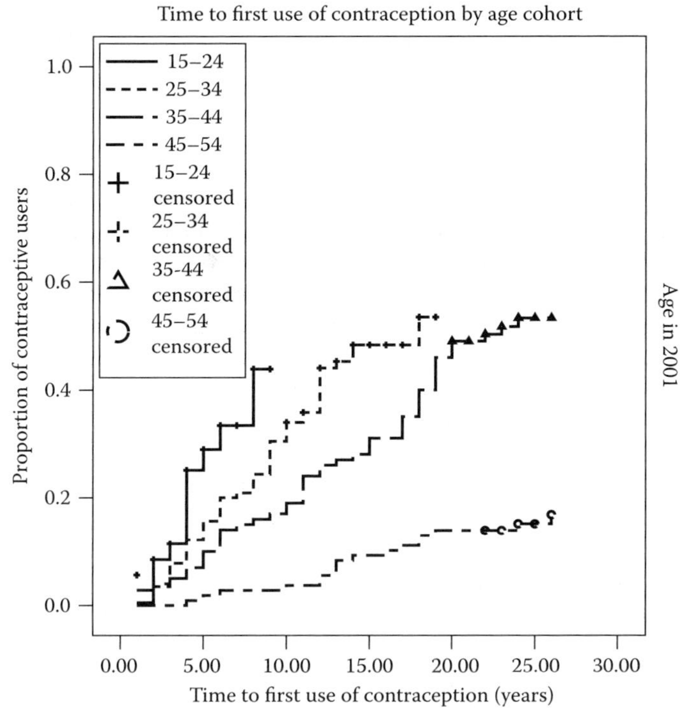


**FIGURE 11.2** The cumulative uptake rate of first use of modern contraception by age by village of residence.

of contraceptives for women in the other villages. Thus, the low uptake rates in Village D are as likely to be due to lack of information as to lack of supply, as has been found in another area of rural Gambia (Luck et al., 2000).

Figures 11.1 and 11.2 show the raw data, which conflate a number of effects. The event history analysis on age at first use allows us to untangle these effects, and results are shown in Table 11.3. The significant influences are age, age squared, parity, village, wealth rank, and date.

The effect of age is broadly negative (although the age term is positive, the larger age-squared term is negative, causing a decline over most of the relevant age range); but this is combined with a strongly positive effect of parity. Because parity increases with age, older women are more likely to be users within cohorts. This pattern is strongly suggestive of strategic use of contraception to space births among those who have a large number of children relative to others of their age.

Only a very few individuals spoke English (an indication of educational level), and this did not significantly influence contraceptive uptake. The effect of wealth was significant, although not in the direction predicted by the notion that wealthy (thus high status) individuals are generally the innovators with respect to contraceptive uptake. The wealthiest families were significantly slower than medium or poor families to first use contraception. This is, however, consistent with the evolutionary ecological perspective, which predicts that wealthy families will have more children because they have more resources with which to successfully raise children.

The effect of village is very significant: the three villages that have access to the same medical center do not have the same rate of uptake. Village A and the smaller Village B follow roughly the same trajectory and time to first use of contraception is not significantly different in these two villages. However, Village C is following a much faster trajectory; nonusers have almost three times the annual probability of first using contraception than in Villages A and B. This is not related to geographic distance from the medical center, which is, in fact, located in Village A, with Villages B, C, and D roughly equidistant (and nearby). However, Villages A and B have much stronger social ties with each other than either do with Village C, most of the residents of which are members of a different clan. This can be illustrated quantitatively by looking at statistics on intermarriage, shown in Table 11.2. Most women in our sample remain in their native village after marriage. Table 11.2 shows that although a few women did migrate between Villages A, B, and D for the purpose of marriage, this was not the case for Village C. No woman born in Village A, B, or D married into Village C and only one girl

**TABLE 11.2**
**Respondents Included in the Analysis, by Village of Residence and Area of Birth,* also Showing Contraceptive Use (Self-Reported)**

| | Village of residence | | | | |
|---|---|---|---|---|---|
| | A | B | C | D | Total |
| Total n | 288 | 87 | 142 | 190 | 707 |
| Village of birth | | | | | |
| A | 238 | 11 | 0 | 4 | 253 |
| B | 2 | 49 | 0 | 9 | 60 |
| C | 1 | 0 | 125 | 0 | 126 |
| D | 2 | 3 | 0 | 158 | 163 |
| Other | 45 | 24 | 17 | 19 | 105 |
| Contraceptive use (self-report) | | | | | |
| Current users | 20 | 9 | 26 | 1 | 56 |
| Ever used | 70 | 19 | 59 | 4 | 152 |

*Moving from birth village is due to marriage.

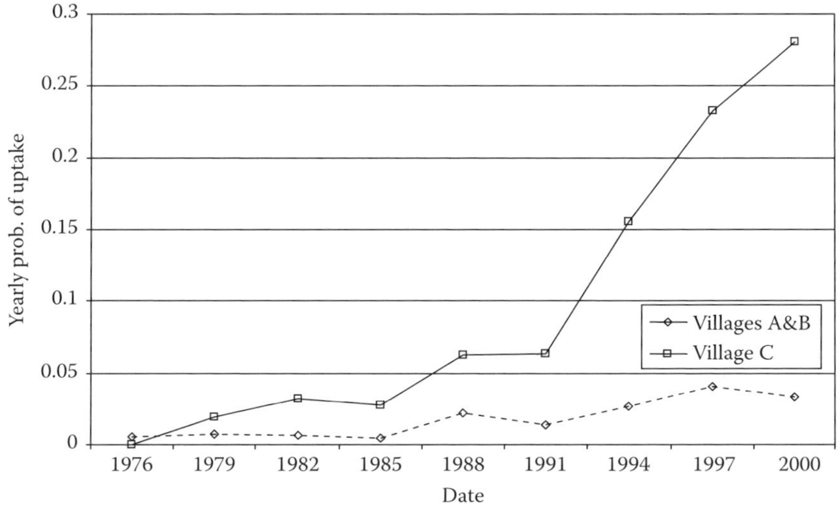


**FIGURE 11.3** Model adjusted uptake rate of first use of modern contraception for villages A and B (combined) and Village C, as a function of date (model output shown for woman aged 30 of parity 3–5).

from Village C married into any other study village (A, and in this case she married a migrant who had moved into Village A for employment). Thus, the similar rates of contraceptive uptake in Villages A and B, and the very different rate of uptake in Village C, seems to reflect the degree of social contact between these villages. Diffusion of infectious disease has already been demonstrated to run along these lines of social contact. In 1961 Villages A, B, and D were afflicted with a measles epidemic that caused high mortality of young children. Village C did not suffer from an outbreak of measles in this year, but did so 5 years later when a village 16 miles away, whose inhabitants do intermarry with Village D, also suffered an outbreak (McGregor, 1976).

Figure 11.3 shows the rates of uptake by village, having controlled for other relevant effects. These results were obtained by running the model again, first only including Villages A and B together (as Table 11.3 shows they did not differ significantly in their rate of uptake), and then just including Village C. After a slow start, it is clear the uptake rates in Village C have escalated — a pattern characteristic of a frequency-dependent effect, i.e., social contact with other users influencing the decision to use contraception for the first time. Uptake rate in Villages A and B is much lower, and although obviously increasing over time, it is hard to tell whether they have yet reached the beginning of any "take-off" in uptake rate. The early sections of an exponentially rising curve are notoriously hard to extrapolate. However, it is clear that, if current rates of uptake continue, within a generation women who have used modern contraception will become the norm and nonusers the minority in Villages A, B, and C.

**TABLE 11.3**
**Determinant of Age at First Use of Contraception**

| Parameter | Estimate | St. error | Odds ratio | P |
|---|---|---|---|---|
| Intercept | –4.773 | 1.4154 | | 0.0007 |
| **Age** | **0.2157** | **0.0942** | **1.241** | **0.0220** |
| **Age squared** | **–0.0062** | **0.0016** | **0.994** | **0.0001** |
| **Date 75–77** | **–1.8436** | **0.5492** | **0.158** | **0.0008** |
| **Date 78–80** | **–1.2040** | **0.4209** | **0.300** | **0.0042** |
| **Date 81–83** | **–1.0023** | **0.4026** | **0.367** | **0.0128** |
| **Date 84–86** | **–1.2364** | **0.4394** | **0.290** | **0.0049** |
| Date 87–89 | — | — | 1.000 | — |
| Date 90–92 | –0.2283 | 0.3355 | 0.796 | 0.4961 |
| Date 93–95 | 0.4997 | 0.2925 | 1.648 | 0.0876 |
| **Date 96–98** | **0.8788** | **0.2881** | **2.408** | **0.0023** |
| **Date 99-01** | **0.8636** | **0.3103** | **2.372** | **0.0054** |
| Village A | — | — | 1.000 | — |
| Village B | –0.2304 | 0.2673 | 0.794 | 0.3888 |
| **Village C** | **1.0800** | **0.1913** | **2.945** | **<.0001** |
| **Village D** | **–2.5749** | **0.5165** | **0.076** | **<.0001** |
| No English | — | — | 1.000 | — |
| Some English | 0.5562 | 0.3462 | 1.744 | 0.1082 |
| **Wealth 1** | **–0.4793** | **0.2291** | **0.619** | **0.0365** |
| Wealth 2 | — | — | 1.000 | — |
| Wealth 3 | –0.1510 | 0.1889 | 0.860 | 0.4240 |
| **Parity 0–2** | **–0.8798** | **0.2766** | **0.415** | **0.0015** |
| Parity 3–5 | — | — | 1.000 | — |
| **Parity 6+** | **1.0223** | **0.2746** | **2.780** | **0.0002** |

*Age, age squared, parity, and date are time varying.
**Village of residence, knowledge of English, and wealth rank are fixed.
***Significant effects are shown in bold.

Table 11.4 shows the fertility (number of live births) and mortality (number of children born alive that have died), corrected for age, of those that have ever used modern contraception and those that have not. It is clear that adopters of modern contraceptives have higher fertility than nonadopters. There is no evidence that infant mortality is significantly different in the two groups. This is consistent with the finding that women with high fertility for their age are more likely to become contraceptive users. However, they are not using contraception for long enough periods to reduce their fertility to a lower level than that of the nonusers of a similar age. Thus, there is no evidence that contraceptive use is necessarily maladaptive, from a Darwinian perspective, in this population.

**TABLE 11.4**
**Reproductive Success of Contraceptive Adopters (Those That Have Ever Used Modern Contraceptives) and Nonadopters, Shown as Number of Currently Living Children and Number of Children Born Alive That Have Died (Assuming Maternal Age of 43)**

| | Living Children | Dead Children |
|---|---|---|
| Adopters | 4.953 (0.166)* | 1.061 (0.113) |
| Nonadopters | 4.196 (0.082) | 1.253 (0.056) |

*Standard error shown in parenthesis.

## 11.4 DISCUSSION

Patterns of contraceptive uptake show several characteristics indicative of an innovation that is spread, in part, by social contact with other users. There is some indication that uptake rates accelerate as the number of users increases over time — an S-shaped adoption curve, with a flat base, is likely to emerge. However, it should be noted that there are a number of models of uptake that could produce curves not dissimilar from those shown in Figure 11.3. More convincing evidence of social transmission of ideas comes from the fact that it is clear that the villages linked through strong social ties, including links through kin, have similar rates of uptake whereas the more socially isolated village is proceeding at a different rate. Figure 11.4 shows a schematic representation of social links and the geographic relationship between the villages (not to scale). It is interesting to note that strong social ties with Village D, where no one is adopting the innovation (because they do not have easy access to it), is likely to increase the relative frequency with which women encounter nonusers; thus, it is interesting to speculate that lack of access to contraception in Village D could be slowing the rate of uptake in Villages A and B. Those sociologists interested in modeling the spread of ideas and of contagious diseases have pointed out that a social structure made up of small groups, who all know each other but some of whom have links with other groups, can lead to the dynamics of spread that are rather different from that of homogeneous, large populations (Watts and Strogatz, 1998). For example, connected small groups can greatly accelerate uptake across populations as "infection" spreads rapidly within small groups with repeated, close contacts and then jumps on to another group via a small number of contacts. Furthermore, groups from which the innovation or infection is blocked influence uptake rates in other groups to which that group is connected (Newman, Watts, and Strogatz, 2002). It seems likely that such models are applicable in this situation. Although we do not have specific data on connectedness to villages outside our sample, it is clear that Village C did not have any more girls marrying in from outside the four villages than did Villages A and B (in fact rather fewer, Table 11.2), whereas their contact with their immediate neighbors was more limited.

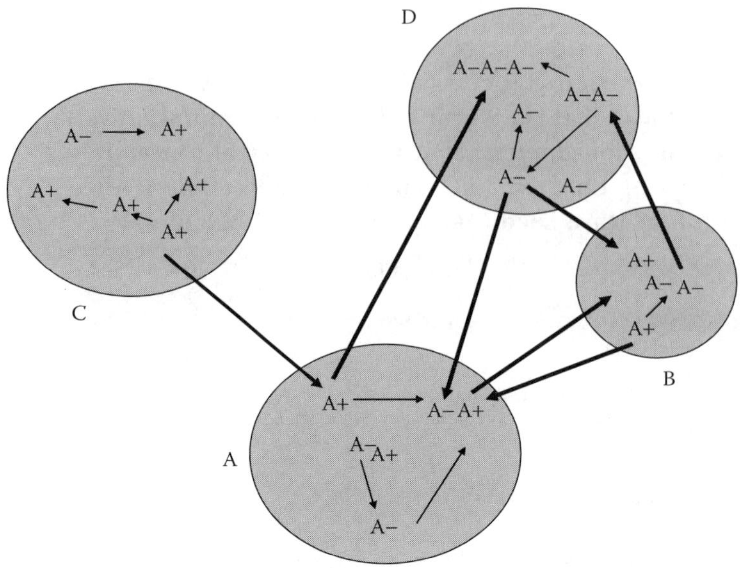


**FIGURE 11.4** Diagram representing the spatial arrangement of the four villages, with the relative balance of A+ to A– indicating the balance of contraceptive adopters and nonadopters in each village, and with arrows representing social ties within and between villages (not to scale).

Hence, once contraception began to be used, contact with other users becomes rapidly more likely in this more isolated community. However, there is no evidence that the faster rate of uptake is due to a founder effect, as the first few cases of contraceptive use were in fact in Village A.

Although analogies with contagion are appropriate at some level, we have also presented evidence that the individual decision to become a contraceptive user is strongly determined by individual sociodemographic status, particularly with regard to fertility. Thus, although the general acceptability of modern contraception appears to be increasing over time, the users are not passive recipients of infection. To use the terminology of cultural evolutionary studies, contraceptive use does not have the characteristics of a "selfish meme" (Dawkins, 1976) that is replicating in individuals as a means of spreading itself, possibly at the expense of their own reproductive interests. Conformism can emerge even when the decision to take up an innovation is based on some level of individual learning. For example, observing other users might help would-be users to evaluate social or health costs that they feared would be associated with contraceptive uptake, which might encourage them to try it for themselves, i.e., uptake rates accelerate when there are more users from which they can learn the true costs and benefits of contraceptive use. Thus, evidence that social contact influences uptake rates does not mean that the decision to use the innovation is not highly strategic for the individuals concerned. Even though contraceptive users may have lower fertility than they might have experienced without contraceptives, they appear to remain, on average, among the higher fertility women in their cohort in their village.

Evolutionary anthropologists have long puzzled over why women limit their fertility and, in particular, the paradox that wealthy populations seem to be of lower fertility than poor populations when they have more resources that could theoretically be used to invest in more children. This pattern seen globally across populations is, to some extent, explicable if one considers that wealthy populations have higher standards of acceptable parental investment per child (Low, 2000; Mace, 1998). Across population, differences may reflect differences in levels of aspiration or in the opportunities available, and, hence, the returns on parental investment will vary in each population. However, within homogeneous groups, where levels of parental investment were comparable, evolutionary ecological theory would predict a positive relationship between wealth and reproductive success. Within this population, there is no evidence that women in wealthy (and thus presumably higher status) families were the first to use contraception. On the contrary, if different, they took longer to first use. Thus, within this homogeneous village setting, it appears to be those for whom resources may be stretched that contraceptive use is more appealing. This fits with the prediction from evolutionary ecology but is at odds with much of the demographic literature on contraceptive uptake, although Bocquet-Appel and Jakobi (1997) found no evidence that the first fertility decline in Victorian Britain was associated with the elite. Cultural evolutionary theorists have also considered "prestige bias" (Boyd and Richerson, 1985) to be an important force in cultural transmission. Here we do not see evidence of prestige bias, if the wealth of the male head of household is what constitutes prestige; however, it is interesting to note that if women can gain prestige through high fertility, it is actually those with high fertility for their age that have the fastest rates of uptake, so they may indeed be the prestigious individuals that others are then copying.

Although there is plenty of evidence that social contact is related in some way to the rate of contraceptive uptake, this is clearly not the only explanation for the rate of spread of this new technology. Twenty-five years is a very long period of exposure to an innovation — so long that it becomes implausible to argue that ideas about contraception had not diffused into all those villages with access to the clinic long before many individuals started to use the technology. It is, therefore, necessary to ask why it took so long for contraceptive use to become more common. Although it has been conventional to evoke social and cultural barriers in this context, both have proved rapidly surmountable obstacles when women find themselves living in conditions favoring low fertility. To take an extreme example, rural Ethiopia has one of the highest birth rates in Africa. Yet in the Ethiopian capital city, Addis Ababa, where a large proportion of the population are actually recent migrants from rural areas, birth rates are below replacement (Gurmu, 2005; Sibanda et al., 2003). Further, in Addis, it is the wealthier individuals who have the highest birth rates, despite the fact that their access to both news media and medical services are likely to be higher than those of the poor (Gurmu, 2005). It seems that competition for resources, and a society that favors high levels of investment per child, are important determinants of receptivity to ideas about limiting family size. In the Gambia, the rural, agrarian economy in these villages did not necessarily provide these ultimate reasons for higher parental investment in fewer children until recently. Since 1975, as infant mortality has declined, the village populations are thought to have nearly doubled without commensurate increase in farmland (indeed opportunities for rice cultivation have actually been reduced since the Sahel droughts of the 1970s). Education in the villages is not readily available beyond the basic level, and outmigration for women to look for work was not a common strategy. However, since the population density in the villages has increased, the reliance on outmigration as a strategy has also increased. Job opportunities outside farming usually require education and/or outmigration. Better nutrition may have been reducing birth intervals when contraception is not used. All these factors have been slowly moving in the direction of providing incentives of limiting family size through contraception over the last 25 years. Ideas may be diffusing through networks in villages, but only when conditions ultimately favor higher parental investment in each child are ideas about lower fertility and contraception converted into actual uptake.

## REFERENCES


Amin, S., Basu, A. M., and Stephenson, R., Spatial variation in contraceptive use in Bangladesh: looking beyond the borders, *Demogr.,* 39, 251–267, 2002.

Billewicz, W. Z., and McGregor, I. A., The demography of two West African (Gambian) villages, 1951–1975, *J. Biosoc. Sci.,* 13, 219–240, 1981.

Bledsoe, C. H., Hill, A. G., D'Alessandro, U., and Langerock, P., Constructing natural fertility: the use of Western contraceptive technologies in rural Gambia, *Popul. Dev. Rev.,* 20, 81–113, 1994.

Bocquet-Appel, J. P., and Jakobi, L., The spatial diffusion of contraception in Great Britain and the origins of the fertility transition, *Popul.,* 52, 977–1004, 1997.

Bongaarts, J., and Watkins, S. C., Social interactions and contemporary fertility transitions, *Popul. Dev. Rev.,* 22, 639–682, 1996.

Borgerhoff Mulder, M., The demographic transition: are we any closer to an evolutionary explanation? *Trend. Ecol. Evol.,* 13, 266–270, 1998.

Boyd, R., and Richerson, P. J., *Culture and the evolutionary process,* University of Chicago Press, Chicago, 1985.

Cleland, J., and Wilson, C., Demand theories of the fertility transition: an iconoclastic view, *Popul. Studies,* 41, 5–30, 1987.

Cohen, B., The emerging fertility transition in sub-Saharan Africa, *World Dev.,* 26, 1431–1461, 1998.

Dawkins, R., *The selfish gene,* Oxford University Press, Oxford, 1976.

Gurmu, E., *Fertility transition driven by poverty: the case of Addis Ababa (Ethiopia).* Unpublished PhD, University College London, London, 2005.

Henrich, J., Cultural transmission and the diffusion of innovations: adoption dynamics indicate that biased cultural transmission is the predominate force in behavioral change, *Am. Anthropol.,* 103, 992–1013, 2001.

Kaplan, H., A theory of fertility and parental investment in traditional and modern human societies, *Yearbk. Phys. Anthropol.,* 39, 91–135, 1996.

Kaplan, H. S., et al., Does observed fertility maximize fitness among New Mexican men — a test of an optimality model and a new theory of parental investment in the embodied capital of offspring, *Hum. Nature,* 6, 325–360, 1995.

Kohler, H. P., Learning in social networks and contraceptive choice. *Demogr.,* 34, 369–383, 1997.

Krebs, J. R., and Davies, N. B., *An introduction to behavioural ecology,* 3rd ed., Blackwells, Oxford, 1993.

Lamb, W. H. et al., Changes in maternal and child mortality rates in three isolated Gambian villages over ten years, *Lancet,* 2, 912–914, 1984.

Lesthaeghe, R., *The decline of Belgian fertility, 1800–1970*, Princeton University Press, Princeton, NJ, 1977.

Low, B. S., Sex, wealth and fertility: old rules, new environments, In *Adaptation and human behaviour: an anthropological perspective*, Cronk, L., Chagnon, N., and Irons, W., Eds., Aldine de Gruyter, New York, 323–343, 2000.

Luck, M. et al., Mobilizing demand for contraception in rural Gambia, *Studies Fam. Plan.,* 31, 325–335, 2000.

Mace, R., The co-evolution of human fertility and wealth inheritance strategies, *Philos. Trans. Roy. Soc.,* 353, 389–397, 1998.

McGregor, I. A., Health and communicable disease in a rural African environment, *Oikos,* 27, 180–192, 1976.

Newman, M. E. J., Watts, D. J., and Strogatz, S. H., Random graph models of social networks, *Proc. Natl. Acad. Sci. USA,* 99, 2566–2572, 2002.

Paz Soldan, V. A., How family planning ideas are spread within social groups in rural Malawi, *Studies Fam. Plan.,* 35, 275–290, 2004.

Rayco-Solon, P. et al., Fifty-year mortality trends in three rural African villages, *Trop. Med. Int. Health,* 9, 1151–1160, 2004.

Rogers, E. M., *Diffusion of innovations*, 5th ed., Free Press, New York, 1995.

Sibanda, A. et al., The proximate determinants of the decline to below-replacement fertility in Addis Ababa, Ethiopia, *Studies Fam. Plan.,* 34, 1–7, 2003.

Tsui, A. O., and Ochoa, L. H., Service proximity as a determinant of contraceptive behaviour: evidence from cross-national studies of survey data, In *Family planning programmes and fertility,* Phillips, J. F., and Ross, J. A., Eds., Oxford University Press, New York, 222–256, 1992.

United Nations. *World Fertility Report 2003.* New York: United Nations Department of Economic and Social Affairs, Population Division, 2003.

Valente, T. W. et al., Social network associations with contraceptive use among Cameroonian women in voluntary associations, *Soc. Sci. Med.,* 45, 677–687, 1997.

Watts, D. J., and Strogatz, S. H., Collective dynamics of 'small-world' networks. *Nature,* 393, 440–442, 1998.

# 12 Sex without Birth or Death: A Comparison of Two International Humanitarian Movements

*John Cleland and Susan Watkins*

## CONTENTS



## ABSTRACT

HIV/AIDS prevention in developing countries has apparently achieved rather little despite massive expenditure on information and services. Increasingly, the effectiveness of a public health approach to the problem is doubted and fundamental changes in poverty, gender inequality, and culture are seen as necessary to stem severe generalized epidemics. We draw upon the much more extensive experience of family planning programs, which share key characteristics with HIV prevention, to argue that this perspective is profoundly mistaken. The family planning literature shows that modern contraception was initially greeted with suspicion and fear. In many countries, initial uptake of contraception was slow and this led many commentators to reiterate the view that poverty, gender inequality, and culture represented intractable barriers to reproductive change. History now shows that these barriers were illusory. We present an explanation for the lagged response to family planning exhortations and services and the speed with which changes in reproductive behavior eventually spread across whole societies. The central reason for the frustrating slowness of response is that the problems, and especially the remedies, were socially constructed in the West and were considered alien in the other regions of the world. Thus, for behavior to change, the definition of the problem and of the solutions must first be domesticated, a process that occurs in the local social networks where international and national messages are evaluated and reinterpreted. In essence, this domestication of the agenda involves the passing of ownership from the domain of officialdom to people themselves. The explanation has direct relevance to the AIDS control movement.

## 12.1 INTRODUCTION

The international population control movement, which started in the mid-1950s and was led by the United States, defined population growth in developing countries as a crisis with global consequences and identified a global solution: sex was to be decoupled from birth through the use of modern methods of family planning. By the turn of the century, sex presumably continued to be practiced and enjoyed, but fertility had declined, or had begun to decline, almost everywhere. Beginning in the mid-1980s, a new global threat appeared. Again, both the problem (AIDS) and the solution (primarily the delinking of sex from death through the use of condoms) were defined in the West and exported to developing countries. The approaches that the two movements took to persuade governments and people in non-Western countries to change their intimate behavior were surprisingly similar: a choreography of measures ranging from the use of international conferences to persuade national leaders to adopt appropriate policies and implement them in standard ways to the emphasis on technology (clinical contraceptive methods in the case of the population movement, condoms in the case of AIDS prevention). The ambitions, assumptions, and implementation of both movements are strikingly similar, and the social processes by which the AIDS crisis is ultimately resolved are likely to be similar to the processes that earlier led to the widespread adoption of fertility control.

The aim of this paper is to identify the lessons of the earlier population movement for AIDS control. We begin at the international level, where problems and solutions were defined, funding raised, and interventions conceived. Although the problems and solutions were perceived as global, the multilateral and bilateral agencies that directed the interventions necessarily had to work through the governments of sovereign states, or at the least with their permission. We then examine the local responses that, in the case of the population movement, led to fertility decline and, we predict, will lead to a downturn in the spread of HIV. In our conclusions, we attempt to distill the key points of these sagas: what have we learned about the process of delinking sex from birth that is relevant for current humanitarian efforts to sever the connection between sex and death?

## 12.2 CREATING INTERNATIONAL AND NATIONAL AGENDAS

Efforts to change the beliefs and behavior of others for their own or society's good are not new. Rather, both the population control and the AIDS prevention movements are on a historical continuum that extends from the earliest efforts to spread Christianity and Islam on through movements to reduce smoking and alcohol consumption and to promote universal suffrage and human rights. What unites population and AIDS movements and sets them apart from other movements to change mass behavior is that both attempt to modify central, but highly sensitive, domains of behavior by breaking the link between sexual intercourse and its undesirable consequences.

At the outset, differences between the two movements should be acknowledged. First, population control emerged earlier and more gradually onto the international agenda in the 1950s and 1960s,[1] than AIDS control: international meetings were held to warn the world about AIDS within a few years of the identification of HIV. Second, the issue of rapid population growth was and remains intellectually divisive. The very existence of a problem was denied by many, including both Marxists and liberal right-wing economists (McQuillan, 1979; Simon, 1982). The proposed solution — the vigorous promotion of modern contraceptives — was attacked by the Vatican and some feminists and was regarded as inherently ineffective by a fair number of demographers (Davis, 1967; Hartman, 1987; Demeny, 1988; Pritchett, 1994). By comparison, denial of the problem of AIDS and its main mode of transmission has been confined to a few mavericks (and one key African leader) though serious disagreements exist about the relative feasibility and effectiveness of condoms, abstinence, and fidelity, the three pillars of prevention. Not surprisingly, death control is intrinsically less controversial than birth control.

Third, the problem of rapid population growth was perceived to be shared by all low income countries but of little relevance to richer industrialized countries that had already achieved moderate birth rates by the middle of the last century. HIV, on the other hand, has the recognized potential to spread in any society; indeed, in the 1980s, there were justified concerns that HIV infections would diffuse from gays and injecting drug users to become endemic in the general populations of Europe and North America, and such fears have not entirely abated. The fact that AIDS in the West was initially associated with socially marginalized groups exerted a profound effect on AIDS control strategies — in particular the emphasis on human rights and the need to address stigma — that have no parallel in the population control movement. Moreover, AIDS cannot be seen as a common problem of all poor countries. So far it has disproportionately affected sub-Saharan Africa, notably the east and southern parts of the subcontinent; all countries with severe generalized epidemics (sero-prevalence at 5% or more among women aged 15–24) are in Africa.

A fourth important difference concerns the target population. Family planning efforts were directed primarily at respectably married women and the mode of action of most modern contraceptives was distanced from the coital act. Discussion of sexual partnerships and sexual practices thus could be largely ignored. In contrast,

AIDS control programs have had to focus on men as well as women, with a greater emphasis on nonmarital than on marital sex and with attention to sexual practices such as anal and "dry" intercourse.

Finally, and most obviously, HIV/AIDS is an infectious disease, whereas childbirth is not. The relationship between behavioral change and fertility is rather straightforward: a strong linear relationship exists between the level of contraceptive use in a society and its fertility level. With AIDS, the relationship is much more complex and still not well understood, partly because the effects of changes in behavior on infections are mediated by many other factors.

The similarities between the two movements, however, are more striking than these differences. In both, the nature of the problem and the solution were defined in the West, with warnings of the extreme consequences of inaction. Large sums of money were mobilized for the development of new biomedical tools (e.g., hormonal contraceptives, microbicides, and vaccines) and for preventive programs. New dedicated international agencies were created to coordinate action (UNFPA and UNAIDS). Skeptical national governments of poor countries were persuaded, cajoled, and even occasionally blackmailed into support (Luke and Watkins, 2002). New government agencies were created to coordinate or implement programs, and rather similar blueprints for action were drawn up at national level with the assistance of foreign experts (Watkins and Hodgson, 1998; Chimbwete, Watkins, and Zulu, 2005). In the following paragraphs, some of these similarities are further outlined.

The literature of the 1950s, 1960s, and 1970s abounds with apocalyptic visions of the future if decisive action to curb population growth was not taken immediately, perhaps best exemplified by the writings of biologists such as Ehrlich and Ehrlich (1970):

> The explosive growth of the human population is the most significant terrestrial event of the past million millennia. No geological event in a billion years … has posed a threat to terrestrial life comparable to that of human over population. (p. 1)

The pioneers of the international population movement drew on several plausible rationales: health benefits to mothers and children, extension of reproduction choice, avoidance of famine, and environmental protection. However, the enduring rationale was provided by an extensive demographic–economic analysis of India that showed that economic and social progress in that country was jeopardized by rapid growth of population (Coale and Hoover, 1958). During the next decade, key leaders in the United States coalesced around this characterization of the problem. In 1969, Robert McNamara, then president of the World Bank, described population growth as "the greatest single obstacle to economic and social advancement of the majority of the peoples in the underdeveloped world" (Symonds and Carder, 1973, p. 17). And Lyndon Johnson famously declared that a dollar spent on family planning was worth one hundred dollars spent on other forms of international assistance. Moreover, the promise that fertility control was a *sina qua non* for development was particularly attractive to developing country governments. With enthusiastic support from Scandinavian governments and more cautious endorsement by the British government, together with the creation of the United Nations Population Fund, the stage was set for the era of mass family planning programs.

Similarly, the AIDS problem was defined internationally, though in this case the single most important catalyst was based in Geneva rather than New York or Washington D.C.: Jonathan Mann, at WHO's Global Program on AIDS. AIDS control had a particularly compelling humanitarian rationale, namely, the prevention of premature death and suffering, with its dire effects on families and children. Though these considerations are still prominent, intriguingly, AIDS, just as population, is increasingly framed as a threat to development and even to international security, thereby making AIDS control more relevant to the World Bank and the UN. Thus, the first two paragraphs, on the impact of AIDS in the UNAIDS 2004 report states:

> In both low- and high-prevalence settings, HIV and AIDS hinders human development. ... In the hardest hit countries, it is erasing decades of health, economic and social progress. (p. 41)

The first sentence of the UK's Call for Action on HIV/AIDS (DFID, 2003) goes even further: "HIV/AIDS is the greatest threat to development in the world today" and this position is endorsed in the preface by the prime minister who characterizes AIDS as the "biggest barrier to tackling poverty" African leaders might justifiably feel bemused by such an abrupt shift in international rhetoric from the vision of too many people and too few productive jobs as a consequence of rapid population growth to a vision of a labor force so diminished by AIDS mortality that agricultural and industrial production is threatened.[2]

The solutions were also defined internationally. Encouraged by survey findings that most couples in Asia and Latin America wanted relatively small families, the leaders of the population movement argued that the remedy was to persuade people to use clinical or surgical methods of contraception — initially the intrauterine device but soon followed by sterilization and the pill. The task ahead was to launch information and educational campaigns and create accessible services initially through static family planning clinics and later through social marketing and community-based schemes.

To the experts who initially led the AIDS prevention movement, the solution also lay in modern technology, primarily the condom but also treatment of curable sexually transmitted diseases and HIV testing. The programmatic blueprint was even more standardized than for population control. In the late 1980s and early 1990s, national AIDS control programs with medium-term action plans proliferated. Treichler's (1999) description seems particularly apt: "Despite local variations, it sometimes seems as though public health campaigns draw ideas and images from some secret central vault, so similarly are they structured over time and space." Underlying both movements was the assumption that people, if properly informed and with ready access to appropriate services, would act rationally to avert the undesirable consequences of sex.

## 12.3 THE FAMILY PLANNING STORY: 1960–2000

State-sponsored family planning programs started in most of Asia and in a few countries of Africa and Arab States in the 1960s and 1970s. In Latin America, the lead was taken by well-funded nongovernmental organizations (NGOs) such as Profamilia, Bemfam, and Mexfam in Colombia, Brazil, and Mexico, respectively.

Despite persistent widespread skepticism among many population scientists, the mood was buoyant, even evangelical. Donald Bogue, the prominent U.S. demographer, characterized the movement as a "crusade," a "Holy war," and "revolutionary" in its aspiration to lower fertility as an early step on the pathway to modernization in distinction to the more patient but passive approach of awaiting the elimination of illiteracy, poverty, and traditionalism (Bogue, 1968, pp. 539–540).

By 1980, the mood had changed. There had been few "success stories." Fertility had clearly declined in islands such as Mauritius, Sri Lanka, and Fiji, in the city states of Hong Kong and Singapore, and in South Korea and Taiwan. But these examples could be dismissed as exceptional. In the larger and poorer countries of Asia — India, Pakistan, Bangladesh, Indonesia, Thailand, and Philippines — there were few trusted signs of widespread falls in fertility. Large changes in reproductive behavior in Latin America appeared to be confined to metropolitan elites. Throughout most of Africa and the Middle East, fertility remained resolutely high.[3] The skeptics reasserted their ascendancy. Events appeared to vindicate the deeply held and time-honored view that human reproduction was firmly embedded in economic, social, and cultural settings and that transformation of these settings was a necessary precondition for fertility decline.

Specifically, three main types of obstacle to fertility change were identified. The first centered on poverty and related insecurity. The idea that poor couples find large families to be more advantageous than small families can be found in the earliest theoretical statements of fertility transition (Notestein, 1945). In poor families, children are cheap to rear and they contribute to the domestic economy from an early age. Costs are kept low by the fact that children of the very poor do not attend school or do so for only a brief period and that aspirations for a better material life are overshadowed by the day-to-day struggle to make ends meet.

Closely related to poverty were considerations of the value of children as a form of security in old age and more generally as mitigators of risk. The argument was marshaled and applied with particular force to Bangladesh by Mead Cain (1978, 1983). In a poor country such as Bangladesh, the lack of state welfare safety nets combined with a high risk environment stemming from lawlessness, a high incidence of premature death and disability, and periodic climatic crises, children provide the only dependable form of assistance to parents. The policy implication was clear. Sustained fertility decline was an unrealistic proposition pending radical improvements in the social fabric.

A second type of obstacle was gender inequality. A common assertion in the voluminous literature on links between the status of women and fertility is that women tend to be less pronatalist than their husbands but lack of decision-making power renders them unable to implement their preference (Dyson and Moore, 1983). Moreover, in societies where women are barred or discouraged from paid employment outside the home, the opportunity costs of childbearing remain low, and male-dominated social systems lead to a preference for sons over daughters, which, in the absence of sex-selective abortion and confidence about child survival, sustains moderate to high fertility. In Kenya, it was argued that men might remain pronatalist because they reap the benefits of children without shouldering the costs (Frank and McNicoll, 1987).

The third type of barrier, though less commonly invoked than the other two, was cultural in nature. The persistence of high fertility in Moslem populations was explained by the pronatalist values of Islam (Leete, 1996; Goldscheider, 1999); the concept of machismo was thought to sustain high fertility in Latin America; and in an influential article, the Caldwells argued that traditional religious beliefs and associated cultural values would act as a serious barrier to fertility decline in Africa (Caldwell and Caldwell, 1987).

By 1990, however, it could be shown that these barriers were far from immutable. Reproductive change had swept across Asia, encompassing Bangladesh and Nepal, among the poorest countries in the world. Fertility decline in Latin America had spread from urban educated elites to less privileged, rural strata. Decline was firmly established in several Arab states and had clearly started in sub-Saharan Africa. Since 1990, this globalization of fertility transition has continued. The level of childbearing is now under three births per woman in developing regions compared with over six 50 years ago. The number of countries where fertility has remained unchanged has shrunk to a few outposts in West and Central Africa.

No consensus exists about the fundamental underlying reason for these fertility declines, but there must surely be a single common cause, or nexus of causes, for such widespread and synchronous reproductive change in such a variety of national settings. But there can be little doubt about the crucial importance of social networks and the diffusion through these networks of new ideas and new information. Evidence from both surveys and ethnographic research suggest that both the very idea of pregnancy control within marriage and the means to achieve it were initially alien and unknown in many developing countries. Majorities of women canvassed in the late 1970s and early 1980s as part of the World Fertility Survey in the least developed and most isolated countries (e.g., Benin, Cameroon, Nepal, and Yemen) denied awareness of any contraceptive method.

Much of the ethnographic evidence is entirely consistent with the impression given by survey results. In North India (Jeffery et al., 1989), Bangladesh (Maloney, Aziz, and Sarker, 1981), Thailand (Knodel et al., 1984), and Mali (van de Walle and van de Walle, 1991), to cite but a few examples, there is convincing testimony that the idea of fertility regulation was unthinkable and/or the means to achieve it unknown. A few quotations illustrate the point:

> "In the old days, no one used medicines to stop having children. They would say it was bad to stick out your leg to trip up God in his work. However many children were in a person's fate, that many would certainly be born." (India)

> "The number of children to be born is indicated in the woman's childbearing tube. Only God knows this, and no one can foretell it." (Bangladesh)

> "We didn't think about how many [children] we should have. [My husband] felt sorry for me when I had lots of children but no one knew what to do." (Thailand)

> "Truly the number I wanted was four, but going on to ten was God's work." (Mali)

The constant reference to religion need not be interpreted literally. Rather these statements probably reflect a distrust of an alien idea that appears unnatural, rather than a carefully considered rejection on the basis of religious doctrine. In societies where religious sentiments permeate every aspect of life, the easiest way to express this distrust is to resort to familiar religious terminology.

In most settings it is also clear that the idea of marital birth control and its methods was initially greeted with suspicion, ambivalence, and fear. Fear of side effects and more serious damage to health were among the most commonly cited reasons for nonuse. Religious objections and fear of social disapproval figure in some settings but not in others. Much of the early evidence is summarized in Bogue (1983). Recent studies document more fully the anxieties, fears, and sometimes even outrage that can be evoked by the introduction of modern contraception into a social system. Simmons et al. (1988) describe the initial uproar caused by the advent of female family planning workers in Bangladesh. Casterline and Sathar (1997) conclude from their detailed study of unmet need in the Punjab province of Pakistan that the three decisive obstacles to contraceptive use are: fear of side effects; concerns about social, cultural, and religious acceptability; and perceptions by wives of husband's disapproval. The most vivid evidence comes from the studies by Watkins and collaborators in South Nyanza, Kenya (e.g., Watkins, 2000). The ambivalence, fear, and anguish with which women viewed modern contraception is unmistakably portrayed.

Thus, contraception was not seamlessly incorporated into reproductive strategies whenever the need arose. On the contrary, in many societies, it encountered serious resistance, the most common expression of which took the form of concerns about health. But this articulation of health concerns probably reflected more profound disquiet about a radical innovation that went to the heart of one of life's central preoccupations.

These feelings naturally provoked frequent discussion. One of the very few detailed accounts of the nature of discussion comes from research in South Nyanza, Kenya. Not surprisingly, perhaps, discussions of contraception among women were embedded in more general debates about family size and modern life. Everyday conversation made no distinction between motives and means. Information exchange went hand in hand with assessment and evaluation. The substance of the diffusion was thus a bundle of interrelated topics: the concept and legitimacy of birth control characteristics of particular methods and ideas about family size. The experiences of close friends, neighbors, and relatives appeared to be of particular importance. What mattered most was the direct testimony of those who had actually tried a method. The deep ambivalence about family planning methods lent credence to alarming stories about babies born with disease or deformity as a result of their mother's contraceptive habits. There was little evidence that women explicitly sought information and advice from high-status or highly educated individuals. Formal health care providers were valued sources of technical information, but they are socially distanced in a way that eroded complete trust (Rutenberg and Watkins, 1997). The metaphor used by Watkins, Rutenberg, and Green (1995, p. 51) aptly sums up the general impression from their work: "Women in these areas are not navigating the domain of uncertainty alone, but rather in flotillas, convoys in which the topics of conversation are relevant, the debates widespread and sometimes intense."

If reproductive control and specific methods of contraception are frequent topics of discussion, it is to be expected that the decisions and behavior of individuals will be influenced by perceived attitudes and behavior of social network members. Cross-sectional studies show that a woman's perception of community approval is predictive of her own contraceptive use, but conclusive evidence requires costly prospective study designs (Retherford and Palmore, 1983; Beckman, 1993; Valente et al., 1997). Complex statistical modeling of temporal-spatial patterns of fertility decline in Taiwan, Costa Rica, and Kenya provided more persuasive evidence that social influences are important (Montgomery and Casterline, 1993; Rosero-Bixby and Casterline 1993, 1994; Kohler, Behrman, and Watkins 2000, 2001). But the most powerful argument in favor of the importance of social influence remains the simple and obvious one of the speed with which new behavior can spread. It is unconvincing to explain rapid marital fertility change solely in terms of increased knowledge or access to contraception. Awareness of contraceptives and supply sources was often well established long before changes in behavior. Nor is it plausible to conclude that couples, independently of each other, performed the same cost–benefit calculations and came to the same conclusion about family size and fertility regulation within a short span of time. The principle of Occam's razor should apply: give the simplest explanation that is consistent with the evidence. The simplest explanation for rapid rises in contraception and declines in fertility is that people were influenced by one another. Not only is it the simplest explanation, it is also the most compelling.

In sum, a social interaction perspective provides a convincing explanation of the lagged behavioral response to programmatic exhortations and services and the speed with which changes in behavior eventually spread across whole societies.[4] Private cost–benefit calculations may suffice for choices of minor significance but, for more important choices such as family size and family planning, collective scrutiny of the empirical evidence provided by the experience of others appears to be necessary.

## 12.4 PARALLELS WITH THE AIDS MOVEMENT

AIDS-control programs started in the late 1980s or early 1990s and, since then, substantial sums have been spent on information and education campaigns and on the distribution and marketing of condoms. As in the first decade or so of family planning programs, the record of success is limited. In only two low or middle income countries (Uganda, Thailand) is there decisive evidence of the reduction in HIV incidence. In most African countries, spread of infection has continued, though at vast differing paces for reasons not fully understood, and new epidemics have started in Asia and Eastern Europe. Even the achievements of specific intensive interventions targeted at small geographical subpopulations or at special groups have been, at best, mixed (Oakley, Fullerton, and Holland, 1995; Oakley et al., 1995; Grunseit et al. 1997; Jemmott and Jemmott, 2000; Stephenson, Imrie, and Sutton, 2000; Speizer, Magnam, and Colvia, 2003). A particularly discouraging example comes from a well-designed and theoretically sophisticated intervention in South Africa, where despite substantial funding, the involvement of a variety of stakeholders (government officials, epidemiologists, social scientists, and a community-based NGO) levels of infection increased over the course of the intervention (Campbell, 2003).

AIDS-control programs appear to be approximately at the same stage reached in 1980 by the family planning movement. Confidence in the effectiveness of the classical public health approach (information and services) has diminished. As in the earlier decades of population control the refrain "they know but they don't change" is repeated in conferences and workshops. Increasing emphasis is given to contextual obstacles to behavioral change: societal factors that severely constrain individual decisions and freedom of action regarding protection against HIV infection. And, just as in the case of family planning, three main types of constraining factor are typically identified: poverty, gender inequality, and culture. Poverty is said to increase vulnerability by dislocating families, through its association with labor migration, and, above all, by forcing women to sell their bodies to survive and provide for their children (Parker, Easton, and Klein, 2000; UNAIDS, 1999). Gender inequality is seen by many as the overridingly important factor and the following typical statement would be widely endorsed: "If we are to contain the HIV epidemic we must tackle its root cause — gender inequality" (Gupta, 2002, p. 184). Enmeshed with gender relations are cultural constructions of sexuality. Men are frequently portrayed as slaves to concepts of masculinity that validate risk-taking, sexual conquest, and prioritization of immediate sexual pleasure over health considerations (Dunckle et al., 2004; Jewkes, Levin, and Penn-Kekera, 2003). Conversely, women are imprisoned by models of desirable femininity that stress innocence, passivity, and acquiescence. The behavioral consequences of this combination of gender stereotypes, it is claimed, are lethal.

Perhaps these analyses are correct. Perhaps radical revolutions in the economic structure of societies and in gender relations and associated cultural constructions are necessary to stem severe generalized epidemics. Nor would we deny for a moment that radical changes are highly desirable in their own right. Nevertheless, we are skeptical of the prevailing pessimism for reasons outlined below.

First, behavioral change is unlikely to occur rapidly. Just as it took a long time to observe and evaluate the advantages and disadvantages of the experiences of friends, relatives, and neighbors who bravely decoupled sex from birth, so also it takes time for infection with HIV to result in weight loss, diarrhea, and death and for the consequences of following admonitions to decouple sex from death through abstinence, fidelity, and condom use — all considered to be inconsistent with the joys of sex — to become evident in a decline in the number of village funerals.

Second, behavioral change is difficult to measure, either directly through surveys or indirectly through declines in HIV prevalence. Surveys respondents are cagey about the details of their sexual activity and probably even more so as HIV prevention programs become increasingly widespread, persistent, and strident. And the current method of tracking changes in HIV infections through annual testing of pregnant women is crude, capturing prevalence rather than the incidence of new cases. Thus, behavioral change may be occurring but overlooked. Indeed, the survey evidence for Africa and Latin America reveals marked upwards trends in condom use among sexually active single people (Ali and Cleland, 2005). Resistance to condom use within marriage and stable relationships is certainly stronger, but there is some evidence of change and that wives are more influential than often assumed (Maharaj and Cleland, 2004).

Third, evidence of behavioral change may be brushed aside by those national and international actors implementing and monitoring prevention programs because, perversely, the incentives favor continued pessimism. Funding for AIDS control may appear stingy to spokespersons for the AIDS control movement. But it supports fiefdoms of bureaucratic power in official agencies and NGOs in Washington, London, and Geneva as well as in other national capitals, and throughout the AIDS-affected countries it provides the best jobs in town, international travel, opportunities for patronage, vehicles, and per diems for workshops and seminars.

Fourth, it is unlikely that those engulfed by severe epidemics would not respond, and, indeed, there is scattered evidence of attempts to navigate the epidemic. In rural Malawi, where the epidemic is mature and generalized, the consequences of its ravages are strikingly evident. At least since 1998, about the time that deaths from AIDS became evident in rural villages, respondents in a longitudinal survey have attended several funerals a month, funerals that are preceded and followed by conversations about those who appear to be on the verge of death, and by verbal autopsies that draw on local knowledge to narrate the successive symptoms and sexual partnerships of the deceased (Watkins, 2004).

The particulars of the deaths provide a basis for domesticating the prevention messages disseminated by donors, the government, and NGOs. Just as modern methods of family planning, initially perceived as suitable only for foreigners, became routine through social interactions among friends, relatives, and neighbors, conversations about local people observed suffering and dying from AIDS lead to generalizations about the causes of the epidemic and to advice on how to avoid death. Thus, the evidence from Malawi refutes the common assertion that people in Africa do not discuss AIDS because of the stigma attached to it. Modified versions of abstinence and fidelity, such as a reduction in the number of sexual partners and a more careful selection of partners, are typically seen as more attractive than condom use, although in some circumstances the moral lesson drawn is one of free will: that those who do not use condoms with bar girls are deliberately choosing, and paying for, death. The evidence for other countries in sub-Saharan Africa with mature and generalized epidemics is less rich, but it suggests that the same process is occurring there as well (Watkins, 2004).

## 12.5 CONCLUSIONS

In this paper we have described the international, national, and local responses to two problems that were defined as threats on a global scale: unprecedented rates of population growth in developing countries and the advent of the AIDS pandemic that has already brought in its wake huge rises in mortality in the worst affected countries of East and Southern Africa and threatens to do the same elsewhere. Solutions to both required at their roots changes in intimate behavior: procreation and sex, respectively. In both instances the international response was similar: mobilization of huge resources, the creation of new dedicated UN agencies and national bureaucracies, immense pressure on governments to take action, and similar blueprints for action with an initial emphasis on the provision of information and services. In both movements, the desired behavioral response was slow to occur, giving rise to much pessimism, even incomprehension that people, when possessed of correct information, did not behave in a rational fashion. In the case of population growth, couples continued to spurn contraception despite stated desires to have no more children. In the case of HIV, individuals continue to have unprotected sex with noncohabiting partners despite full knowledge of the risks involved.

In both movements, the slowness of the behavioral response led to new conceptualizations of the problem in terms of intransigent barriers to change posed by poverty, gender inequality, and culture. It was compellingly argued that the problems of excessive fertility and continued heterosexual spread of HIV would never be overcome unless poverty and income inequality were reduced, a revolution in gender relations was achieved, or deep-rooted cultural beliefs overcome. As a result, the agenda for action widened far beyond the mere provision of information and services to address these newly identified barriers.[5] In the case of AIDS control it is too early to assess the validity of the redefinitions of the problem and the new action agenda. HIV incidence has fallen in only a handful of countries and their circumstances diverge so widely that the identification of general lessons is premature. But because the family planning movement started some 40 year earlier, it is more feasible to use what has been learned about the timing and determinants of fertility declines to draw conclusions about the role of different types of policies and programs. We contend that these lessons have direct relevance to HIV control.

Perhaps the most important lesson from the family planning experience is that ideational and behavioral adaptations to new threats, particularly to threats that relate to core concerns of sex and procreation, take time to emerge. The central reason for the apparent slowness of response is that the problem, and especially the remedies, were socially constructed in the West and were thus considered alien in other regions of the world. Thus, the definition of the problem and of the solutions must be domesticated, a process that occurs in local social networks where international and national messages are evaluated and reinterpreted. In essence this domestication of the agenda involves the passing of ownership from the domain of officialdom to the people themselves. A convincing body of family planning evidence indicates this phase is a necessary precursor to widespread changes in ideas and behavior. The process is one of discussion and debate within networks of relatives, friends, and associates, in which stories of the personal experience of those known to the network members is of special importance. This process takes time, and even carefully designed and theoretically driven interventions have proved largely ineffective at accelerating the pace of change in the general population.

The central importance of this process of domestication and local ownership, however, does not imply that actions of donors, governments, and NGOs are irrelevant. First, information and education campaigns provoke conversational networks in local communities. Although it is not possible to quantify the contribution of knowledge dissemination, it is one obvious way in which change in ideas and behavior can be facilitated. Second, donors, governments, and NGOs can provide services that are appropriate, affordable, and accessible. The experience of the provision of modern contraceptives suggests that a wide array of approaches to meet a diversity of needs is the most effective strategy.

Third, vigorous information programs and the speedy creation of services is most easily effected when political will is favorable to the cause. As we have discussed, donors expended much money and deftly choreographed efforts to persuade governments to publicly promote the international agenda; however, whether or not government programs hastened fertility change is a topic of controversy among experts. Our view is that although strong government support is certainly conducive, it has not proved necessary for mass uptake of contraception and widespread decline of fertility to occur.

One of the great surprises of the past 40 years has been the mass uptake of modern contraception in some of the poorest and least literate countries of the world, such as Bangladesh, Nepal, and Malawi. Many predicted that this would not happen, postulating a nexus between low income and large family size. But it proved to be the case that the poor were just as interested in smaller families as were their richer neighbors. Similarly, countries such as Brazil that are characterized by huge disparities of income have now achieved levels of contraceptive prevalence that are found in much more egalitarian countries such as Costa Rica.

A frequent refrain in strategies to enhance the use of modern contraception has been the need to address huge gender inequalities as a *sine qua non*. Men have constantly been portrayed as more pronatalist than their wives and more opposed to the adoption of modern methods; as a corollary, women were constantly portrayed as under their man's thumb. The scientific evidence for this stereotype is slender, to say the least. Although the vast bulk of contraception is provided to women by women, it is obvious that most men either supported or did not oppose the use of family planning by women. And the substantial minority of women who used despite their husband's objection suggests more autonomy than had been attributed to them. Huge strides in women's emancipation did not precede the fertility transition. Rather, the historical record shows the reverse sequence of events to be more common. Much has been written about supposedly immutable cultural barriers to the uptake of contraception. But, as Frank Notestein remarked long ago, "the cake of custom crumbles rapidly." Many observers have mistaken the initial turbulence in the early stages of the spread of an innovation for a more profound resistance to change. The extent and nature of the resistance to the advent of modern contraception has been very context specific: in Taiwan and Thailand, for instance, modern contraception was welcomed with little reserve, whereas in Bangladesh and Kenya the resistance was more enduring. In our view an initial resistance is an almost inevitable feature of the dynamics of change in intimate areas of the human experience.

Despite differences outlined earlier between population control and AIDS prevention, we believe there are lessons from the past for AIDS. The primary one is that the formulation and deployment of effective strategies of AIDS prevention by the communities at risk takes time but we believe it will occur, just as pessimism over high fertility was overtaken by events. Just as the domestication of the population agenda occurred worldwide, the domestication of the HIV prevention agenda appears to be occurring in Malawi, and there is reason to believe a similar process is occurring in other heavily afflicted areas where partner reduction and selection together with greater recourse to condoms have been documented. Sooner would certainly have been better, and it seems at best insensitive and at worst immoral to recommend patience in the face of so many predictable deaths and so many predictable orphans. However, if we are correct that the appreciation of risk and the domestication of prevention strategies must be done by those who themselves face that risk, it is unlikely that interventions to promote foreign prescriptions of behavior change can do much to hasten local responses. The evidence reviewed in this paper supports this view. Most instances of successful behavioral interventions relate to groups with special characteristics that facilitate change, such as the collective identity and organization of western gay men and the structural features of sex work in some Asian societies.

Our diagnosis gives great grounds for optimism, for it suggests that supposedly insurmountable barriers to ideational and behavioral change that are described in the literature will not be insurmountable in stemming the AIDS pandemic, as they were not insurmountable in preventing rapid and pervasive fertility declines. For instance, it seems to us predictable that the high premium placed on "flesh to flesh" sex, the emphasis on men's uncontrollable sexual urges (also a feature of the portrayal of men that justified noncoital contraceptive methods), and other cultural barriers to change will prove to be adaptable to the new circumstances. Nor will poverty or inequitable distribution of resources stand in the way of people's desire to prevent an untimely death. Most importantly, we think that women are more capable of protecting themselves than is commonly believed and that the characterization of men as indifferent to their own death is ridiculous. No doubt, behavioral change promotion projects to address presumed cultural barriers, structural obstacles, and gender inequality will continue to proliferate under the justification that these are the only ways to stem epidemics. We are not denying that such projects are driven by compassion, nor do we wish to deny the magnitude of problems of poverty and gender inequality. However, there are sound grounds for skepticism that they will be effective in stemming HIV epidemics, and the danger exists that they may divert funds and energies from the more prosaic actions that really matter: the dissemination of basic information and appropriate services.

## NOTES

1. For detailed accounts, see Donaldson (1990), Harkavy (1995), Hodgson and Watkins (1997).
2. Although it is true that, in the longer term, unchecked population growth and unchecked spread of AIDS spell economic disaster, the empirical evidence for serious macroeconomic impacts in the short or median term is rather weak. See Cassen (1994); Birdsall, Kelley, and Sinding (2001); Ainsworth, Transen, and Over (1998).
3. This perception of unchanging fertility was in part illusory, a reflection of lack of sufficient evidence and distrust of the available evidence. It is now clear that declines in fertility prior to 1980 were more widespread than believed at the time (Chesnais, 1992).
4. For the classic exposition of the innovation-diffusion framework, see Rogers (1983). For a more recent assessment of the evidence see Casterline (2001).
5. In the case of family planning, large sums were not diverted to address these barriers, but such interventions are increasingly common for AIDS prevention.

## REFERENCES

Ainsworth, M., Transen, L., and Over, M., Eds., *Confronting AIDS: evidence from the developing world*, The European Commission, Brussels, 1998.

Ali, M., and Cleland, J., Sexual and reproductive behavior among single women aged 15–24 in eight Latin American countries: a comparative analysis, *Soc. Sci. Med.,* 60, 1175–1185, 2005.

Beckman, L.J., Communication, power and the influence of social networks in couple decisions on fertility, In *Determinants of fertility in developing countries, Vol. 2*, Bulatao, R.A., and Lee, R.D., Eds., Academic Press, New York, 415–443, 1993.

Birdsall, N., Kelley, A.C., and Sinding S.W., Eds., *Population matters,* Oxford University Press, Oxford, 2001.

Bogue, D., Normative and psychic costs of contraception, In *Determinants of fertility in developing countries, Vol. 2*, Bulatao, R.A., and Lee, R.D., Eds, Academic Press, New York, 151–192, 1983.

Bogue, D.J., Progress and problems of world fertility control, *Demography*, 5, 539–540, 1968.

Cain, M.T., The household life cycles and economic mobility in rural Bangladesh, *Popul. Dev. Rev.,* 4, 421–438, 1978.

Cain, M.T., Fertility as an adjustment to risk, *Popul. Dev. Rev.*, 9, 688–702, 1983.

Caldwell, J.C., and Caldwell, P., The cultural-context of high fertility in sub-Saharan Africa, *Popul. Dev. Rev.*, 13, 409–437, 1987.

Campbell, C., *'Letting them die': why HIV/AIDS prevention programmes fail*, James Currey, Oxford, 2003.

Cassen, R., Ed., *Population and development: old debates, new conclusions*, Transaction Publishers, Oxford, 1994.

Casterline, J., Ed., *Diffusion processes and fertility transition,* National Academy Press, Washington, DC, 2001.

Casterline, J., and Sathar, Z., *The gap between reproductive intention and behavior*, Population Council, Islamabad, Pakistan, 1997.

Chesnais, J.-C., *The demographic transition*, Oxford University Press, Oxford, 1992.

Chimbwete, C., Watkins, S.C., and Zulu, E.M., The evolution of population policies in Kenya and Malawi, *Popul. Res. Policy Rev.,* 24, 85–106, 2005.

Cleland, J., and Ali, M., The use of demographic and health surveys to monitor reproductive behaviour in single women, Paper presented at International Quetelet Seminar, Nov. 17–20, Louvain-La-Neuve, Belgium, 2004.

Coale, A.J., and Hoover, E.M., *Population growth and economic development in low-income countries: a case study of India's prospects*, Princeton University Press, Princeton, NJ, 1958.

Davis, K., Population policy: will family planning programs succeed? *Science*, 158, 730–739, 1967.

Demeny, P., Social science and population policy, *Popul. Dev. Rev.*, 14, 451–479, 1988.

DFID (UK Department for International Development), *Call for action on HIV/AIDS*. London, 2003.

Donaldson, P.J., *Nature against us: the United States and the world population crisis, 1965–1980*, University of North Carolina Press, Chapel Hill, 1990.

Dow, T.E., and Werner, L.H., Prospects for fertility decline in rural Kenya, *Popul. Dev. Rev.,* 9, 77–97, 1983.

Dunckle, K.L., et al., Gender-based violence, relationship power, and the risk of HIV infection in women attending antenatal clinics in South Africa, *Lancet*, 363, 1415–1420, 2004.

Dyson, T., and Moore, M. On kinship structure, female autonomy, and demographic behavior in India. *Popul. Dev. Rev.,* 9, 35–60, 1983.

Ehrlich, P.R., and Ehrlich, A.H., *Population resources, environment: issues in human ecology,* W.H. Freeman and Co, San Francisco, 1970.

Frank, O., and McNicoll, G., An interpretation of fertility and population-policy in Kenya. *Popul. Dev. Rev.,* 13, 209–243, 1987.

Goldscheider, C., Religious values, dependencies, and fertility: evidence and implications from Israel, In *Dynamics of values in fertility change,* Leete, R., Ed., Oxford University Press, Oxford, 310–330, 1999.

Green, M., The evolution of United States international population policy, 1965–92 — A chronological account, *Popul. Dev. Rev.,* 19, 303–321, 1993.

Grunseit, A. et al., Sexuality education and young people's sexual behavior: a review of studies, *J. Adolescent Res.*, 12, 421–453, 1997.

Gupta, G.R., How men's power over women fuels the HIV epidemic, *Br. Med. J.*, 324, 183–184, 2002.

Harkavy, O., *Curbing population growth: an insider's perspective on the population movement,* Plenum Press, New York, 1995.

Hartman, B., *Reproductive rights and wrongs: the global politics of population control and contraceptive advice*, Harper and Row, New York, 1987.

Hodgson, D., and Watkins, S.C., Feminists and neo-malthusians: past and present alliances. *Popul. Dev. Rev.,* 23, 469–523, 1997.

Jeffery, P., Jeffery, R., and Lyon, A., *Labour pains and labour power,* Zed Books, London, 1989.

Jemmott, J.B., and Jemmott, L.S., HIV risk reduction behavioral interventions with heterosexual adolescents, *AIDS,* 14 (Suppl. 2), S40-S52, 2000.

Jewkes, R., Levin J.B., and Penn-Kekera, L.A., Gender inequalities, intimate partner violence and HIV prevention practices: findings of a South Africa cross-sectional study, *Soc. Sci. Med.*, 56, 125–134, 2003.

Knodel, J., Havanon, N., and Pramualratana, A., Fertility transition in Thailand: a qualitative analysis. *Popul. Dev. Rev.,* 10, 297–328, 1984.

Kohler, H.P., Behrman, J.R., and Watkins, S.C., Empirical assessments of social networks, fertility and family planning programs: nonlinearities and their implications, *Demographic Research*, v.3, http://www.demographicresearch.org/volumes/vol, 2000.

Kohler, H.P., Behrman, J.R., and Watkins, S.C., The density of social networks and fertility decisions: evidence from South Nyanza District, Kenya, *Demography*, 38, 43–58, 2001.

Leete, R., *Malaysian's demographic transition: rapid development, culture and politics*, Oxford University Press, Kuala Lumpur, 1996.

Luke, N., and Watkins, S.C., Reactions of developing-country elites to international population policy. *Popul. Dev. Rev.* 28, 707–733, 2002.

Maharaj, P., and Cleland, J., Condom use within marital and cohabiting partnerships in KwaZulu-Natal, South Africa, *Studies Fam. Planning*, 35, 116–124, 2004.

Maloney, C., Aziz, K.M.A., and Sarker, P.C., *Beliefs and fertility in Bangladesh,* Monograph No. 2, International Centre for Diarrhoeal Disease Research, Dhaka, 1981.

McQuillan, K. Common themes in Catholic and Marxist thought on population and development. *Popul. Dev. Rev.*, 4, 689–698, 1979.

Montgomery, M.R., and Casterline, J.B., The diffusion of fertility control in Taiwan: evidence from pooled cross-section time-series models. *Popul. Studies* 47, 457–479, 1993.

Notestein, F., Population — the long view, In *Food for the world*, Shultz, T.W., Ed., University of Chicago Press, Chicago, 36–57, 1945.

Oakley, A., Fullerton, D., and Holland, J., Behavioural interventions for HIV/AIDS prevention, *AIDS,* 9, 479–486, 1995.

Oakley, A. et al., Sexual health education interventions for young people, *Br. Med. J.,* 310, 158–162, 1995.

Parker, R.G., Easton, D., and Klein, C.H., Structural barriers and facilitators in HIV prevention: a review of international research, *AIDS*, 14 (Suppl. 1), S22-S32, 2000.

Pritchett, L., Desired fertility and the impact of population policies, *Popul. Dev. Rev.* 20, 1–55, 1994.

Retherford, R.D., and Palmore, J.A., Diffusion processes affecting fertility regulation, In *Determinants of fertility in developing countries, Vol. 2*, Bulatao, R.A., and Lee, R.D., Eds., Academic Press, New York, 295–339, 1983.

Rogers, E.M., *Diffusion of innovations*, Free Press, New York, 1983.

Rosero-Bixby, L., and Casterline, J.B., Modelling diffusion effects in fertility transition, *Popul. Studies* 47, 147–167, 1993.

Rosero-Bixby, L., and Casterline, J.B., Interaction diffusion and fertility transition in Costa Rica, *Soc. Forces,* 73, 435–462, 1994.

Rutenberg, N., and Watkins, S.C., The buzz outside the clinics: conversations and contraception in Kenya, *Studies Fam. Planning*, 28, 290–307, 1997.

Shisana, O., and Davids, A., Correcting gender inequalities is central to controlling HIV/AIDS, *Bulletin W.H.O.*, 82, 812, 2004.

Simmons, R. et al., Beyond supply: the importance of female family planning workers in rural Bangladesh, *Studies Fam. Planning,* 19, 29–38, 1988.

Simon, J., *The ultimate resource*, Princeton University Press, Princeton, NJ, 1982.

Speizer, I.S., Magnam, R.J, and Colvia, C.E., The effectiveness of adolescent reproductive health interventions in developing countries: a review of the evidence, *J. Adolescent Health,* 33, 324–348, 2003.

Stephenson, J.M., Imrie, J., and Sutton, S.R., Rigorous trials of sexual behaviour interventions in STD/HIV prevention: what can we learn from them? *AIDS,* 14 (Suppl. 3), S115-S124, 2000.

Symonds, R., and Carder M., *The UN and the population question 1945–1970*, McGaw-Hill, New York, 1973.

Treichler, P.A., *How to have a theory in an epidemic*, University of North Carolina Press, Durham, North Carolina, 1999.

UNAIDS, *Gender and HIV/AIDS: taking stock of research and programmes*, Geneva, 1999.

UNAIDS, *2004 report on the global AIDS epidemic,* Geneva, 2004.

Valente, T.W. et al., Social network associations with contraceptive use among Cameroonian women in voluntary associations, *Soc. Sci. Med.,* 45, 677–687, 1997.

van de Walle, E., and van de Walle, F., Breastfeeding and popular aetiology in the Sahel, *Health Transition Rev.,* 1, 69–81, 1991.

Watkins, S.C., Local and foreign models of reproduction in Nyanza Province, Kenya, *Popul. Dev. Rev.,* 26, 725–759, 2000.

Watkins, S.C., Navigating the AIDS epidemic in rural Malawi, *Popul. Dev. Rev.*, 30, 673–705, 2004.

Watkins, S.C., and Hodgson, D., From mercantilists to neo-Malthusians: the international population movement and the transformation of population ideology in Kenya, Paper presented at a Workshop on Social Processes Underlying Fertility Change in Developing Countries, Committee on Population, National Academy of Sciences, Washington, DC, 29–30 January 1998.

Watkins, S.C., Rutenberg, R., and Green, S., Diffusion and debate: reproductive change in South Nyanza Province, Kenya, Paper presented at the annual meeting of the Population Association of America, April 5–8, San Francisco, CA, 1995.

# 13 Smoking and the New Health Education in Britain, 1950s–1970s

*Virginia Berridge and Kelly Loughlin*

## CONTENTS



## ABSTRACT

Advertising is a key site of engagement for public health in contemporary Britain. The control or prohibition of mass advertising detrimental to health is a central objective. Using mass advertising has also been a more general public health strategy, for example, during the initial government responses to HIV/AIDS in the 1980s. This paper traces the initial significance of mass advertising in public health in Britain to the postwar decades up to the 1970s. Smoking is identified as the key issue that helped to define this new approach. The factors involved were: road safety and drunk driving models, U.S. advertising theory, the relocation of health education within central government, the arrival of mass consumption, and the rise of the "new public health" agenda.

(We gratefully acknowledge permission to reproduce this article from the *American Journal of Public Health* (*Am. J. Public Health*) 95, 956–964, 2005).

## 13.1 INTRODUCTION

Advertising is a key site of engagement for contemporary British public health. Control of advertising deemed detrimental to population health is an important strategy, recently demonstrated in the success of efforts to prohibit tobacco advertising

(Guardian, 2003a, 2003b).[1] Mass advertising has a paradoxical dual function in public health. A central cause for concern, it is also a central resource, a strategy used by public health interests and by government. Health campaigns use striking visual and verbal imagery and the full resources of the mass media. The Conservative government in the 1980s was notable for its use of mass media initiatives. The British AIDS campaigns, especially the national campaign in early 1987, were illustrations (Berridge, 1996; Dorn, 1986).[2,3] The campaigns in the 1980s for the privatization of state utilities such as British Gas and British Telecom also relied on mass advertising: the government became the largest advertiser in the UK by 1990 (Williams, 1998).

We focus on developments in the UK to argue that a mass media style of health education had its origin in the redefinition of smoking as a health issue in the period from the 1950s to the 1970s. Redefining smoking was also part of a broader move within public health: the rise of a new ideology that stressed individual responsibility for good health, "lifestyle," and behavior. The health agenda that grew out of this redefinition laid particular stress on the visual, on techniques of mass persuasion. Its roots lay in transatlantic influence; in the emergence of mass consumption in the aftermath of wartime restrictions on consumer goods and promotions; and in structural changes in responsibility for health, the central/local tension that has characterized much of British health policy.

## 13.2 HEALTH PROPAGANDA IN THE INTERWAR AND WARTIME YEARS

Before World War Two, the government recognized the value of publicity but had been reluctant to take it on. The newly established Ministry of Health was given responsibility in 1919 for the dissemination of information about health. Sir George Newman, the Ministry's chief medical officer, recognized that better health would be established through mass education. But publicity was not accepted as a legitimate responsibility of central government until the late 1930s (Grant, 1994). The Ministry's approach depended on part funding voluntary organizations such as the Central Council for Health Education (CCHE) and working through local government rather than taking action centrally, a tension that has continued to mark efforts to the present day. Commercial methods were discussed but little use was made of them. This approach changed, and greater centralization was introduced with the establishment of the Ministry of Information (MoI) during the war. But tensions remained. The early wartime Ministry of Information has been criticized by historians for its "clumsy and condescending" publicity arrangements (Grant, 1994, p. 245). Most of its officials were civil servants, who had little experience of publicity work (Taylor, 1980, p. 1981).

## 13.3 THE 1950S: LOW KEY RESPONSES AND LOCALIZED HEALTH EDUCATION

By the time the *British Medical Journal* published the connection between smoking and lung cancer in 1950, responsibility for health education had changed back to the local level. The initial response to the "discovery" of the connection between smoking and lung cancer in the UK was low key (Webster, 1984). But this was not just the result of procrastination. There were certainly links between the Treasury's reliance on tobacco revenue and the tobacco industry, and the latter funded research through the umbrella of the Medical Research Council (MRC) (Berridge, 1998). But this period of negotiation was also one of paradigm shift, the growing acceptability of the epidemiological rather than the biomedical, laboratory-based mode of proof (Burnham, 1989; Brandt, 1990). It was a period of conflict between different statistical traditions, between the UK's dominant genetic and hereditarian tradition and the new style that emphasized relative risk (Berlivet, 2005a; Parascandola, 2001). Civil servants in the Ministry of Health debated what form of proof they were dealing with. Was it really conclusive? What forms of health education would be appropriate? A Ministry of Health statement in May 1956 explained why central publicity would not be the right thing.

> The considerations on publicity concerning smoking and lung cancer differ slightly from those on cancer publicity generally in that the special point — that people might give up smoking — is not a matter of reporting symptoms. It does however concern an individual decision which involves others to a very much smaller extent than the subjects of past central public health campaigns. (National Archives MH55/2220a)

Smoking, it argued, was not a disease in the way cancer, or indeed infectious disease, was. It might lead to disease, but not for many years. The notion of long term risk was not part of public health in the 1950s. Publicity would be asking people to curtail a habit that was at this stage deeply embedded in everyday culture. It might also raise public fear about cancer, which the Ministry had been concerned to damp down. Unfounded cancer phobia might generate a demand for services at a time when NHS costs were becoming a political issue (Cantor, 2002; Webster, 1984). Nor did the Ministry want to resume central funding of the main health education body, the CCHE. After the war, funding responsibility had been passed back to the local authorities and away from central government. Central government had no wish to resume financial responsibility, as discussions at the time make clear (National Archives MH55/2203). There were thus practical, structural, and theoretical reasons for lack of action.

The message that came across in public education at this time was equivocal. A 1957 pamphlet issued by the Central Council dealt with the adventures of the fictional Wisdom family under the title "What — no smoking?" (Figure 13.1). In this comic strip, a boy and his mother drew the attention of the smoking father to the risks he was running. Worried, he goes to see his general practitioner, Dr. Brain, who presents the facts. One in every 300 smokers gets lung cancer. If he gives up, he is three times as likely to get it; if he continues, he is seven times as likely. Dr. Brain's advice is measured and calm. "… it still does not sound as if the risk is very great, so there's no need to get in a panic, whatever you decide to do." The idea of outlining specific courses of action to take was anathema to a society that associated "propaganda" with wartime central direction and with earlier Nazi propaganda. Health education placed its faith in the citizenship and responsibility of its recipients.

The scientific message about smoking had not "hardened" at this stage. It could be described as a period of dissent and negotiation before the diffusion of an

**FIGURE 13.1** Pamphlet issued by the Central Council for Health Education in 1957: "The Adventures of the Wisdom family: What — no smoking?" Health Education Authority leaflet archive.

ostensibly stable consensus suitable for public dissemination.[4] In the late 1950s, publicity on smoking was the responsibility of the local authorities and of the Medical Officer of Health (MoH), the local government public health official. There was no lack of interest. After a Ministry of Health circular went out to the 129 English local authorities in 1958, of the 127 who replied, 118 accepted the need for local action and only 9 did not (National Archives MH55/2225). However, with the exception of Edinburgh, only modest campaigns of action were initiated. The attitude and personality of the local MoH appeared to have influenced local action as much as anything else. Detailed reports on the local response showed a variety, ranging from a prompt response to those where the relevant committee discussed antismoking publicity in a cloud of cigarette smoke, or where the Director of Education was a heavy smoker and it was recorded that there would be no health education in schools (National Archives MH55/2228).

Within the Ministry of Health, there were the beginnings of a change of stance. An internal paper from the spring of 1960 commented presciently on possible future directions.

> It now seems apparent that local health authorities are not likely to be the most effective major agencies for conveying to the adult population information on smoking and lung cancer. Newspapers, magazines, radio and television are the main instruments for informing the public and these naturally look for their sources of news on this subject either to Government announcements or to scientific papers written by researchers in the field. (National Archives MH55/2226).

The publication of the Royal College of Physicians (RCP) 1962 report, *Smoking and Health*, led initially to only small signs of change. An Advisory Group on Publicity was set up in the Ministry of Health, and a circular was issued to local authorities offering free publicity. Posters showing coffins and graphs of rising deaths were prominent; and there were also posters showing a teenage boy and girl hesitating before they started smoking. A range of material was produced by other organizations, including the British Temperance Society. A record, "No Smoking" was produced by Transatlantic records of Edgware. "A Scottish psychologist outlines colloquially and effectively the dangers of smoking…." (National Archives MH82/205). The economic dimension of smoking also came increasingly to the fore (see Figure 13.2).

> "So she said she was giving up smoking as she could save money and I said don't be daft you won't save a penny and my Norman agrees with me; But the next thing you know she's got herself one of these cut out dresses and a trouser suit and pair of those white boots on top of which she's got rid of that cough and what's more my Norman's dating her up. Honestly you can't trust some people can you!"

The Central Council (CCHE) ran a van campaign in 1962–1963 that disseminated antismoking propaganda throughout the country (National Archives MH82/205-209). Four nonsmoking male graduates were recruited to disseminate the message, finding a readier acceptance in schools than in youth clubs or factories. "The girls enjoyed it, they were such charming young men!" wrote the head of one school in 1963. There were reports of a lecture to 3,000 schoolchildren in September 1962 at the new Gallery Evangelical Centre in London's Regent Street. But the focus was still on the local authorities, where some MoHs took up the cause enthusiastically. However Devon education committee refused to allow leaflets supported by the pop stars Cliff Richard and Frankie Vaughan to be distributed in schools. They were in "beatnik language" objected a teachers' representative, and quite contrary to what was being taught. Here is some of the language reported: "Always puffin" a fag — squares,

> Never snuffin' the habit — squares,
>
> Drop it, doll, be smart, be sharp!
>
> Cool cats wise,
>
> And cats remain,
>
> Nonsmokers, doll, in this campaign… (National Archives MH151/18)

**FIGURE 13.2** Poster prepared by Reginald Mount for the Central Office of Information in the 1960s. Wellcome Library for the History and Understanding of Medicine. Iconographic Collection. Photo no. L24904.

## 13.4 HEALTH EDUCATION BEGINS TO CHANGE IN THE 1960S

The 1960s began to bring a change in tempo — toward a mass media focused, slicker, pretested by market research, advertising agency product. At the same time, the nature and content of the message changed — away from neutral information presentation, away from reduction of risk, toward more direct advice and an absolutist line. This shift was initially evident outside the health field as such. A nationwide campaign for the Ministry of Transport over the Christmas period in 1964, mounted by the government's Central Office of Information, sought to change public attitudes to drinking and driving, informing people of the dangers and penalties. This short-lived media blitz (six weeks) used press, television, and poster advertising and was supported by research into public attitudes and responses before, during, and after the campaign.[5] Drunk driving was in some sense a model for the later smoking and public health media model and one can trace a process of "policy seepage" as well as one of "policy transfer" from the U.S. Here, an initiative developed in one government department, transport, influenced the model subsequently developed in another, health.

Key milestones for that in the 1960s were the 1962 RCP report, the 1964 Cohen report on health education, and the replacement of the CCHE by the Health Education Council (HEC) in 1968–1969. The 1962 RCP report placed great emphasis on the media presentation of its conclusions in a deliberate move to appear modern and to lay out possible policy initiatives to government (Royal College of Physicians, 1960). Unlike previous reports from the Royal College, it was aimed at the public rather than the medical profession. The production of the Report was part of a wider reorientation of medicine and the College itself in the 1960s, which saw moves to make its role more relevant and less distant from society. The secretary of the committee, Dr. Charles Fletcher, was well known as a pioneer of the media presentation of medicine; and Dr. Jerry Morris, another committee member, also placed great emphasis on the importance of the discussion of health in the media (Booth, 1991; Loughlin, 2000). In 2000, at his 90th birthday conference at the London School of Hygiene and Tropical Medicine, the leading epidemiologist Sir Michael Marmot remarked that "Jerry has always told me I should watch more television rather than less."[6] The College hired a public relations consultant to manage the report's launch and held one of its first press conferences.[7] The role of the media and of advertising was also given some prominence in the report's actual content, which recommended the restriction and not the banning of advertising (Royal College of Physicians, 1962). The government set up a Cabinet committee after the report was issued, and this also discussed mass media campaigns. The role of the media began to seep into the government agenda on smoking. The industry agreed to a voluntary ban on television cigarette advertising that could appeal to young people, on advertising before 9 p.m., and restrictions on press and poster advertising.[8] These indicated a new focus at this time on the national media as well as on local action.[9]

The 1964 Cohen committee took the new tendencies further (Ministry of Health, 1964). Its origin lay in lobbying from health educators who sought greater national organization and coordination (Blythe, 1987). They formed an Institute of Health Education in 1962 and gave evidence to Cohen, claiming that health education techniques were a synthesis of methods used in "formal education, advertising and PR" (Ministry of Health, 1964, p. 95). The advertising and the PR sections of commercial organizations were presented as a model of effective action and expenditure throughout the Report. The image was one of up-to-date, media savvy professionals who knew their market. The 1960s were a boom time for advertising and market research in the UK. Membership of the Market Research Society grew from 23 to 2,000 in the years between 1947 and 1972 (Williams, 1998, p. 217). These activities had been curtailed during the war and hampered by the rationing of goods and newsprint into the 1950s. With the birth of commercial television in 1955 and the end of rationing in 1956, Britain entered an era of mass consumption. The Cohen committee itself had a strong media membership. Its deputy chair came from the Consumers Association and from a BBC background, while there was also an advertising agency representative and the health editor of *Woman* magazine. The report wanted to drop the traditional health education focus that had been on individual advice to mothers on specific action like vaccination and immunization.

It considered that more was needed on human relationships — sex education, mental health, the risks of smoking and being overweight, the need for physical exercise. These were difficult areas, where “self-discipline” was required.

A strong emphasis on individual risk avoidance mingled moral and medical imperatives. Diseases such as chronic bronchitis could be prevented if individuals would modify habits such as smoking and if the government would accelerate its health-promoting campaigns. What was needed was a greater degree of central publicity, using habit changing campaigns and social surveys, as well as strengthening the new profession of health educators. The models came from American social psychology. The new breed was to be trained in journalism, publicity, the behavioral sciences, and teaching methods. Health education would involve both imparting knowledge and inculcating self-discipline — a telling phrase. The new health educator was to be part salesman, persuading people to take appropriate action. Just knowing about the risks of cigarette smoking was not enough: Cohen called tobacco advertising “propaganda” and it had to be countered in the same way (Ministry of Health, 1964, pp. 14 and 46). This was a major change from the evenhanded response of health education in the1950s. Persuasion was now the key.

The Cohen report emphasized the role of the mass media in health education — one TV program could reach five million people, whereas it would take 250,000 group discussions of 20 people each to do the same. The Report led to the establishment of the central government funded Health Education Council in 1968, reconstituted in the early 1970s.[10] By the late 1960s, under the impact of these changes, the approach was very different. In October 1969, a major antismoking poster campaign was launched. There was research input, pretesting, and market evaluation. The campaign was pretested on a statistically selected group of subjects and based on Ministry of Health research published in 1967. An advertising agency was used and the “look” of the advertisements was quite different.

## 13.5 THE 1970S: MEDIA PERSUASION

Expenditure almost doubled on the HEC’s smoking campaign in the early 1970s, after a further RCP report in 1971, *Smoking and Health Now*, and a reconstitution of the HEC itself. TV campaigns and press advertising were areas of growth. £413,899 was spent in 1972/3; £702,292 in 1973/4 (Health Education Authority, c. 1975). Research also became more important in the design of campaigns and their evaluation. The quantitative sociologist Ann Cartwright had evaluated a campaign carried out in Edinburgh in 1959. There had been another evaluation by the sociologist Margot Jeffreys in Hertfordshire in the early 1960s. In the 1960s, this style of research expanded. There was OPCS (Office of Population Censuses and Surveys) research on smoking, research surveys by the educationalist John Bynner on boys smoking, and research studies by the public health researcher Walter Holland on the impact of health education on schoolchildren.[11] The social survey found a role in health through the smoking issue.

The new style of advertising campaign was mounted for the HEC by the advertising agency Saatchi and Saatchi. Saatchi’s emerged originally out of the reconfiguration of advertising in the UK in the 1960s under the influence of developments in the U.S. Vance Packard’s *The Hidden Persuaders* was published in 1957, and Charles Saatchi visited the U.S. in the early 1960s (Packard, 1957; Fendley, 1995). He recalled how the press advertising people dominated the UK advertising scene, with little knowledge of the possibility of TV. The new style agencies changed the image of advertising and started calling in academics to help — here was a more professional scientific approach also using typically a combination of humor and hard sell. Cigarette advertisements were among the first to use it, as for example in the Gold Box Benson and Hedges advertisements created by the agency Collett Dickinson and Pearce.

The HEC account was Saatchi's first big break and marked the importation of the new advertising style into the ranks of the public health opposition. At first the work was confined to posters and brochures, but later came full scale advertising. Saatchi produced a number of advertisements early in 1970 with the lines "The tar and discharge that collect in the lungs of the average smoker" and "You can't scrub your lungs clean" (Figure 13.3). Images like these generated an antiesthetic around

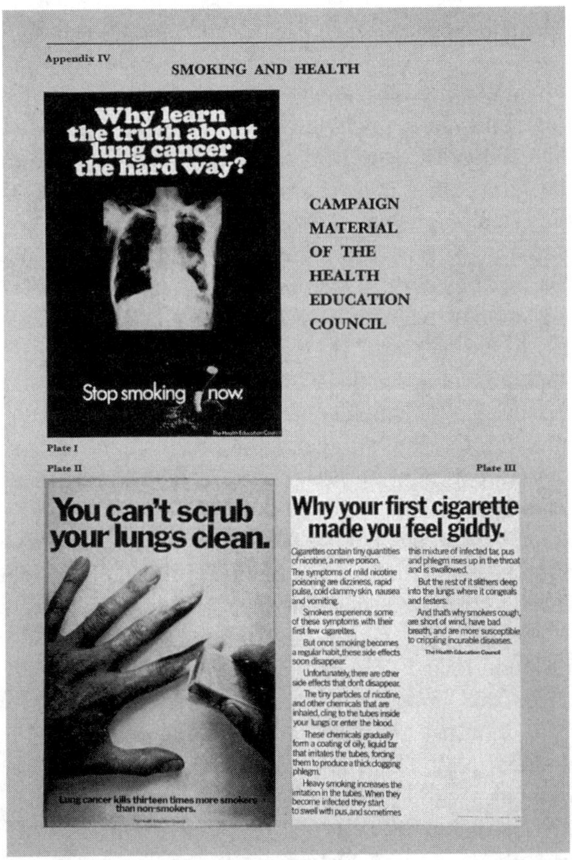


**FIGURE 13.3** Advertisements placed by Saatchi's for the Health Education Council in 1970. Wellcome Library for the History and Understanding of Medicine, ASH archive, SA/ASH Box 77, R. 14 William Norman collection, A. Mackie file, Antismoking in pregnancy campaign. Pre & post campaign study (Communication Research Ltd, May 1974).

smoking, visible effects, such as the unsightly nicotine stains on fingers, provided a visual cue to deeper more significant damage. The rest of the media started to become interested. The *Sun* newspaper wrote about the antismoking campaign, noting how dynamic and brutally effective the copywriting was. Earlier hesitations about generating public fears concerning cancer were swept aside. Graphic images of diseased lungs were featured in posters asking, "Why learn the truth about lung cancer the hard way?"

In 1970, the Saatchi brothers formed the agency Saatchi and Saatchi and Charles Saatchi brought the HEC account with him. For the first time the antismoking campaign was extended to TV. Advertisements in 1971 showed smokers crossing London's Waterloo Bridge intercut with film of lemmings throwing themselves off a cliff. A voice over said,

> "There's a strange Arctic rodent called a lemming which every year throws itself off a cliff. It's as though it wanted to die. Every year in Britain thousands of men and women smoke cigarettes. It's as though they want to die...." (Fendley, 1995, p. 35).

There was a change of focus to target women — and pregnant women especially. In 1957, the fact that education would have to be directed to men, as the main group of smokers, was considered by the Ministry of Health to be another reason not to make smoking the subject of a major campaign; it just did not fit the public health and health education stereotype (National Archives MH55/2220b). Women re-emerged in the 1970s as a major focus of new style health education for smoking, as part of the focus on individual behavior, which reproduced the concern for women shown by public health as social hygiene 70 years earlier and also added new concerns about reproduction in the "pill era" (Lewis, 1980; Marks, 2001). The most striking image from a campaign run in 1973–1974 was a naked, pregnant, smoking woman. "Is it fair to force your baby to smoke cigarettes?" it asked (Figure 13.4). There was a clothed version of the woman but evaluation concluded that it was less effective as a campaign tool (Department of Health, 1977). The Council's main preparatory research for the campaign had been based on a clothed model and had not used the nude option. Alistair Mackie, head of the HEC, later explained that the nude emerged out of a conversation he had with his chief medical officer, "I can remember thinking in a crude way what a tremendous topic this was for public relations work" (Jacobson, 1981). Public images of female nudity were not common in the advertising culture of the time, and a nude image of a pregnant woman had an obvious shock value. The poster enhances this impact through its portrayal of the woman as serene, unconcerned, or possibly just unaware. Indeed, if the image is reduced and we see the head and shoulders only, the smoking woman would not be out of place in a cigarette advertisement.

The campaign cost £160,000 and took up nearly two-thirds of the HEC's anti-smoking budget for the year. Two further campaigns were planned and over 20% of the HEC's antismoking budget was to go to smoking in pregnancy. Some critics disputed its effect. Surveys commissioned by the HEC before and after the first campaign showed that the proportion of smokers among pregnant women fell from 39 to 20%. This seemed significant, but the research was flawed because it had not

**FIGURE 13.4** The naked smoking mother poster produced by Saatchi's in the 1970s. ASH Archive, Wellcome Library, William Norman collection, SA/ASH Box 77, R. 19, Alistair Mackie file.

compared the women who watched the advertisements with a control group of pregnant women who did not. Similar HEC research into later campaigns showed no overall impact and also showed that 15% of women stopped smoking spontaneously anyway when they became pregnant (Jacobson, 1981). Women were the focus; but there was no doubt that the at-risk fetus was male. The HEC press release for the second pregnancy campaign said, "Mums-to-be will be told that smoking can restrict the baby's growth, make *him* underdeveloped and underweight at birth and even kill *him*" (emphasis added). There was also a focus on teenage children and on keeping them from starting to smoke. The pregnant woman campaign paralleled Saatchi's campaign for the HEC on contraception, where a picture of a doleful pregnant man was captioned, "Would you be more careful if it was you that got pregnant?" The problem for smoking was defined in the 1970s as women in their reproductive role although there was little evidence that this was in fact the major issue. Smoking was failing to decrease among women and young girls in general, not just pregnant women (Berridge, 2001a).

Even though the subject matter was traditional in its "women as mothers" emphasis, the style and nature of these advertisements, their use of television as well as traditional poster campaigns were all new departures. The message of the advertisements also took a new, harder, line. The 1950s Wisdom leaflet left it up to the individual to decide. Others advised a harm reduction approach through switching to safer forms of smoking like pipes and cigars. By the end of the 1970s, the HEC began to promote a definite style of behavior change. In 1978, the Council responded to a report produced by the government's Independent Scientific Committee on Smoking and Health (ISCSH). The committee's report was on product modification and lower risk (sometimes called "safer") cigarettes. The Council's advertisement on the topic stated that switching to a tobacco substitute cigarette was like jumping from the 36th rather than the 39th floor of a building. This was a graphic illustration of risk that signaled the end of a more liberal policy line that had been followed since the 1950s (Health Education Council, 1977–1978).

This change of health education "view" can be linked to two developments, one smoking specific, the other related to overall changes in public health ideology and practice. As far as smoking was concerned, the 1970s had been occupied at the governmental level with moves to work with industry to develop tobacco substitutes or to remove the harmful components of tobacco and produce a "safer cigarette." The publication of the 1978 report led to a barrage of criticism; seasoned smoking researchers argued that its proposals for reduced tar and nicotine cigarettes could lead to "compensatory smoking," with the smoker inhaling more harmful products rather than less. The harm reduction agenda for smoking was under threat (Berridge, 1998; Berridge and Starns, 2005).

The wider public health agenda was also changing by the end of the 1970s. Smoking formed part of this redefinition and epitomized it. At the international level, documents such as the Lalonde Report (1974) had stimulated new thinking about public health, while in the UK this agenda was underlined by the government publication of *Prevention and Health: Everybody's Business* at the end of the 1970s (Petersen and Lupton, 1996). "Everybody's business" implied individual and community responsibility, rather than the previous emphasis on the role of state intervention or clinical facilities and services. In the UK this came not long after the removal of medical public health practitioners (MoHs) from their local authority bases and into clinical health services (Lewis, 1986). The old local health education model, dominated by the MoH, was undermined at the structural level at the same time that the ideology of public health itself was changing. Despite the Cohen committee's emphasis on the new breed of health educators, there were initially few at the local level, around 50 in the mid-1960s (Blythe, 1986). The new policy agendas — for smoking, but also for a host of other health issues (diet and heart disease was another) — were typically underpinned by epidemiological research and the social survey and focused on the reformation of behavior, on the culpable role of industry, and on the use of taxation or other economic incentives as a health tool. Above all, they had a media agenda — with a dual focus. Advertising was a key site of engagement with the industrial "opposition."

The media began to suffuse a new style of health activism. ASH (Action on Smoking and Health), established in 1971, exemplified this approach. This led to some initial problems for the new organization. In its launch leaflet, it had envisaged that it would undertake an advertising campaign itself. It wanted to drop the "Black

Widow" approach of the road accident campaigns, which had aimed to shock drivers into responsible behavior. Instead, ASH envisaged marketing social acceptability. Financial and material incentives were to be encouraged, with group therapy on the lines of Weight Watchers and a focus on children and young people. "Primarily the campaign will attempt to take the social cachet that surrounds smoking and turn it on its head" (National Archives MH154/169). But these proposals ruffled feathers because these ideas infringed on the territory of the Health Education Council. The HEC was annoyed at the overlap with its role. ASH subsequently defined its public role round effective media "spin" in a way that did not involve so much of a direct marketing role (Berridge, 2005). But the fact that both it and the HEC had reached similar conclusions about health marketing in the early 1970s is a striking illustration of the overall change toward media persuasion at this time.

"Hidden persuasion" was used from the feminist perspective to characterize women as "victims" of mass advertising. Bobbie Jacobson, Deputy Director of ASH, in *The Ladykillers. Why Smoking Is a Feminist Issue*, an influential public health text of the early 1980s, characterized women as the passive "dupes" of mass media messages about smoking (Jacobsen, 1981). This implied a role for the media that drew on the Frankfurt school of media sociology, which had stressed the role of the population as receptors for the influence of the mass media (Curran and Seaton, 1991).

In the UK, American commercial techniques of education and persuasion affected public health professionals at a time of structural change, in terms of their location and role. In France and Switzerland, too, the role of psychological and mass media models grew in importance; France also set up a centralized agency in the late 1970s (Berlivet, 2005b; Lengwiler, 2005). In Switzerland the route initially came via advertising in the occupational health field and through road safety; in Britain, too, road safety seems to have been the precursor. The mass media were "modern" and so, for public health in the UK, they were part of its modernizing project.[12] Their use was also part of a redefined consumerism that was linked with developments in health advertising. The Consumers Association was represented on Cohen and it was that Association that published the first report in 1971 on tobacco labeling. Stacey has argued that the health consumer as a concept originated in the 1970s (Stacey, 1976). Three decades later, mass communication remains a widely used strategy. Social marketing in health promotion attests to the importance of consumerism and persuasion in health. For the tobacco field, advertising retains its symbolic status, and controversy over "effect" has had little impact on that. Research into the effectiveness of mass communication strategies in health promotion has discounted older "injection" models of media effect and concludes that production of behavior change is unproven. However, media coverage can define the framework of a public agenda (for recent discussion of the effect of the mass media, see Reid, 1996; Ellis and Grey, 2004; Wellings and Macdowall, 2000). Recent UK policy commentary has criticized the concentration on mass advertising techniques as of little proven impact, although they retain political utility (see, for example, Edwards et al., 1994). In commercial advertising the pendulum is now swinging toward individual, personalized advice, and it remains to be seen whether health promotion marketing will follow suit.[13] Understanding the historical rationale for the emergence of mass advertising in public health during the 1970s could help inform the reassessment of strategies in the present.

## ACKNOWLEDGMENTS

We are grateful to the Wellcome Trust, which has funded the historical work on which this work is based. Our thanks are also due to Wendy MacDowall of the Sexual Health program at the London School of Hygiene and Tropical Medicine, who has advised on current views of "media persuasion" in health promotion.

## NOTES

1. Control or prohibition of candy advertising during children's programs on television is currently under discussion.
2. They used "shock–horror" tactics — "AIDS, Don't Die of Ignorance," extensive use of television, and advertising agencies and market research companies were involved.
3. An antiheroin campaign mounted in 1985–1986 had been a precursor of the later AIDS campaign. It proved controversial with researchers, who argued that mass media might be counterproductive and that the campaign evaluation by market research had not been rigorous.
4. On the processes through which public statements about scientific advice is performed for nonscientific audiences, see Hilgartner, 2000.
5. The research was carried out by the Road Research Laboratory and the campaign organized by the Advertising Division of the Central Office of Information, which drew on the services of commercial advertising companies. See Clark, 1970.
6. Transcript of witness seminar for 90th Birthday of Jerry Morris, 21st July, 2000, London School of Hygiene and Tropical Medicine. See also Berridge, 2001b; Loughlin, 2001.
7. Interview with public relations consultant by Virginia Berridge, 1995.
8. These were agreed on through the office of the Postmaster General, which was charged with regulating television advertising.
9. By the 1960s television had joined the press as a key media player in Britain. The one BBC channel in the 1950s was joined by a commercial channel in 1955. The BBC gained a second noncommercial channel in 1963. By 1958 income from television advertising exceeded that from the press for the first time. See Williams, 1998, pp. 217–219.
10. The World War Two and postwar history of health education up to the 1980s is surveyed in Smith, 1987; also Blythe, 1987.
11. Walter Holland, interview by Michael Ashley Miller, Oxford Brookes University/Royal College of Physicians video interviews, May-December 1996. Walter Holland, interview by Virginia Berridge, March 1997. Also Bynner, 1967, 1969.
12. Royal College of Physicians, minutes of the committee on smoking and atmospheric pollution, fifth meeting 4 January 1961. "It was agreed that the report should include a section on the use of advertising against smoking: modern methods should be employed to combat modern methods."
13. The alcohol industry is currently experimenting with personal recommendation as a marketing technique, whereby promoters befriend and advise partygoers on what drink to buy.

## REFERENCES


Berlivet, L., Association or causation? The debate on the scientific status of risk factor epidemiology, 1947 - c. 1965, In *Making health policy: networks in research and policy since 1945,* Berridge, V., Ed., Rodopi, Amsterdam, 2005a.

Berlivet, L., Uneasy prevention: the problematic modernization of health education in France after 1975, In *Medicine, the market and the mass media: producing health in the twentieth century,* Berridge, V., and Loughlin, K., Eds., Routledge, London, 2005b.

Berridge, V., *AIDS in the U.K.: the making of policy, 1981–1994*, Oxford University Press, Oxford, 1996.

Berridge, V., Science and policy: the case of post war British smoking policy, In Ashes to ashes: the history of smoking and health, Lock, S., Reynolds, L., and Tansey, E.M., Eds., Rodopi, Amsterdam, 143–163, 1998.

Berridge, V., Jerry Morris, *Int. J. Epidemiol.*, 30, 1145–1145, 2001a.

Berridge, V., Constructing women and smoking as a public health problem in Britain, 1950–1990s, *Gender Hist.,* 13, 328–348, 2001b.

Berridge, V., New social movement or government funded voluntary sector? ASH (Action on Smoking and Health) science and anti tobacco activism in the 1970s, In *The practice of reform in health, medicine and science, 1500–2000,* Pelling, M., and Mandelbrote, S., Eds., Ashgate, London, 2005.

Berridge, V., and Starns, P., The "invisible industrialist" and public health: the rise and fall of 'safer smoking' in the 1970s, In *Medicine, the market and the mass media: producing health in the twentieth century,* Berridge, V., and Loughlin, K., Eds., Routledge, London, 2005.

Blythe, G.M., *A history of the central council for health education, 1927–1968,* Unpublished M.Litt. thesis, Oxford University, 1987.

Blythe, M., A century of health education, *Health Hygiene*, 7, 105–115, 1986.

Booth, C.C., Smoking and the gold headed cane, In *Balancing act: essays to honour Stephen Lock,* Booth, C., Ed., Keynes Press, London, 1991.

Brandt, A., The cigarette, risk and American culture, *Daedalus,* 119, 155–176, 1990.

Burnham, J., American physicians and tobacco use: two surgeons general, 1929 and 1964, *Bull. Hist. Med.,* 63, 1–31, 1989.

Bynner, J. M., *Medical students' attitudes towards smoking. A report on a survey carried out for the Ministry of Health,* HMSO, London, 1967.

Bynner, J. M., *The young smoker: a study of smoking among school boys carried out for the Ministry of Health (Government social survey, 383),* HMSO, London, 1969.

Cantor, D., Representing 'the public': medicine, charity and emotion in twentieth-century Britain, In *Medicine, health and the public sphere in Britain, 1600–2000,* Sturdy, S., Ed., Routledge, London, 145–168, 2002.

Clark, F., *The central office of information,* George Allen & Unwin, London, 1970.

Curran, J., and Seaton, J., *Power without responsibility: the press and broadcasting in Britain,* 4th ed., Routledge, London, 1991.

Department of Health, Witnesses to the Select Committee on Preventive Medicine, 1976. *First Report from the Expenditure Committee. 1976–1977 session, Preventive Medicine, Vol. II.*, Qs 2137–2138, 1977.

Dorn, N., Media campaigns, *Druglink,* 1, 8–9, 1986.

Edwards, G. et al., *Alcohol policy and the public good,* Oxford Medical Publications, Oxford, 1994.

Ellis, S., and Grey, A., *Prevention of sexually transmitted infections (STIs): a review of reviews into the effectiveness of non clinical interventions,* Health Development Agency, London, 2004.

Fendley, A., *Commercial break. The inside story of Saatchi and Saatchi,* Hamish Hamilton, London, 1995.

Grant, M., *Propaganda and the role of the state in inter-war Britain,* Clarendon Press, Oxford, 1994.

*Guardian*, Campaigners hail tobacco ad ban, 14 February, 2003a.

*Guardian,* Curb on junk food adverts to combat child obesity, 1 December, 2003b.

Health Education Authority leaflet archive (now Health Development Agency), The Adventures of the Wisdom family, "What — no smoking?" Central Council for Health Education leaflet, 1957.

Health Education Council, *Annual Report, 1977–1978.*

Hilgartner, S., *Science on stage: expert advice as public drama*, Stanford University Press, Stanford, CA, 2000.

Jacobson, B., *The ladykillers: why smoking is a feminist issue*, Pluto Press, London, 1981.

Lengwiler, M., Between war propaganda and advertising: the visual style of accident prevention as a precursor to post war health education in Switzerland, In *Medicine, the market and the mass media: producing health in the twentieth century,* Berridge, V., and Loughlin, K., Eds., Routledge, London, 2005.

Lewis, J., The politics of motherhood. Child and maternal welfare in England, 1900–1939, Croom Helm, London, 1980.

Lewis, J., *What price community medicine? The philosophy, practice and politics of public health since 1919,* Wheatsheaf, Brighton, United Kingdom, 1986.

Loughlin, K., "Your Life in Their Hands": the context of a medical-media controversy, *Media Hist.*, 6, 177–188, 2000.

Loughlin, K., Epidemiology, social medicine and public health: a celebration of the 90th birthday of Professor J. N. Morris, *Int. J. Epidemiol.*, 30, 1198–1199, 2001.

Marks, L., Sexual chemistry: a history of the contraceptive pill, Yale University Press, New Haven, CT, 2001.

Ministry of Health, The Central Health Services Council, The Scottish Health Services Council, *Health Education. Report of a joint committee of the Central and Scottish Health Services Councils,* HMSO, London, 1964.

National Archives, Ministry of Health papers, Public health propaganda; smoking and lung cancer: publicity policy, 1957–1960, MH55/2203.

National Archives, Ministry of Health papers, Tobacco smoking and cancer of the lung, Brief for adjournment debate 1 March 1957, MH55/2220a.

National Archives, Ministry of Health papers, Publicity: statement by the MRC and Ministry of Health, 1955–1958, Minute to Mr. Pater, 1 April, 1957, MH55/2220b.

National Archives, Ministry of Health papers, Summary of local health authority replies to circular 17/58, MH 55/2225.

National Archives, Ministry of Health papers, Paper to Mr. Galbraith on smoking and lung cancer, 15 March 1960, MH55/2226.

National Archives, Ministry of Health papers, Publicity-action taken by local authorities 1957–1958, MH55/2228.

National Archives, Ministry of Health papers, Files on the mobile unit campaign, MH 82/205–209.

National Archives, Ministry of Health papers, Smoking and health campaign policy, *Daily Mail* report 22 February 1963, MH 151/18.

National Archives, Ministry of Health papers, MH 154/169, Letter from Dr. John Dunwoody to Dr. Julia Dawkins, 5 January 1971.

Packard, V., *The hidden persuaders,* Longmans, Green and Co., London, 1957.

Parascandola, M., Cigarettes and the US Public Health Service in the 1950s, *Am. J Public Health,* 91, 196–205, 2001

Petersen, A., and Lupton, D., *The new public health: health and self in the age of risk,* Sage, London, 1996.

Reid, D., How effective is health education via mass communications? *Health Ed. J.,* 55, 332–344, 1996.

Royal College of Physicians, Committee to report on smoking and atmospheric pollution, Minutes of fourth meeting 17 March, 1960.

Royal College of Physicians, *Smoking and health,* Pitman, London, 1962.

Smith, R., *Working paper no 66. The national politics of alcohol education: a review,* School of Advanced Urban Studies, Bristol, United Kingdom, 1–21, 1987.

Stacey, M., The health service consumer: a sociological misconception, *Sociological Review Monograph,* 22, The Sociology of the National Health Service, University of Keele, United Kingdom, 194–200, 1976.

Taylor, P.M., Techniques of persuasion: basic ground rules of British propaganda during the second world war, *Hist. J. Film Radio Television,* 1, 57–67, 1980.

Taylor, P.M., *The projection of Britain,* Cambridge University Press, Cambridge, 1981.

Webster, C., Tobacco smoking addiction: a challenge to the National Health Service, *Br. J. Addict.,* 79, 8–16, 1984.

Wellings, K., and Macdowall, W., Evaluating mass media approaches, In *Evaluating health promotion: practice and methods,* Thorogood, M., and Coombes, Y., Eds., Oxford University Press, Oxford, 113–128, 2000.

Williams, K., *Get me a murder a day! A history of mass communication in Britain*, Arnold, London, 1998.

# 14 The Demographic and Health Impact of the One Child Family Policy

*Therese Hesketh*

## CONTENTS



## ABSTRACT

The One Child Policy has had a profound effect on the lives of nearly one-quarter of the world's population for a quarter of a century. The Policy was introduced in 1979 because population containment was seen as essential to economic growth. It consists of regulations governing family size, with one child allowed in urban areas and two children in most rural areas. The demographic impact includes: 1) a reduction in total fertility rate from 2.9 to 1.7, with 250–300 million births prevented; 2) an increase in the male:female sex ratio, from 1.06 in 1970 to 1.17 in 2000; and 3) an increase in the population over 65 years from 5% in 1982 to 7.5% in 2004. Women have benefited from fewer pregnancies and safe abortion, but contraceptive choice is limited and unapproved pregnancies have a higher mortality. There is little empirical evidence for harm to psychosocial adjustment and development. The Policy was planned for the short term, with a goal to create a small family culture, and evidence suggests this has been achieved. This, together with the huge societal changes, the unpopularity of the policy, and problems with enforcement, suggests that China is ready for a Two Child Policy.

## 14.1 INTRODUCTION

For a quarter of a century the One Child Family Policy has had a major impact on the lives of nearly one-quarter of the world's population. From an evolutionary perspective, this represents a powerful example of niche construction in a social context, whereby the manipulation of human behavior through institutional social information transmission has profound effects on demographic and health outcomes.

This paper will examine the demographic and health impact of the One Child Policy in China. After describing the background to the Policy and how it works, the paper will examine the demographic impact in terms of population growth, sex ratio, and the age structure of the population. It will then discuss the impact of the policy on the health of women and children. Finally, it will examine the relevance of the Policy in contemporary China and consider whether the dramatic societal changes that have been witnessed over the past 25 years mean that the time has come for the Policy to be lifted.

## 14.2 THE EARLY DAYS OF THE POLICY

Mao Ze Dong famously said there could never be too many Chinese, because human resources would be China's greatest defense in the widely predicted Third World War. But the population increase from 540 million in 1950 to 850 million by 1970 was regarded as too rapid and led to the introduction of the "late, long, few" policy in 1970. This was a voluntary family planning program, calling for later child bearing, longer interbirth spacing, and fewer children. As a result the Total Fertility Rate (TFR), the mean number of children born to each woman, dropped dramatically from 5.9 in 1970 to 2.7 in 1979 (Yao and Yin, 1994). But this figure was still too high for the new leader, Deng Xiao Ping, who was embarking on an ambitious program of economic reform after the economic stagnation of the Cultural Revolution. Around two-thirds of the population were under 30 and predictions for population growth remained high (Zhu, 2003). Strict population containment was seen as essential to economic expansion, so the One Child Policy was born.

By 1984 the TFR was reported to have dropped to 1.94. The apparent demographic success of the policy together with its unpopularity, especially in the countryside, led to considerable relaxation, including allowing two children in rural areas. (It was later discovered that there had been considerable under-registration of births in rural areas and that the actual fertility rate was probably closer to 2.5.) Since 1984 there have been alternate relaxations and tightenings according to population projections, but the policy has never returned to the stringency of the early 1980s (Hesketh and Zhu, 1997).

## 14.3 HOW DOES THE POLICY WORK?

The One Child Family Policy is not a law; it consists of a set of rules and regulations governing the approved size of Chinese families. The State Family Planning Bureau, which is separate from the Ministry of Health, sets the overall targets and policy direction. Family planning committees at provincial and county level devise local strategies for implementation. These include restrictions on family size, late marriage and child-bearing, and spacing where second children are permitted, together with easy access to contraception and abortion. Despite its name, the one child rule applies to a minority of the population: mainly urban residents and government employees. For urban residents, the Policy is strictly enforced with few exceptions to the one child rule. These include: where the first child has a disability; in the case of remarriage if one or the other partner has no previous children; for certain groups of workers in high risk occupations such as mining; and finally, in some areas if both partners are themselves from one child families (Hesketh and Zhu, 1997). In rural areas, where around 70% of the population lives, a second child is generally allowed after five years, but in some areas, this is only if the first is a girl, clearly acknowledging the traditional preference for boys (Short and Zhai, 1998). Third children are allowed for some ethnic minorities and in remote underpopulated areas. Marriage is not usually permitted until age 24 for men and 22 for women in cities, or ages 22 and 20, respectively, in rural areas, and the target for spacing between first and second birth is fixed at a minimum of five years.

The whole Policy is underpinned by a system of rewards and penalties. Since enforcement is the responsibility of local officials, interpretation of the regulations and the administration of the system of rewards and penalties varies widely even from village to village (Short and Zhai, 1998). Rewards consist mainly of financial incentives and sometimes access to better schools for the only child and help with medical insurance. Fines typically amount to around 10%–20% of a family's annual income for 3–5 years, but cases of much higher fines have been reported (Hardee-Cleaveland and Banister, 1988). When families are unable to pay, personal belongings are sometimes confiscated instead. For government employees noncompliance is punished with loss of employment.

In general, communities have allocated birth quotas. In each community a woman, usually a representative of the Women's Federation, has responsibility for family planning issues and ensuring that birth quotas are not exceeded. A couple normally needs to apply for permission to have a child. There is considerable social pressure to comply with the rules. In the past if an unapproved birth pushed the community above its birth quota, then other couples would have to wait to be granted permission to have a child, though this occurs less commonly now (Hardee-Cleaveland and Banister, 1988). The general pattern is that most women get married soon after they are permitted. They give birth a median of 15 months after this (Short and Zhai, 1998), an intrauterine device (IUD) is inserted after the first birth and is removed to allow for a second birth, if this is permitted. After the second birth an IUD is generally used again, and then one or the other partner is sterilized, usually several years after, when the child has reached an age when mortality risk is low.

To improve compliance, the Policy has been promoted through a massive health education campaign, focusing on the societal dangers of overpopulation and the personal material benefits of having only one child (Green, 1988). Slogans that have been widely used include: "With two children you can afford a 14 inch TV, with one child you can afford a 21 inch TV" and "The One Child Policy can guarantee that children will be better cared for and educated."

## 14.4 THE IMPACT OF THE POLICY ON POPULATION GROWTH

A major caveat when considering population issues in China is the difficulty with obtaining accurate figures. Routine population statistics in China are known to be subject to manipulation to conform with family planning regulations (Merli and Raftery, 2000), and surveys are notoriously prone to bias because of motivation to hide higher order births; family planning officials are often unwilling to uncover deliberate breaking of the rules (Pan, 2003). One source that is regarded as relatively reliable is the National Family Planning and Reproductive Health Survey (NFPRHS), which was carried out in 2001 (Theses Collection 2001). This survey was carried out among a nationally representative sample of 39,500 women (aged 15–49), the generation that has been the target of the Policy. For this very comprehensive survey, interviewers were specifically instructed to be nonjudgmental and respectful and to ensure confidentiality and impunity. Data from this source will be used for much of the demographic information in this paper.

The Policy has substantially contributed to a reduction in population growth over the past 25 years. When the Policy was introduced, a target population of around 1.2 billion by the Year 2000 was set (Greenhalgh and Bongaarts, 1987). The total population reported for that year was 1.27 billion, close enough for the Policy to be deemed a demographic success (China Population Information and Research Centre [CPIRC], 2001) despite the view of some demographers that this was an underestimate (Wang, 2003). The authorities claim that the Policy has prevented 250 to 300 million births, has reduced the natural increase from 11.6 to 8 per thousand (Festini and de Martino, 2004), and has helped to lift over 200 million people out of poverty (Anon., 1997). The median age of first birth rose by two years between 1979 and 1985 but has remained unchanged for the past 20 years at 24.1 years. Spacing between first and second children increased from a median of 2.5 years in 1981 to 5.3 in 2001, achieving the Policy target of at least five years (Zheng and Wang, 2003).

A unique demographic picture has emerged of mostly (around 83%) one child urban families, and mostly (around 70%) two child rural families, with third and higher order births accounting for around 12% of all births, though with almost all of these in rural areas.

The mean TFR across the country is around 1.7 (slightly higher than in the UK) with a rate of 1.25 in urban areas and 1.98 in rural, with a range from 1.0 in the eastern cities, such as Beijing and Shanghai, to 3.1 in Tibet (Wang, 2003).

But an important question is how much of the reduction in birth rate seen over the past 25 years can be attributed to the Policy. It is probably only partially responsible. The most dramatic fall in TFR occurred *before* the Policy was imposed (see Figure 14.1). As already noted, between 1970 and 1979 the voluntary "late, long, few" policy had resulted in a reduction in TFR from 5.8 to 2.9. After the One Child Policy started, there was a much more gradual fall in TFR until 1995, and it has more or less stabilized at around 1.7 since then (US Central Intelligence Agency, 2004). This begs the question of whether the downward trajectory of the TFR would have continued without imposition of the Policy. Many countries have experienced major declines in fertility over the past 25 years and neighboring East Asian countries

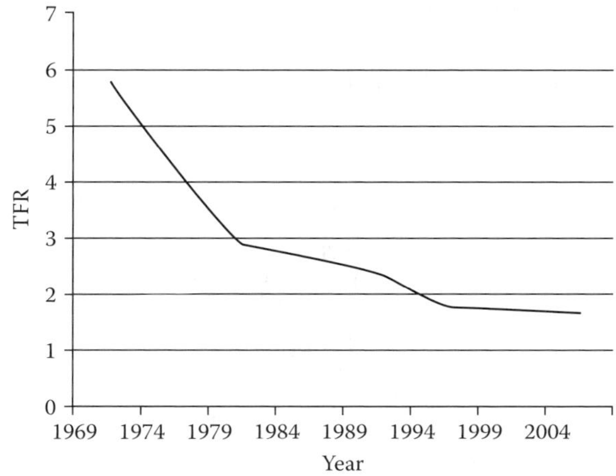


**FIGURE 14.1** Total Fertility Rate in China from 1969 to 2004.

have some of the lowest TFRs in the world: 0.91 in Hong Kong, 1.04 in Singapore, 1.38 in Japan (US Central Intelligence Agency, 2004). Even allowing for the fact that these are more developed, urbanized countries, the parallels are hard to ignore. Therefore, it is reasonable to speculate that there would have been a further decline in China's TFR had the voluntary policy of the 1970s continued. This is the conclusion of a study examining fertility trends in 288 villages in rural Hebei Province. This showed consistently declining fertility from the Revolution in 1949 onward. The fertility rate was actually slightly higher for women born after 1960, at whom the Policy was targeted, than those born in the 20 years previously: 2.4 compared with 2.1 (Anson and Anson, 2003).

## 14.5 IMPACT ON THE SEX RATIO

The impact of the Policy on the sex ratio has become an area of great concern. The sex ratio is usually defined as the proportion of male live births to female live births, although in the Chinese context the proportion of registered male to female births is probably a more appropriate definition, since that is what is reflected in official figures. The normal sex ratio at birth ranges between 1.03 and 1.07 (Davis et al., 1998). Since the onset of the Policy, there has been a steady increase in the reported sex ratio from 1.06 in the late 1970s to 1.11 in the late 1980s to 1.17 in the late 1990s (Festini and de Martino, 2004). There are marked and well-documented local differences with ratios of up to 1.3 in rural Anhui, Guangdong, and Qinghai. Recent data from the National Family Planning and Reproductive Health Survey have highlighted the problem (see Table 14.1). What they show clearly is that the increased sex ratio is a phenomenon of urban and rural China, and not predominantly rural China as has been suggested elsewhere (Festini and de Martino, 2004). There is a very marked gradient across birth order: in rural areas sex ratio for the first birth is

**TABLE 14.1**
**Sex Ratio by Birth Order (n = 56,830)**

| | Total | Urban | Rural |
|---|---|---|---|
| First $n = 29{,}310$ | 1.06 | 1.13 | 1.05 |
| Second $n = 16{,}448$ | 1.24 | 1.30 | 1.23 |
| Third n = 7,285 | 1.28 | 1.19 | 1.29 |
| Fourth $n = 3{,}787$ | 1.31 | 1.19 | 1.32 |
| | 1.15 | 1.16 | 1.15 |

*Source:* Adapted from: Kang, C., and Wang, Y., Sex ratio at birth, *Theses Collection of 2001 National Family Planning and Reproductive Health Survey*, China Population Publishing House, Beijing, 2003.

1.06 within normal limits, but it is 1.13 for first urban births rising steeply in both urban and rural areas with birth order (Kang and Wang, 2003).

What happens to all these "missing" girls is a matter of speculation. Sex-selective abortion undoubtedly accounts for a large proportion (Sen, 2003) though actual figures are impossible to obtain, because it is strictly illegal. However, it is known to be widely carried out, helped by a burgeoning private sector. There is also evidence that nonregistration of female births is making a contribution. A household survey in six counties in three provinces carried out in 1995 found a normal sex ratio in the under-14 age group, with the number of girls exceeding the number registered by 22% (Bogg, 1998). While female infanticide is probably extremely rare now, less aggressive management of sick infant girls is known to occur (Peabody et al., 1992), and abandonment of newborn girls greatly exceeds that of boys, as the gender balance in Chinese orphanages clearly demonstrates (Kandela, 1996).

The picture that emerges is that some urban Chinese, knowing they are allowed only one child, make the choice to sex-select. In rural areas there is no pressure with the first birth, since most are permitted to go on to have a second child. If the first child is female and the second pregnancy is female then again the child "disappears" either through sex-selective abortion, abandonment, or simply nonregistration. This means that a couple would be allowed to go on to have another child.

The Chinese government has acknowledged the potentially disastrous social consequences of this sex imbalance. For men these concerns center on the risk of mental health problems, socially disruptive behavior, and reduced marriageability; for women concerns have grown about kidnapping and trafficking and the selling of women as brides (Tuljapurkar et al., 1995). There are already increased reports of these problems and they could represent a real threat to China's stability in the future (Hudson and Den Boer, 2002).

Although the Policy has been blamed for the unusual sex ratio, it is probably just one contributory factor. A high sex ratio was reported in China in the 1930s and 1940s, then it declined in the post-Revolution Years after 1949 (Gu and Roy,

1995). Many other Asian countries with declining fertility rates and traditional preference for males are seeing gender imbalances: Taiwan 1.19, Singapore 1.18, South Korea 1.12, and parts of northern India 1.20. These are thought to be largely a result of sex-selective abortion (Wu et al., 2003; Bogg, 1998). It is also of note that the marked increase in sex ratio between the 1980s and 1990s in China coincided with much easier access to cheap ultrasound in rural and suburban areas, and not with any change in enforcement of the Policy. It is likely that even in the absence of the Policy, sex-selective abortion would continue, though its frequency would certainly reduce. A sustainable solution will only come with a change in attitudes toward female offspring.

## 14.6 THE IMPACT ON THE AGING POPULATION

The other demographic concern is the increase in the elderly population. The percentage of the population over 65 was 5% in 1982 and now stands at 7.5%, but is expected to rise to over 15% by the Year 2025 (CPIRC, 2001). Although these figures are much lower than in most industrialized countries (in the UK 16% of the population is already over 65, in Japan 20%), lack of adequate pension coverage in China means that financial dependence on offspring is still necessary for around 70% of elderly people (Sun, 1998). This means that the ratio of working people to dependents will be very low. At a family level this problem has been designated the "4:2:1" phenomenon, meaning that increasing numbers of couples will be solely responsible for the care of one child and four parents. This is of particular concern to Chinese policy makers (Peng, 1998). A specific measure has been introduced to address this. As already mentioned, one of the exceptions to the one child rule in most cities is where both partners are only children themselves. As the cohorts of urban-only children reach reproductive age many couples will meet these criteria, though how many avail themselves of the opportunity to have more than one child remains to be seen. In addition the government is trying to extend social welfare benefits to more people. Present coverage is available only to those employed in the formal sector.

## 14.7 THE IMPACT ON WOMEN'S HEALTH

The Policy was bound to have a massive demographic impact. The health impacts are somewhat less obvious and confounded by general improvements in socioeconomic status and sophistication of health care over the past 25 years. Nevertheless, there have been some clear health benefits for women. With small families women are freed from the continuous burden of pregnancy and its associated morbidity and mortality. Mothers also have greater freedom to work outside the home and to acquire skills and training, with all the associated benefits (Hesketh and Zhu, 1997). The One Child Policy has also resulted in ready access to safe, legal abortion; abortion is an extremely rare cause of death in China (Young, 1990). Early and hence relatively safe medical abortion is greatly encouraged (Xiao, 1997).

However, there are many negative consequences for women's health. Abortion rates are high and the government itself regards the rate as unacceptably high, with around 20% of all women over the age of 30 having had an abortion and with around 2% having had three or more (Qian, 2004). The overwhelming majority (over 90%) of pregnancies before marriage end in therapeutic abortion because babies born outside marriage are unapproved by definition. Forced abortion and sterilization certainly did take place during the early very stringent years of the Policy, but they are probably extremely rare now (Hesketh and Zhu, 1997; Qian, 2004).

A further direct consequence of the policy has been the reluctance of women with unapproved pregnancies to seek antenatal and obstetric care. An eight province study showed women with unapproved pregnancies were 75% less likely both to seek antenatal care and to deliver in a modern medical facility (Doherty et al., 2001). The reasons are the fines incurred for unapproved pregnancies and fears that pressure will be applied to terminate the pregnancy. A study carried out in the large central province of Sichuan between 1989 and 1991 showed that the risk of maternal death was twice as high for unapproved pregnancies because many of the unapproved deliveries occurred at home without trained personnel (Ni and Rossignol, 1994).

The Policy also means there is very little individual choice in method of contraception. Eighty percent of the women interviewed for the National Family Planning and Reproductive Health Survey said they had no choice about contraceptive method; they simply agreed to use the method recommended by family planning workers (Yin et al., 2003). Trends in the use of contraception among married women over the past 10 years are shown in Table 14.2. There remains a great reliance on long-term contraceptive measures. Over the past 10 years the use of sterilization has declined, although female sterilization is carried out over four times more often than male sterilization. The use of IUDs is on the increase. This is despite the fact that of the 22% of women who said that they were dissatisfied with their contraceptive method, 66% of them were IUD users (Yin et al., 2003). IUDs are commonly inserted

**TABLE 14.2**
**Trends in Contraception Use of Married Chinese**

| | 1992 | 1997 | 2001 |
|---|---|---|---|
| Male sterilization | 12 | 9.2 | 7.9 |
| Female sterilization | 42 | 40 | 37 |
| IUD | 40 | 43 | 46 |
| Condom | 1.8 | 4.0 | 5.6 |
| OCP | 3.8 | 2.1 | 2.5 |
| Other | 0.8 | 1.3 | 1.3 |

*Source:* Adapted from: Yin, Q., Shuai, Y., Zong, Z., and Xu, P., Choice of contraceptive methods among women of child bearing age and influencing factors, Theses Collection of 2001 National Family Planning and Reproductive Health Survey, China Population Publishing House, Beijing, 2003.

immediately after delivery. This technique has been found to be convenient, safe, and effective, though with slightly higher expulsion rates than with interval insertion (Grimes et al., 2003).

However, there are moves to encourage more choice, and in some carefully selected counties programs of informed choice have now been introduced. The family planning workers in these counties have been trained to provide women with the information they need to make appropriate choices about contraception rather than simply accept advice.

## 14.8 HEALTH OF CHILDREN

The impact of the policy on the health of children is more complex. There are clearly some direct economic benefits for children who benefit from the increased resources devoted to them. Girl children may particularly benefit because families place all their investment in them with no competition from sons (Hesketh et al., 1997).

When the policy was first instigated, psychologists voiced serious concerns about a generation of only children. In China anecdotal evidence about only children as overindulged, overweight "Little Emperors" abounds (Jing et al., 1987; Jiao et al., 1986; Wan et al., 1994), but hard evidence for the ill effects is hard to find. The literature has concentrated largely on issues of psychological adjustment and development, and the results are equivocal. For example, a study of 697 preschool children in Nanjing found slightly higher scores for temper, moodiness, and depression in only girls compared with those with siblings, but no such differences in boys (Tseng et al., 1988). In a four province study of 1,000 8–17-year-olds no evidence was found that only children had "undesirable personalities" as judged by self-report, teachers, and mothers, though only children demonstrated slightly higher academic achievement (Poston and Falbo, 1990; Falbo and Poston, 1993), a finding also of studies in the West (Falbo and Polit, 1986). In the current climate of fierce academic competition, there have been particular concerns that with parental expectations and aspirations focused on just one child, these pressures are leading to the high rates of anxiety and depression documented in Chinese adolescents (Liu et al., 1999). But a study carried out in 300 children in Beijing and Tianjin found higher levels of anxiety and depression in sibling children (Yang et al., 1995). A study carried out in three areas of Zhejiang Province examined a wide range of health, nutritional, and social indicators in only and sibling children. After adjustment for sex and socioeconomic status, only two differences remained between the two groups: only children were less likely to be bullied and were more likely to confide in parents (Hesketh et al., 2003).

One of the methodological difficulties with comparing only children with sibling children is that only children live in wealthier urban environments, where only children are the norm, and sibling children come from poorer rural families where sibling families are the norm. It is difficult to adjust for such major confounding. The Zhejiang study did attempt to control for these confounders by analyzing a subsample that compared 759 only children with 629 sibling children in one socio-economically homogeneous rural area where there were almost equal numbers of only and sibling families. The results are shown in Table 14.3.

**TABLE 14.3**
**Selected Indicators for Xiaoshan only by Family Size and by Family Position in Sibling Families: Percentages and Adjusted Odds Ratios**

| | 1-child % n=759 | 2+ children n=629 | Adjusted OR* 95% CI | P |
|---|---|---|---|---|
| | **Nutrition** | | | |
| Overweight | 0.26 | 0.19 | 0.6 (0.2,1.2) | 0.2 |
| Underweight | 15 | 17 | 0.9 (0.6,1.4) | 0.5 |
| Anemic | 36 | 38 | 0.9 (0.7,1.3) | 0.5 |
| | **Health care seeking behavior** | | | |
| Visit to doctor in last year | 62 | 58 | 0.8 (0.6,1.1) | 0.08 |
| Visit to dentist in last year | 12 | 6 | 0.7 (0.3,1) | 0.06 |
| | **Risk behaviors** | | | |
| Ever smoking | 20 | 12 | 0.6 (0.3,1.3) | 0.5 |
| Ever drinking alcohol | 38 | 30 | 0.7 (0.6, 1) | 0.04 |
| Ever been drunk | 16 | 11 | 0.7 (0.5,1) | 0.04 |
| | **Mental health** | | | |
| Self-esteem: high | 30 | 28 | 0.9 (0.6,1.2) | 0.7 |
| Anxiety: severe/moderate | 35 | 40 | 1.2 (0.9,1.4) | 0.06 |
| Depression: severe/moderate | 26 | 39 | 1.4 (1.1,1.8) | 0.03 |
| Have contemplated suicide | 14 | 15 | 1.1 (0.5,1.4) | 0.9 |
| Have tried to commit suicide | 10 | 9 | 0.9 (0.7,1.3) | 0.9 |
| | **Social** | | | |
| Bullied sometimes/often | 33 | 48 | 0.7 (0.5,0.8) | 0.001 |
| Generally liked by peers | 74 | 65 | 0.6 (0.5,0.8) | 0.005 |
| Readily confide in friends | 68 | 64 | 0.8 (0.6,1.0) | 0.07 |
| Readily confide in parents/other relatives | 66 | 38 | 0.3 (0.2,0.6) | 0.006 |
| Self-reported academic record: excellent/good | 50 | 39 | 0.5 (0.3,0.7) | 0.002 |

* adjusted for sex and parental education.

The conclusions were that being an only child seems to confer a marginal protective effect: only children were found to be consistently better off in terms of social adjustment, popularity, and willingness to confide in others, and they are less likely to be the victims of bullying though their self-reported academic record was reported as lower. This suggests that in the absence of siblings only children rely more on friends for social interaction and support.

## 14.9 THE FUTURE

The One Child Policy has clearly had a very major impact on many aspects of the health of Chinese people. The authorities have managed to win large scale acceptance of quite drastic birth limitation. This is perhaps all the more remarkable in view of the Chinese traditional love of children, the dependence of parents on their children in old age, and of course the thorny problem of the traditional preference for boys. When the One Child Policy was instigated in 1979 the authorities claimed it was for the short term. The aim was to move toward a small family culture. To what degree has that happened and how relevant is the Policy in contemporary China? There are a number of factors to consider.

First, what size of family would Chinese couples choose today if there were *no* Policy? In 1985 the China In Depth Fertility Survey found that the preferred number of children was 1.8 in urban Shanghai and 3.1 in poor rural Shaanxi (China In Depth Fertility Survey, 1985). Most recent data from the National Family Planning and Reproductive Health Survey show that, overall, 35% of the women questioned would prefer one child, 57% would prefer two, but importantly, very few women (5.8%) would like more than two (Lin, 2003), (see Table 14.4). The mean number of children preferred is lower among younger, urban, more educated women. Even in rural areas 30% of the women questioned claimed to want just one child, with 61% wanting two. Concerns about the division of diminishing plots of land among children are cited as one important reason (Zhu, 2003). However, in urban areas, where (with very few exceptions) only one child is currently allowed, 43% would still prefer

**TABLE 14.4**
**Preference for Family Size by Age, Residence, and Education by Percentage (n = 39,344)**

| No. of children preferred | | 0 | 1 | 2 | 3 | No preference | Mean no. of children preferred |
|---|---|---|---|---|---|---|---|
| Total | | 1.1 | 35 | 57 | 5.8 | 1.1 | 1.7 |
| Age | 15–24 | 1.9 | 50 | 44 | 2.1 | 2.3 | 1.5 |
| | 25–34 | 0.8 | 37 | 57 | 4.5 | 0.7 | 1.7 |
| | 35–44 | 0.9 | 29 | 61 | 8.1 | 0.9 | 1.8 |
| | 45+ | 0.9 | 24 | 64 | 9.9 | 1.2 | 1.8 |
| Type | Rural | 0.4 | 30 | 61 | 7.5 | 1.2 | 1.8 |
| | Urban | 3.1 | 52 | 43 | 1.5 | 0.9 | 1.4 |
| Education | Illiterate/semiliterate | 0.4 | 17 | 67 | 14 | 1.6 | 2.0 |
| | Primary school | 0.3 | 25 | 65 | 9.5 | 0.8 | 1.8 |
| | Middle/high | 2.1 | 47 | 46 | 3.9 | 1.1 | 1.5 |
| | College | 4.0 | 49 | 45 | 3.3 | 1.0 | 1.4 |

*Source:* Adapted from: Lin, B., Fertility desires of women of childbearing age and influencing factors. *Theses Collection of 2001 National Family Planning and Reproductive Health Survey*, China Population Publishing House, Beijing, 2003.

**TABLE 14.5**
**Gender Preferences (n = 39,344)**

| No. of boys | No. of girls | n | % |
|---|---|---|---|
| 0 | 1 | 2,323 | 5.9 |
| 0 | 2 | 267 | 0.7 |
| 1 | 0 | 2,218 | 5.6 |
| 1 | 1 | 17,882 | 45 |
| 1 | 2 | 254 | 0.6 |
| 2 | 0 | 274 | 0.7 |
| 2 | 1 | 1,123 | 2.8 |
| 2 | 2 | 566 | 1.4 |
| No gender preference | | 14,437 | 37 |

*Source:* Adapted from: Lin, B., Fertility desires of women of childbearing age and influencing factors, *Theses Collection of 2001 National Family Planning and Reproductive Health Survey*, China Population Publishing House, Beijing, 2003.

two, so the one child restriction remains unacceptable for nearly a half of urban Chinese women. Gender preference confirms this: though 36% state they have no gender preference, 45% of women want a boy *and* a girl (see Table 14.5). Of note is the fact that of those women preferring one or two girl children, 90% are young urbanites, perhaps indicating awareness of the problems that will be faced by men if the high sex ratio persists. Most surprising in the Chinese context, where traditional values still influence strong pronatalist pressure, is the finding that 10% of all Shanghainese 19–24-year-olds say they want no children at all. This all strongly suggests that China *is* becoming a small family culture.

Second, the China of today is very different from the China of 1979; the economy has grown at an average of 9% per year, the disposable income of most people has risen dramatically. Along with this has come much greater freedom and higher expectations and aspirations. Improved socioeconomic status has been associated with falls in fertility in many populations (Omran, 1971). In the Chinese context increased wealth makes it much harder for the authorities to enforce the Policy through economic disincentives. Many couples can now afford the fines. Furthermore, with the new greatly increased freedom of movement and a mobile population of casual workers of around 120 million, it is much more difficult for family planning authorities to track individuals who choose to flout the regulations.

The evidence against the Policy is mounting up: concerns about the sex ratio, the aging population, the effect on maternal mortality for unapproved pregnancies, the lack of reproductive choice, the unfair differential in the policy between urban and rural areas, and concerns about the unnatural character of cities full of only children. All this combines with the evidence for preference for family size and the difficulty with enforcing the Policy in the current economic climate. The government is also not immune from the criticism of the international community that the Policy is a violation of the basic human right to reproduction. In 1979 China was still a pariah in the world community. Now that China is a major global player, the government recognizes that a relaxation of the Policy would be desirable both for the Chinese people and for China's standing in the world community.

A number of different options for the future have been suggested. One possibility is that everyone should be allowed to have up to two, but *only* two children with a space of at least five years between them (Greenhalgh and Bongaarts, 1987). It is predicted that that this option would give a fertility rate of 1.7 over the next two decades, the same as it is now. This option would help to normalize the sex ratio, would reduce the 4:2:1 phenomenon and the financial burden of the ageing population, and would be acceptable to the overwhelming majority of the people.

However, the government has recently announced that there will be no fundamental change in the policy in the near future (Anon., 2002). The basic rules and regulations still apply, but the implementation and enforcement of the rules is relaxing, albeit slowly, in many areas. More counties are attempting to allow women informed choice in contraception and are not insisting that formal permission be obtained for first children. An increasing number of cities are now allowing couples consisting of only children to have two children. Notably Shanghai has amended its local policy to include this measure, whereas Beijing has explicitly blocked its introduction. One of the concerns is that this measure may be very divisive, with friends and neighbors allowed different numbers of children. To allow everyone to have two would seem to be a much easier and fairer solution.

The Policy has been highly successful on many levels and the demographic impact has been enormous. From a broader perspective, the policy illustrates how dramatically the institutional regulation of behavior can affect the primary evolutionary currencies of offspring quantity and quality. But the costs have begun to outweigh the benefits. The evidence is compelling: the Chinese should seriously consider moving toward a Two Child Policy in the near future.

## REFERENCES

Anon., An integration programme of poverty alleviation and development with family planning, *China Popul. Today,*14, 6, 1997.

Anon., No relaxation of Chinese one couple, one child policy, *People's Daily Beijing*, www.english.peopledaily.com.cn, 2 Sep, 2002.

Anson, O., and Anson, J., Fertility trends in rural Hebei province, People's Republic of China. *Popul. Res. Policy Rev.,* 22, 399–409, 2003.

Bogg, L., Family planning in China: out of control? *Am. J. Public Health*, 88, 649–651, 1998.

China In Depth Fertility Survey, 1985, www.sscnet.ucla.edu/issr/da/index/techinfo.

China Population Information and Research Centre (CPIRC), Basic population data of China 1949–2000 and major figures of the 2000 population census, CPIRC, Beijing, 2001.

Davis, D.L., Gottlieb, M.B., and Stampnitsky, J.R., Reduced ratio of male to female births in several industrial countries, *JAMA*, 279, 1018–1023, 1998.

Doherty, J.P., Norton, E.C., and Veney, J.E., China's one-child policy: the economic choices faced by pregnant women, *Soc. Sci. Med.*, 52, 745–761, 2001.

Falbo, T., and Polit, D.F., A quantitative review of the only child literature: research evidence and theory development, *Psychol. Bull.*, 100, 176–189, 1986.

Falbo, T., and Poston, D.L., The academic, personality and physical outcomes of only children in China. *Child Dev.*, 64, 18–35, 1993.

Festini, F., and de Martino, M., Twenty-five years of the one child family policy in China, *J. Epidemiol. Commun. Health*, 58, 358–360, 2004.

Green, L., Promoting the one child policy in China, *J. Pub. Health. Policy*, 6, 273–278, 1988.

Greenhalgh, S., and Bongaarts, J., Fertility policy in China: future options, *Science*, 235, 1167–1172, 1987.

Grimes, D. et al., Immediate post-partum insertion of intrauterine devices. *Cochrane Database Syst. Rev.*, 1, CD003036, 2003.

Gu, B., and Roy, K., Sex ratio at birth in China, with reference to other areas in East Asia; what we know. *Asia Pac. Popul. J.*, 10, 17–42, 1995.

Hardee-Cleaveland, K., and Banister, J., Fertility policy and implementation in China 1986–1988, *Popul. Dev. Rev.*, 14, 245–286, 1988.

Hesketh, T., Qu, J.D., and Tomkins, A., The health effects of family size: cross sectional survey in Chinese adolescents. *Arch. Dis. Child.*, 88, 467–471, 2003.

Hesketh, T., and Zhu, W.X., The one child family policy; the good, the bad and the ugly, *BMJ*, 314, 1685–1687, 1997.

Hudson, V., and Den Boer, A.M., A surplus of men, a deficit of peace : security and sex ratios in Asia's largest states, *Int. Security,* 4, 5–38, 2002.

Jiao, S., Ji, G., and Jing, Q., Comparative study of behavioural qualities of only children and sibling children, *Child Dev.*, 57, 357–361, 1986.

Jing, Q., Wan, C., and Over, R., Single child family in China: psychological perspectives. *Int. J. Psychol.*, 22, 127–138, 1987.

Kandela, P.. Exposing the abuse of Chinese orphans, *BMJ*, 24, 495–496, 1996.

Kang, C., and Wang, Y., Sex ratio at birth, *Theses Collection of 2001 National Family Planning and Reproductive Health Survey*, China Population Publishing House, Beijing, 2003.

Lin, B., Fertility desires of women of childbearing age and influencing factors. *Theses Collection of 2001 National Family Planning and Reproductive Health Survey*, China Population Publishing House, Beijing, 2003.

Liu, X.C. et al., Self-reported depressive symptoms among Chinese adolescents. *Soc. Psychiatr. Pychiatr. Epidemiol.*, 34, 44–47, 1999.

Merli, M.G., and Raftery, A.E., Are births underreported in rural China? Manipulation of statistical records in response to China's population policies, *Demography*, 37, 109–126, 2000.

Ni, H., and Rossignol, A.M., Maternal deaths among women with pregnancies outside of family planning in Sichuan, China, *Epidemiology*, 5, 490–494, 1994.

Omran, A.R., The epidemiological transition. A theory of the epidemiology of population change, *Milbank Memorial Fund Quarterly*, 49, 509–538, 1971.

Pan, J.H., Analysis of outcomes of pregnancies and fertility level of married Chinese women. *Theses Collection of 2001 National Family Planning and Reproductive Health Survey*, China Population Publishing House, Beijing, 2003.

Peabody, J.W., Hesketh, T., and Kattwinkel, J., Creation of a neonatology facility in a developing country: Experience from a 5-year project in China, *Am. J. Perinatol.*, 9, 401–408, 1992.

Peng, P., Causes and consequences of fertility decline in China, *China Popul. Today,* 15, 5–6, 1998.

Poston, D.L., and Falbo, T., Academic performance and personality traits of Chinese children: onlies versus others, *Am. J. Sociol.*, 96, 433–451, 1990.

Qian, X., Qian, X., Tang, S., and Garner, P., Unintended pregnancy and induced abortion among unmarried women in China: a systematic review, *BMC Health Serv. Res.*, 4, 1–6, 2004.

Sen, A., Missing women — revisited, *BMJ*, 327, 1297–1298, 2003.

Short, S.E., and Zhai, F.Y., Looking locally at China's one-child policy, *Studies in Family Planning*, 29, 373–387, 1998.

Sun, F., Ageing of the population in China: trends and implications, *Asia Pac. Popul.*, 13, 75–92, 1998.

*Theses Collection of 2001 National Family Planning and Reproductive Health Survey*, China Population Publishing House, Beijing, 2003.

Tseng, W.S. et al., Family planning and child mental health in China: the Nanjing Survey, *Am. J. Psychiatr.*, 145, 1396–1403, 1988.

Tuljapurkar, S., Li, N., and Feldman, M.W., High sex ratios in China's future, *Science*, 267, 874–876, 1995.

US Central Intelligence Agency, Total fertility rates: 2004 Estimates, www.cia.gov/cia/publications/worldfactbook.

Wan, C. et al., Comparison of personality traits of only children and sibling children in Beijing, *J. Genet. Psychol.*, 155, 377–378, 1994.

Wang, J.Y., Evaluation of the fertility of Chinese women during 1990–2000, *Theses Collection of 2001 National Family Planning and Reproductive Health Survey,* China Population Publishing House, Beijing, 2003.

Wu, Z.C. et al., Perinatal mortality in rural China: retrospective cohort study, *BMJ*, 327, 1319–1322, 2003.

Xiao, B., Abortion and emergency contraception: Chinese experience, *Chin. Med. J.*, 110, 36–42, 1997.

Yang, B. et al., Only children and children with siblings in the People's Republic of China: levels of fear, anxiety and depression, *Child Dev.*, 66, 1301–1311, 1995.

Yao, X.W., and Yin, H., Basic data of China's population. China Population Publishing House, Beijing, 1994.

Yin, Q., Shuai, Y., Zong, Z., and Xu, P., Choice of contraceptive methods among women of child bearing age and influencing factors, *Theses Collection of 2001 National Family Planning and Reproductive Health Survey,* China Population Publishing House, Beijing, 2003.

Young, M.E., Maternal health in China: challenges of the next decade. *Health Policy,* 14, 87–125, 1990.

Zheng, Z.Z., and Wang, Z.L., A preliminary study of the timing of marriage and child bearing of Chinese women since 1980, *Theses Collection of 2001 National Family Planning and Reproductive Health Survey,* China Population Publishing House, Beijing, 2003.

Zhu, W.X., The one child family policy, *Arch. Dis., Child.*, 88, 463–464, 2003.

# 15 Social Trends and Psychopathology

*Roland Littlewood*

## CONTENTS



## ABSTRACT

All societies undergo change, and all perceive the speed and extent of change as disorientating, possibly as pathological. In a now more evidently globalized world, models for identity and pathology come from a bewildering variety of sources. And in this, an emblematic new illness in Euro-American societies is multiple personality disorder which collects together and exemplifies the creolising post-modern self.

All societies face transition from a familiar present to a more or less uncertain future. And each, doubtless, has seen its predicament as unique, has taken its accustomed patterns of experience and power as natural or at least as somehow given, their threatened transformation as forced and arbitrary. Is the world of Christianity's third millennium faced with anything so very different? Is the pace perhaps more giddy? But then each European generation in the modern period has attributed its psychological ills to a disorientating acceleration of change. Does the latest wave of globalization just present us with more of the same, with a simultaneous multiplicity of potential yet not altogether unfamiliar identities — or perhaps something qualitatively rather different? Is the individual of cultural psychiatry less an embodied linear accretion of knowledge than a dramatic series of historically situated dialogues or performances?

Will some more variegated self seek embodiment in *civis mundi* (the world citizen), freely selecting from a global market of alternative goods, roles, values, appropriations, and practices: identities now as "lifestyles?" Shall we take this pluralism as itself "a culture," one that perhaps transcends many of those characteristics we had previously identified as "modern," in particular modernity's notion of a developing self-consciousness and instrumental rationality, the mind withdrawing from its anchoring in the material world to become the object of its own inquiry, the self-sufficient individual as the apparently unchanging locus and source of awareness and meaning, where self-aware human agency became the recognized goal of human history (Gellner, 1988)? Those local patterns in non-European societies that have been of interest to cultural psychiatrists in the 20th century have generally been interpreted as determined by, or resolving, "deeper" processes that underlay apparent human volition — whether we took these processes as unconscious mentation, as evolutionary imperatives or procreative stratagems, as the expression of underlying conflicts for power between the classes, sexes, or generations, or even as some standardized resolution of such conflicts that were represented in a pivotal individual (Littlewood, 1996).

Such an approach, characteristic of modernity itself, we might term rather generally the *structural*; for, as Lévi-Strauss put it, we made an attempt "to reduce apparently arbitrary data to some kind of order, and to attain a level at which a kind of necessity becomes apparent, underlying the illusions of liberty" (Lévi-Strauss, 1970). (And these unmasking attempts carried with them certain assumptions about representation, symbolization, and so on: less our contemporary preference for an arbitrary correspondence between fairly autonomous registers of biology, psychology, and culture than the necessary enactment of one "real" level in individual agency, and thus in our whole historical process.) The culture-bound syndromes that provided the essential problematic for comparative psychiatry were commonly seen as strategies for the resolution of underlying social tensions, strategies that could be said to work only through participants' disavowal of any motivation: agency being objectified and attributed to an external power such as a spirit, a disease, or an overwhelming addiction (Littlewood and Lipsedge, 1987). Whether the participant was to be considered as genuinely motivating the pattern depended on whether agency was taken as a valid objective characteristic or as a locally ascribed subjectivity (Bourdieu, 1977).

Recent postmodern trends in medical anthropology and cultural psychiatry, following the abandonment of the grand metanarratives in the human sciences, have reacted against the unmasking procedures of such structuralism to affirm the volitional individual as authentically making choices to construct their own world (howbeit still, as Marx had remarked, not exactly under conditions of their choosing). Ethnicity and identity are now recognized, not as intrinsic properties, but as actively recreated in each generation (Hall, 1991). Is this the final achievement of unfettered agency — a potentially empirical fact — or just a new academic model? Kirmayer (1992) reminds us that our currently fashionable personalism (hysterics are simple people simulating hysteria) simply reproduces modern Euro-American assumptions of the centrality of a directing self. And, thus, cultural psychiatry's models do seem to have come closer to contemporary Western ideology and folk psychology. All therapies are always part of the biosocial processes they identify and seek to resolve, and I do not propose to answer here the commensurability of causal versus voluntarist modes of thought (mad? bad? disease? malingering?) but rather to anticipate briefly some transformations of personal agency as a cultural assumption in the process of globalization. Should we anticipate new pathologies of the self (Kristeva, 1995)? And how will we understand them? Will we even recognize mental pathology or rather pursue some even less naturalistic model?

Transformations of the self have been a subject of medical anthropological interest since Marcel Mauss's influential paper to the Royal Anthropological Institute (Mauss, 1938). Particularly under the influence of psychoanalysis, we have favored a model of some sort of development or progression, whether evaluative or otherwise, in which our Western experience of the self has come even closer to some single self-aware unity distinguished from other such selves, living or dead, and from an external world. To what Clifford Geertz, no sympathizer with this view, described as "a bounded, unique, more or less integrated motivational and cognitive universe, a dynamic center of awareness, emotion, judgment and action organized into a distinctive whole and set constructively both against other such wholes and against a social and natural background ..." (Geertz, 1983). The modern self had gradually withdrawn from its mystical participation in the natural world or in the norms of others, in a secularization from the Christian soul through Descartes, Kant, and Fichte, to the contemporary idea of a unique consciousness generated by a single body and that body's biography. Psychology, particularly in the United States, became the master human discipline of the individualized modern world.

And along with this, there had been a progression in the nature of distress and restitution, away from socially embedded ritual to an apparently more explicit manifestation of suffering or dislocation now internalized as an individual psychology. Pierre Janet (1925) noted a historical progression from demonic possession to spiritualism and hysteria and thence to anxiety; as his contemporary psychological and psychodynamic idioms assumed the origin of all of these was psychological, the idiom of distress was presumed to have developed in ever more rational stages, finally approaching a recognition of the real nature of human dilemmas, the self-sufficient individual assuming ever greater personal agency and responsibility in the course of Western civilization. In his introduction to the 1978 edition of Morton Prince's classic account of 19th-century hysteria and double consciousness, the psychoanalyst Charles Rycroft was thus able to describe the book as the fascinating document of a bygone age, the increasing self-awareness of a unitary self rendering such pathologies increasingly unlikely (Rycroft, 1978).

But something rather odd has happened. Contrary to all expectation, Western society is currently in the midst of an epidemic of multiple personality, even demonic possession, and which has appeared in the fulcrum of modernity, North America (although now increasingly reported in Europe and Latin America). And this time round, not as the modest emergence of a related secondary or tertiary personality but of dozens, sometimes hundreds, of variant identities associated with a single physical body — children, spirits, selves from earlier lives, shamans, aliens, and extraterrestrials (Ross, 1994). Are we to take this current epidemic (which its therapists argue accounts for a tenth of all Western psychiatric illness) as some atavistic throwback or does it represent a fundamental change in a now globalizing Western self? The pathology of postmodernity, the incorporation of the Other (or *alter* as each variant new identity is expressly called), a pluralism of the self? If the idioms of 19th-century hysteria were those of the representation of neurological disorder (or capital accumulation) as "psychic energy" (Littlewood, 1996) the idioms of global postmodernity, as Gilles Bibeau (1997) reminds us, are those of cultural simulation, appropriation, and multiplicity.

Bibeau suggests that the immigrant is our exemplary global citizen, her creolized identity ours. And it is perhaps no accident that multiple personality disorder — the problematization of multiple identities — has emerged in the United States, a country in which all are immigrants, a physical core incorporating multiple peripheries. North America has always taken itself as the locus for strategic self-fashionings by which diverse immigrant groups realize themselves as Americans as they move to higher status jobs, improve their education, and change residence, neighborhood, profession, friends, spouses, political affiliation, leisure activities, voluntary associations, and even their name and religion, with associated changes of presentation in mannerism, comportment, language, idiom, and dress. With 19th-century dedication to manuals of kinesthesis, style, communication, and handwriting, earlier selves and practices become private or even forgotten. The new identity was achieved in the very opportunities of transformation, in fulfilling some apparently inherent potential, both as the normative expectation of what it is to be genuinely American and as a practical possibility; articulated for men in manuals of self-help and entrepreneurial psychology in the sort of positive transformation (eliding religious conversion, self-knowledge, and managerial presentation) advocated by Dale Carnegie and Vincent Peale; for women through remodeling their bodies by dieting, orthodontics, and plastic surgery; actively realizing one's "personality" in the marketing of it, through self-help groups and civil associations that now emphasize the attainment of a "real" identity articulated through a politicotherapeutic language of communication, growth, space, and authenticity. Yet the achievement of the new self is endlessly deferred (Falk, 1994), producing an American society uncertain whether it should possess cultural and linguistic norms or an educational canon, its emphasis on pluralism and personal self-transformation, the concealment of class conflict, its institutions maintained only through periodic moral panics and social dramas, its citizens unable to agree on what constitutes reality without recourse to legal and medical expertise.

Has the Judeo-Christian self, then, finally fragmented; or is our cultural multiplicity perhaps just the latest rerun of a previous European *fin-de-siecle* hysteria, yet another pause and consolidation on the route to an enhanced and unitary self-awareness, now incorporating other identities serially accessed, as it transcends the local contexts in which they took their meaning? Heelas (1981), like Haraway (1991), argues from locus of control experiments that where societies progressively emphasize an autonomous self, there is a concomitant tendency not only to see the body as other but to attribute undesirable deviations to some discrete agency external to the self, whether a human conspiracy or a malevolent spirit. With the collapse of the socialist economies and their communitarian suppositions in the last 15 years, we can identify a significant political enhancement of autonomous agency in North America and Europe (and also in the postcolonial countries through both the World Bank and renascent Islam [Stirrat, 1993]): toward individual responsibility for sickness, unemployment, and poverty, toward the rights (and responsibilities) of the mentally ill (Etzioni, 1994), with diminished space in a contracted natural world remade by self-sufficient individuals for "accidents" — now experienced as violating traumata occasioned by the deliberate abuse or negligence of others whom we hold morally and legally accountable. Paranoia (double bookkeeping or overinterpretation) and competitive individualism (whether capitalism or monotheism) are inevitable bedfellows.

Yet, linear and determinist theory does seem generally in decline, whether biography, text, or lived biography, in favor of fragmentation, multiplicity, iteration, and indeterminacy; a loosely integrated but pervasive cultural impetus which, in the Western university, includes evolutionary epistemology, biosociality, critical dialogics, game theory, nonlinear chaos theory, and fractal mathematics. Philosophers of mind are attracted to multiple personality because it fits well with the questions they are now raising about personal identity and agency, memory and consciousness; and the multiple personality disorder (MPD) movement in turn deploys their arguments against an invariant single self. Daniel Dennett (1987) argues that MPD follows relatively easily from the multiple functions of the mind for there never was a unitary center given in nature, no Cartesian or Kantian transcendent self, material or otherwise, but rather a competition between a number of multiple drafts. And, thus, there can be no single canonical experience or representation of the outside world (or of our own internal processes) at a particular point in space or time. Rather than enacting a genetic specification of a fully structured human, the central nervous system emerges in the competition between groups of cells that colonize each other (Edelman, 1992) while nucleotide sequencing argues that our mitochondria, cilia, and other organelles are the descendents of immigrant prokaryotic bacteria and spirochetes incorporated by our early ancestors (Margulis, 1982).

The notion of the self as simply a factitious node in a flexible neurosocial network appears not only in cognitive philosophy and in psychology (state-specific memory, cognitive modularity [Pinker, 1994]) in the new cybernetics of systemic group therapy and dissipative systems theory, in the control systems of managerialism (human resource and matrix management), but, significantly for immediate experience rather than for cultural psychiatric metaphor, in the increasingly pervasive human potential therapies, which, as practiced clinical procedures or popular manuals for self-transformation, promote the realization of all our "selves" recognized under an appropriate multiplicity of terms — *selves*, *subselves*, *subpersonalities*, *ego states*, *scripts*, *prototypes*, and so on. Akin to the earlier American manuals of self-perfecting and managerial efficiency, such human technologies commodify not only a popularized psychological idiom of untapped levels of consciousness but of potential new identities that can be serially accessed through our no longer self-sufficient body, in which our variant identities are neither transitory masks nor imaginary novelties but rather the realization of something that is authentically there. We might argue that such a therapy downplays the radical novelty of our new self while at the same time encouraging us to acquire it — less the Puritan struggle for arduous refashioning than freedom for what was already there to assume its natural place. Slippery and elusive though a new personae may be, it works when taken by individual and others for a conventional embodied self of everyday experience. Whether individualism remains the appropriate term for this latest stage of consciousness is questionable. For there is a Hegelian paradox: the more the individual becomes the single center of its own awareness, the less enmeshed in local contingencies through its universalization, the more fragile it becomes. To have a hundred personalities is perhaps to have none.

Michael Kenny (1986) argues that "the current popularity of multiple personality is the product of the disorder of our times." Overinterpretation, as Umberto Eco calls it, is however all too easy, the besetting vice of postmodernism itself. Cultural psychiatry and anthropology in particular have proposed many imaginative analogies between social structure and personal being, the incorporation of fragmentations of the body political read through those of the body physical (Littlewood, 1996). The coherence of such models is in part our need to construct an appealing cultural coherence, but also, I suspect, frequently because of our inadequate fieldwork so that we fail to appreciate the instability and plurality of any local society. As Bibeau (1997) reminds us, in pursuit of coherent meaning we have generally failed to explore the pragmatic networks of power by which experts, whether local healers or psychiatrists, through countless everyday decisions, maintain their influence and clientele by proposing to reconcile a society's core dilemmas, their claims to be — in La Barre's term — culture healers. In the current epidemic of MPD we need to trace the surprising alliances between the American women's movement and the Christian Right, between UFO enthusiasts and the psychiatrists of the human potential movement, between compensation lawyers and MPD therapists, not to mention an enduring "primitivism" currently represented by deep ecology. And these coalescences may be fairly short lived as yet newer social formations of experts in turn accuse the MPD movement of inducing "false memories." Bearing in mind that cultural idioms are only psychiatrically significant in as much as they enter individual subjectivity, can we however argue any broader but noncausal parallels between multiple personality disorder and the incipiently globalized, European, as if with the "self as a reflexive project" (Giddens, 1992), late modernity can no longer sustain the Enlightenment attempt to unify, place, and action, yet cannot cope with experiential flux alone, and retreats to diminished simulacra of an embodied individual, a simulation akin to what Guattari (1992) terms the "reterritorializations" of the nascent nationalisms of our postimperial and post-Soviet world? Or the fragile essentialisms of "identity politics?"

Twenty years ago Bolter (1986) suggested that the computer, like the medieval mechanical clock, had become late modernity's defining technology, a model for the self that is now experienced through an idiom of inputs and outputs, the body as hardware with consciousness as its software, now established as a global commerce of electronic mail, multidimensional user interfaces, satellite television shopping, and hypertext: a world of information rather than material production, and one that questions individual time, proximity, work, and ownership and that radically resituates the location of our taken-for-granted physical experience and volition in a postindustrial and postcorporeal global space, embodied experience fragmented and commodified. As do the replaceable parts available in the biomedical body shop — sperm, egg, and embryo banks, cerebral implants of fetal tissue and computer chips, life support machines to harvest organs from the brain dead, the international market in transplant-ready tissue-typed kidneys from Turkey, India, and China, electronic prostheses and gene splicing. Evolutionary epistemology and deep ecology, like the not dissimilar "hard AI" (Artificial Intelligence) and Artificial Life arguments, propose a virtual but potentially immortal self that has no inevitable locus in the physical body, yet is as real as any other self. "Cyberspace becomes another venue for consciousness itself … Animism is not only possible, it is implicit … To the body in cyberspace we are the mind. By a strange reversal of our cultural expectations, however, it is the body in cyberspace that is immortal, while the animating soul, housed in a body outside cyberspace, faced mortality" (Benedict, 1991). Artificial Life, neopaganism, holistic medicine, and deep ecology collapse our Enlightenment categories of matter, life, volition, and artifact, the whole world becoming a "cellular automaton": "There is no 'I' for a person, for a beehive, for a corporation, for an animal, for a nation, for any living thing" (Kelly, 1994).

Much of our current debate on globalization takes it as a gradual encroachment by Western cultural values, goods, and advertising; and it may be objected that I have privileged here recent technological procedures and their popularly accessible idioms in an essentially Euro-American world. Yet the electronic information economy brings together global images and global virtual embodiments, identities increasingly shared, appropriated, transvalorized, and remade in multiple creolizations. Globalization of culture is not a one-way flow from the First to the Third World: both indigenize alien knowledges and practices (Littlewood, 1993). South Asian meditation practices become domesticated and organized as akin to a Protestant sect; traditional ritual healing is packaged as group therapy or managerial training; Amerindian shamanism, Greek Orthodox fire-walking, and Senoi dream practices are incorporated as Western psychotherapies (Danforth, 1989). Possessing spirits in Samoa takes on the attributes of characters in an American television soap opera, while Trinidadian sects fuse Yoruba orishas with colonial history, deep ecology, and Italian cinema. The *yagé* visions of Columbian shamans reframe the terror of colonial conquest in therapies to which Europeans are themselves now having recourse (Taussig, 1987), while pharmaceuticals became appropriated as traditional medical understandings (Van der Geest and Whyte, 1989). The technovisionaries offer cyberspace ("wherever electricity runs with intelligence" [Benedict, 1991]) as populated not only by Western technology's nascent cyborgs and clones but by not altogether unfamiliar shades, zoomorphic hybrids, and orishas, a parallel world entered not only through virtual reality, computer conferencing, and the Internet, but through dreams, Amerindian psychoactive plants, and spirit possession (Tomas, 1989). Its global matrix subsumes all world culture, the world of Dawkins's memes, Popper's World 3 (Benedict, 1991).

Fragmentation and reconstruction of the volitional self though multiple appropriations and realizations may be nothing so very new, but the pace and extent do appear more salient. As James Clifford (1987) puts it, "we are all Caribbeans now," hybrid and parodic, polyphonic, and self-creating, endlessly recursive. Brazilian doctors trained in the spiritualist tradition of Kardecism (itself reframing from French homeopathy) treat multiple personality disorder by encouraging the patient to become a spirit medium (Krippner, 1987). Canadian psychiatrists sympathetic to the empirical existence of ultrahuman agencies propose coercive clinical exorcism for MPD (Ross, 1994). Christian therapists accept MPD's alters (secondary personalities) as the vengeful spirits of fetuses aborted by feminists. MPD activists and self-help groups argue for the legal recognition of every *alter* as a separate person; therapy is to consist simply in facilitating communication between independent beings in what has increasingly come to resemble systemic family therapy or Ndembu healing ritual. Other American patients with MPD market their skills as television personalities or spirit mediums (trance channelers) (Ross, 1994). Half of the United States population accepts the physical reality of angels; a third anticipate personal rapture into space before the end time tribulation (Strozier, 1994). Twenty-eight percent of American psychotherapists recognize that their patients can recall memories of themselves as a previous life form (Crews, 1994). A recent textbook written by lecturers at London University to train counselors of the "survivors" of a widespread demonic conspiracy for child sacrifice and cannibalism decries "a war-time atrocity in peace-time England" and proposes the establishment of special sanctuaries where survivors can be safely treated in extended sessions away from the powerful jurists and psychiatrists who ridicule the existence of diabolism only to practice it the better in secret (Sinason, 1994). A Harvard University clinic counsels women who have been abducted by flying saucers (Mack, 1994) to be impregnated by alien seed, their hybrid offspring being harvested nine months later (Schnabel, 1994). Public opinion polls indicate that over 15 million American citizens have been abducted in this way (Schnabel, 1994).

As Richard Rorty has proposed, diversity goes along with universality. In an increasingly global culture, it seems we are not altogether alone.

## ACKNOWLEDGMENTS

This chapter uses an argument from my paper "Agency and Its Vicissitudes" in *Transcultural Psychiatry*, 34, 78–90, 1997.

## REFERENCES

Benedict, M., Ed., *Cyberspace: first steps*, MIT Press, Cambridge, MA, 124, 140–141, 1991.

Bibeau, G., Cultural psychiatry in a creolising world: questions for a new research agenda, *Transcult. Psychiatr.*, 34, 1997

Bolter, J.D., *Turing's man: Western culture in the computer age*, Penguin, Harmondsworth, England, 1986.

Bourdieu, P., *Outline of a theory of practice*, Cambridge University Press, Cambridge, 1977.

Clifford, J., *The predicament of culture: twentieth century ethnography, literature and art*, Harvard University Press, Cambridge, MA, 175, 1987.

Crews, F., The revenge of the repressed: 2, *New York Review of Books*, December 1, 49, 1994.

Danforth, L.M., *Firewalking and religious healing: the Anastenaria of Greece and the American Firewalking Movement*, Princeton University Press, Princeton, NJ, 1989.

Dennett, D.C., *The intentional stance*, MIT Press, Cambridge, MA, 1987.

Edelman, G.M., *Bright air, brilliant sun: on the matter of the mind*, Basic Books, New York, 1992.

Etzioni, A., T*he spirit of the community: rights, responsibilities and the communitarian agenda*, Crown, New York, 1994.

Falk, P., *The consuming body*, Routledge, London, 1994.

Geertz, C., 'From the native's point of view': on the nature of anthropological understanding, In *Local knowledge*, Geertz, C., Ed., Basic Books, New York, 1983.

Gellner, E., Plough, sword and book: the structure of human history, Cambridge University Press, Cambridge, 1988.

Giddens, A., *The transformation of intimacy: sexuality, love and eroticism in modern societies*, Polity, Cambridge, 1992.

Guattari, F., Regimes, pathways, subjects, In *Incorporations*, Crary, J., and Kwinter, S., Eds., Zone, New York, 1992.

Hall, S., Old and new identities, old and new ethnicities, In *Culture, globalisation and the world-system*, King, A.D., Ed., Macmillan, London, 1991.

Haraway, D., *Simians, cyborgs and women: the reinvention of nature*, Free Association Books, London, 1991.

Heelas, P., The model applied: anthropology and indigenous psychologies, In *Indigenous psychologies: the anthropology of the self*, Heelas, P., and Lock, A., Eds., Academic Press, London, 1981.

Janet, P., *Psychological healing*, Allen and Unwin, London, 1925.

Kelly, K., *Out of control: the new biology of machines*, Fourth Estate, London, 1994.

Kenny, M., *The passion of Ansel Bourne: multiple personality in American culture*, Smithsonian Institution Press, Washington, 180, 1986.

Kirmayer, L., Social constructions of hypnosis, *Int. J. Clin. Exp. Hypnosis*, 11, 276, 1992.

Krippner, S., Cross-cultural approaches to multiple personality disorder: practices in Brazilian spiritism, *Ethos*, 15, 273, 1987.

Kristeva, J., *New maladies of the soul*, Columbia University Press, New York, 1995.

Lévi-Strauss, C., *The raw and the cooked*, Cape, London, 1970.

Littlewood, R., *Pathology and identity: the work of Mother Earth in Trinidad,* Cambridge University Press, Cambridge, 1993.

Littlewood, R., Reason and necessity in the specification of the multiple self, Royal Anthropological Institute, London, 1996.

Littlewood, R., and Lipsedge, M., The butterfly and the serpent, *Cult. Med. Psychiatr.*, 11, 289, 1987.

Mack, J., *Abductions: human encounters with aliens*, Simon and Schuster, New York, 1994.

Margulis, L., *Symbiosis in cell evolution*, Freeman, San Francisco, 1982.

Mauss, M., A category of the human mind: the notion of person, the notion of 'self', *J. Roy. Anthropol. Inst.*, 68, 263, 1938.

Pinker, S., *The language instinct: how the mind creates language*, Harper Collins, New York, 1994.

Ross, C.A., *The Osiris complex: case studies in multiple personality disorder*, University of Toronto Press, Toronto, 1994.

Rycroft, C., Introduction, In *The dissociation of a personality* (orig. pub. 1905), Prince, M., Ed., Oxford University Press, New York, 1978.

Schnabel, J., *Dark white: aliens, abductions and the UFO obsessions*, Hamish Hamilton, London, 1994.

Sinason, V., Ed., *Treating survivors of satanist abuse*, Routledge, London, 8, 204, 280, 1994.

Stirrat, R.L., Developing Africa? *Afr. Affairs*, 92, 294, 1993.

Strozier, C.B., Apocalypse: on the psychology of fundamentalism in America, Beacon Press, Boston 1994.

Taussig, M., *Shamanism, colonialism and the wild man: a study in terror and healing*, Chicago University Press, Chicago, 1987.

Tomas, D., The technophilic body: on technicity in William Gibson's cyborg culture, *New Formations*, 8, 111, 1989.

Van Der Geest, S., and Whyte, S.R., *The context of medicines in developing countries: studies in pharmaceutical anthropology*, Kluwer, Dordrecht, The Netherlands, 1989.

# 16 Epilogue: Memory, Tradition, and Teleology

*Simon Strickland*

## CONTENTS



## 16.1 INTRODUCTION

The papers in this book range widely and deeply across many issues. They touch on gene-culture coevolutionary processes; models of cultural evolution; conformity and innovation; personal identity and the sense of self; purposive behavior and chance events; modes of individual and mass communication; and the role of the state as a regulator, a source of information, and an agent in changing behavior. On this broad canvas, these studies address the implications of cultural variation and transmission for the evolution and development of human biological characteristics. In doing so, they make more explicit some of the many links between biological sciences and work in the field of social anthropology. In this epilogue, I reflect on three themes that surfaced from time to time during the symposium and that suggest some ideas for further research. These link social information transmission to memory and recollection, tradition and the heritability of culture, and the wider role of teleological argument in human biological and social sciences.

## 16.2 MEMORY AND RECOLLECTION

Aristotle distinguished "memory or remembering" as the perception of an image, from "recollection" as the deliberative inference or recovery of some prior knowledge or experience. He suggested that although many animals possessed memory, man was unique in being able to recollect and argued that recollection was facilitated by orderly arrangement (*On memory*). Aristotle's influence remains pervasive but no longer exclusive to the human sciences (e.g., Menzel, 2005). Some studies on juvenile nonhuman primates indicate that memory for friendships and dominance hierarchies characterizes early ages (Jolly, 1988), while evidence suggesting deliberative recollection can also be interpreted in terms of inventive and predictive problem solving (Hauser, 1988). Experimental cognitive studies point to the importance among apes of multiple past experiences in creating accumulated knowledge and flexible interpretative rules for solving novel problems in the search for food (Call, 2005). If the functional importance of memory and recollection is inversely correlated with exploratory physical activity or foraging energy expenditure in primates (Jolly, 1988), an idea that is also loosely reminiscent of Aristotle's thinking, then a case might be made that this relationship has influenced the evolution of primate brain size (Andrews and Martin, 1991).

Spatial and temporal predictability of primary resources such as fruits would on this argument be associated with (and have exerted selective pressures for) behavioral flexibility, learning, and the ability to retain information for future use (Milton, 1988). In this volume, the significance of this complex for nonhuman primates is illustrated by Whiten's account of Chimpanzee cultures with tool use for coping with seasonal food insecurity, sex differences in foraging, and self–medication. However, comparison between the foraging of contemporary hunter–gatherers and nonhuman primates suggests that humans have exploited far larger ranges and diets of far greater diversity, consistent with their large body size and brain, greater longevity, and extended development (Foley, 1987, p. 138). Recent studies suggest that humans have been able to make an unstable natural environment more predictable in part by evolving unique attributes of memory.

One hypothesis is that an "episodic memory" for mental time travel has evolved as a development of the more general semantic or declarative memory that is present in many mammals as "knowing", and that episodic memory enables a conscious awareness of and thought about both past time ("remembering") and the future (Tulving, 2005). The argument that this is probably unique to humans (resembling that of Aristotle) remains contested, but there is evidence from studies of ontogeny and a small number of individuals with brain injuries. The concept adds a useful dimension to the analysis of social information transmission and use over time.

The relationship between memory or recollection and energy expenditure or physical activity level has not yet been investigated systematically across human groups with varying subsistence patterns. Contemporary reports of average human adult energy expenditure do not suggest marked differences between general categories of subsistence patterns, but seasonal extremes tend to be higher in cereal farming systems than in nonseasonal hunter–gatherer or horticultural systems (Strickland, 1990). Evidence in such groups of the relative significance attributed to memory or recollection does not appear to exist in any systematic form. However, in many societies people use spatial mapping of known geographical features in ways that enable them to transform the landscape and to explore and regulate social relationships among the living and memories or recollections of the dead (see the studies by Feld, 1982; Brosius,

1986; Vitebsky, 1993; and Gaenszle, 2002). Acquisition of such notions and beliefs is therefore integrally linked to (but not necessarily teleologically explained by) the practicalities of living in and depending on these social and physical environments. It is also related to the importance attached to an exactly retentive memory and to powers of recollection, and in association with multiple learned skills can be presumed to enhance efficiency of resource use.

Reports of "collective" genealogical memories are also relatively numerous. Studies of largely nonliterate hunter–gatherer–horticulturalists like the Penan of Central Borneo (Brosius, 1986) suggest that genealogical depth in recalled lineages may span six generations. Studies of generally sedentary communities like the Gurungs of Nepal (Macfarlane, 1976), who maintained limited written records of land holdings before registration was introduced, suggest an interest in lineage genealogical depth of from four to nine generations. There is no systematic evidence to suggest that mobile populations lacking well-defined inheritance of rights to fixed resources are less likely than sedentary peoples to maintain strong spatial memories. Nevertheless, on the scale of a human lifetime or generation, socially significant genealogical information about lines of descent and inheritance is often deliberately retained for its value in securing rights to scarce resources whether movable, fixed, or knowledge-based.

Attention to innovative methods for retaining information might, then, testify to greater energy investment in scarce resources rather than in highly mobile and more risky exploratory behavior. However, sedentarization also incurs risks. There is skeletal evidence of energetic and other forms of nutritional stress during the Neolithic around 5000 years BP (Angel, 1984). This period coincides broadly with one of increasing social inequalities (Cohen, 1998), and with the development of the earliest writing systems that were used in marking or listing property and possibly for more intellectually autonomous purposes (Goody, 1977).

Selective preservation of information is subject to capacity but also to social values and power structures (Durham, 1991, pp. 198–202). An earlier chapter in this volume discusses niche construction theory with a focus on the biological consequences of animal and plant domestication and on the development of language (Odling-Smee). The biological consequences of early writing systems remain unclear, though as noted elsewhere in this volume (Aunger) representational artifacts play an important role in the dynamics of contemporary cultural evolution. In cognitive terms, they may have enabled abstraction and clarity of thought (Goody, 1977). One can only speculate that such systems may also have facilitated a sharpening of distinctions between subjectivity and objectivity, and a developing emphasis on the individuated personal identity arguably characteristic of Christian cultures (Mauss, 1938). A tendency to think of memory as interior to the individual mind recalls Aristotle's analysis. As noted elsewhere in this volume (Littlewood), an emphasis on this together with a strong sense of individuation appears to be distinctively European. By contrast, many peoples locate their recollections of the dead, of dreams, and of personalities in the exterior world (Vitebsky, 1993).

Vitebsky's penetrating analysis of memory and the sense of self in an Indian "tribal" society — the Sora — suggests that as people mature they do not reach a high degree of individuation but become progressively more deeply involved in complex social relationships with others. To some degree, persons then become understood or defined in terms of each other rather than as individuals. Recollections of the dead (memories) can be both intimate and public through the techniques of shamanic séance. What might, to a European, be termed *subjective* can then seem to move along on a largely undifferentiated continuum of reality ranging from the subjective to the objective. In this analysis, *agency* is a critical conceptual variable expressed through subtle changes in linguistic and emotional tone as the process of adjusting to bereavement runs its course from the initial experience through to a greater emotional distance. In such settings, there may be some strategic use of such concepts to manipulate social and economic relationships in competition for scarce resources and for power over other people, perhaps reflecting the prioritization of learning, important in the past (Vitebsky, 1993, pp. 196–197).

It would be helpful to extend niche construction analysis to the more complex neurological and cognitive forms of feedback linking innovative methods for representing and storing information in material form to changes in brain structures and functions. In Odling-Smee's terms (this volume), such changes could be inceptive or counteractive perturbations. For example, neuroimaging studies comparing sighted readers with developmentally and congenitally blind subjects able to read Braille suggest that a multimodal area of the left basal posterior temporal lobe (BA37) integrates visual and tactile sensory information with linguistic knowledge; and this finding is independent of the imageability or abstractness of words (Büchel et al., 1998). Other recent neuroimaging studies (Siok et al., 2004) suggest that alphabetic systems, which graphically represent phonemes as in English, and logographic systems, which graphically represent semantic morphemes as in Chinese, activate contrasting neural circuits. In fluent Chinese readers the left middle front gyrus of Brodmann's area appears to coordinate and integrate processes of verbal and spatial memories. This anatomical region of the brain is larger in Chinese-speaking Asians than in English-speaking Caucasians and is dysfunctional in Chinese dyslexics. By contrast, in English dyslexics the left temporo-parietal region is functionally impaired. These findings suggest that different writing system designs interact with developing neurological characteristics and may therefore result in selective pressures that depend on the wider social functions and correlates of writing.

In summary, information transmission among humans has come to be associated with the apparently unique features of episodic memory in individuals and with varied collective storage methods including systems of representation in material form. In each case, these methods are likely to have promoted cultural and behavioral continuity and flexibility in an unpredictable setting.

## 16.3 TRADITION AND THE HERITABILITY OF CULTURE

The relationship of memory to "tradition" is a complex one and bears on the extent to which the extended phenotype is heritable. Reflecting on the works of Herbert Spencer, Medawar once defined tradition broadly as "the transfer of information through nongenetic channels from one generation to the next" (Medawar, 1967, pp. 47–48). In this context he described cultural or psychosocial evolution as "the secular accumulation of fact and fancy, knowledge and know-how, rules and rites, that is mediated through tradition"; and he noted that this "accumulation" was both Lamarckian in style and reversible in the sense that it could in principle be wholly lost within a single generation. By contrast, a more recent definition centers on shared behavioral practice that is enduring and depends quantifiably on social contributions to learning (Fragaszy, 2003).

These definitions imply differing views of the heritability of culture. On the one hand, one might query the idea of progressive "accumulation" in a species in which techniques for deliberately recording ideas with a credible degree of permanence across generations are a recent evolutionary novelty. Earlier forms of representation would appear transient and limited in their reliability. Compared to Lamarckian thinking, current theories about fetal and postnatal influences on the developing metabolism and risk of obesity (Prentice et al., 2005) offer a better biological analogy with the plastic development and transfer of cultural tradition across generations. On the other hand, predictive and innovative forms of behavior are arguably a more satisfactory index of human cognitive and social intelligence than is the ability to learn and repeat actions through social or operant conditioning (Dennett, 1988). Small group size might raise the risk that innovations would be lost by chance alone or be reinvented. Ability to accumulate novel traditions is, therefore, also likely to be linked to increases in collective memories dependent on population size, density, social networks, and resilience to risks.

The epithet *traditional* applied to societies may also blind observers to those social networks that foster deliberately inventive risk taking. These are well exemplified by the careful experimental test of novel food plant seeds (Thiele, 1999). For example, in Sierra Leone, rice farmers may approach agricultural problems with inventiveness and experimentally, and show a flexible regard for complex ecotones; on the other hand, top-down economic development policies informed by a dominant Western perspective have often seemed unnecessarily inconsistent with these practices (Richards, 1986). This has only begun to change substantively over the past two decades or so with recognition of the importance of innovative risk management.

As noted elsewhere in this volume (Nettle), however, language to some degree allows cultures to acquire cumulative and sophisticated adaptations particularly where they develop external means to store information thought to be of current or future importance. Pertinent here are empirical studies of the degree of creative "fecundity" and of the stability or "replicative copying fidelity" achieved by orally transmitted traditions. Their findings bear upon the idea that some qualities might have greater survival value than others (Laland and Brown, 2002 and this volume). Conventional or sophisticated statistical methods have not generally been used in these studies; and there has long been debate about the appropriate units for quantitative analysis (Finnegan, 1977, pp. 73–87) as also over the utility of the concept of *meme* (Aunger this volume). However, counts of words or phrases used in oral performances by storytellers and their pupils suggest that each new generation might expect to retain from 40% to 80% of the verbal content of their predecessors' creations. This applies whether in a context of entertainment (Phillips, 1981) or of ritual (Strickland, 1983). Sometimes simple permutable design features such as repetition of sound, structure, and meaning seem to play a major role in the survival value of these traditions.

The critical questions are surely then what determines preference for creative innovation or for strict fidelity; to what degree these preferences are independent of broader aesthetic aims but result from within- or between-group socioeconomic pressures; how far semantic content can be separated from communicative mode; and in what ways a competitive element influences the outcome. As yet there are no general answers to these questions. In some cases, an aptitude to vary expression through skill in composition can be highly valued in the expert narrator, whose inexperienced pupil by contrast relies more heavily on memory and repetition. This is not to make a virtue of necessity. In others, correct ritual replication facilitated by joint performances, abstruse language, and authoritative intervention appears largely attainable within a given "school." In still others, an oral art form may in effect facilitate communication through decorative verbal redundancy but be valued less for communicative efficiency than for extending the listener's aesthetic pleasure (Strickland, 1995, pp. 45–47). In contrast to some modeling assumptions (Henrich and Boyd, 1998), it cannot be assumed *a priori* that there are genetically variable "propensities" to imitate or to conform unless one accepts a radically deterministic view of gene-culture coevolution. However, biological variation in memory and the capacity to recollect suggests that the ability to imitate will be plastic and may vary with experience in early life.

Even when there is intentional variety in orally transmitted culture, there may remain strong continuities in geographical content that link social or mythical events with the known landscape, resources, settlement pattern, ethnic identity, and a wide array of practices and imagined worlds. In some cases, the landscape resembles a large-scale mnemonic device (Vitebsky, 1993). Visual memory, based on the relative strength of the primate sense of sight, may appear more prominent in such oral traditions than more generally aural and other sensory impressions including synesthetic experiences. This is not to advocate a specific localist theory of the cognitive primacy of spatial over nonspatial expressions represented in language (Lyons, 1977, 718–724). However, checked against a written record such localizing preferences may have facilitated recollection and the training of memory in ways that were familiar to classical and subsequent thinkers. Historically, these mnemo-techniques gradually became less popular following the advent of printed books (Yates, 1966), indicating a divergence in the heritability of oral and written traditions.

In partially or wholly nonliterate settings, the expression of complex and sophisticated traditions linked to customs of pilgrimage or other forms of ritual journey would have combined repetitive exposure with clarity and memorability. A remarkable instance of this means of transmitting social information, reinforced by its representation in stone, is provided by the arrangement of narrative bas-reliefs on the ninth century Buddhist monument of Barabudur in Java (see Figure 16.1). Mus described these reliefs as "cinematographic" (Mus, 1935/1998, p. 269). Their probable didactic functions seem clear (de Casparis, 1981). The monument probably also testified to the now long-forgotten major dynastic achievements of the founder and to his reputation (Macdonald, 1952/1975).

More recently, photography too can be understood both as an instrument of memory and in a social context perhaps also as a means to invent or to replace memories (Sontag, 1973, 164–166). In this volume, Berridge and Loughlin show how strong visual techniques of mass persuasion have used cues for the biological effects of

(a)

(b)

(c)

**FIGURE 16.1** The ninth century AD monument of Barabudur in Java and two illustrative bas-reliefs (photographed by the author). (a.) The monument of Barabudur, which comprises concentric galleries containing visible panels stretching a distance totalling about 3.1 km. Other panels at the base were for unknown reasons concealed beneath a wide stone terrace at around the time of construction; and a short sequence has been exposed at one corner of the monument as shown in this photograph (see Miksic, 1990, p. 40). (b.) This panel depicts a scene from a story in which the god Indra arrives disguised as a Brahman. The future Buddha is a rabbit that teaches the importance of generosity to his friends — a jackal, an otter, and an ape. These animals all bring Indra food but the rabbit is unable to do this (left panel). The rabbit therefore jumps into the cooking fire himself (right panel) (see Miksic, 1990, pp.72–73). (c.) This panel depicts the Prince Sakyamuni, who was to become the Buddha, shooting an arrow through seven trees to prove his merit at archery. He therefore wins the hand of his wife, the daughter of King Suddhodhana (see Miksic, 1990, p.107).

smoking in attempts to alter behavior and promote public health. As Wolf (1994) noted, both memory "codes" and their analogical use need investigating in parallel if the changing formation of human cultures is to be understood; but this alone would not explain why such cultures come to exist. These characteristics and uses of visual images are central to the social transmission of information, yet are under-researched from a comparative and evolutionary viewpoint.

In summary, human social information transmission shows strong innovative and preservative tendencies, with oral and visual media enabling both. The human ability to capture cultural elements permanently is novel and exceptional. The heritability of culture is therefore itself dynamic.

## 16.4 TELEOLOGY AND HUMAN PURPOSE

Biologists have often reasoned teleologically and this method of argument perhaps comes especially readily to human social intelligence (Jolly, 1988). Monod (1972) held that all structures, performances, and activities that contribute to successful species' preservation and reproduction are "teleonomic" and that the teleonomic level of a species corresponds to the average amount of information needed to assure invariant reproduction. In considering properties of the human capacity for memory and recollection from this perspective, one might adopt the theory that consciousness, self-awareness, foresight, and purpose "represent a system for competing with other humans for status, resources, and eventually reproductive success" (Alexander, 1989, pp. 112–113). One might then use this theory to account for the wider intentional characteristics associated with episodic memory. On the other hand, the competitive use of these mental characteristics does not entail that this is only for a single ultimate teleonomic purpose. Elsewhere in this volume a case is made for human biology to embrace a wider array of "ultimate causes" than reproductive success (Wells and Strickland).

Aristotle distinguished between a primary or ultimate cause as "that for the sake of which" and a secondary or immediate cause as "whence the beginning of motion comes" (*Parts of animals*); and proposed that "where there is an end, all the preceding steps are for the sake of that", and that "It is plain that nature is a cause, a cause that operates for a purpose" (*Physics*). However, Monod (1972) argued that the biosphere is essentially unpredictable; and there is then arguably an "epistemological contradiction" between denying Aristotelian final causes on the one hand, and on the other recognizing the "teleonomic character" of living organisms that decide on and pursue a given purpose. He formulated and sought to resolve this dilemma for humans by arguing that their ability to comprehend and foresee events, enriched by learning, enables behavior to orientate selective pressures. On this approach, one could be forgiven for suggesting that some elements of learning and belief *happen by chance* to survive, to proliferate, and to demand a measure of social conformity (Nettle, 2001, p. 31).

Advocating a less reductionist approach, others have observed that subtle feedback processes enable phenotypes themselves to influence the pressures to which they will become subject and, therefore, to more robustly moderate the unpredictability of their environment (Thorpe, 1978, p. 33). Nonrandom evolutionary variants are then the matter for inquiry. The theory of niche construction provides a framework for investigating these. In this volume, for example, Mace and others argue that competition for resources and a social context favoring high parental investment per child are conditions that promote sustained contraceptive use and ultimately lower fertility in the short term. The fitness implications of contraception in a given context of resources and social networks may therefore contribute to explaining its use. Further, the chapters by Hesketh, Berridge and Loughlin, and Cleland and Watkins illustrate how, at a broader level than the face-to-face community, social institutions such as the state can play a critical role in determining this context and in shaping human behavior on a large scale. If natural selection acts on strategies rather than on their outcomes, as argued in this volume by McNamara and Houston, then one might predict a high degree of plasticity in human behavior at all levels of society.

In a human setting there can be no escape from difficult philosophical questions including those surrounding the nature of belief (Beattie, 1984). Further issues surround the relationship between biological inequalities and social mobility mediated through links between IQ, educational opportunity and attainment, occupation, physical size and shape, and through other biosocially differentiating features of societies (Strickland and Shetty, 1998). The central issue here is again the extent to which intentionality or explicitly purposive behavior can be accommodated by the biological sciences in general and by the kind of niche construction processes discussed elsewhere in this volume.

In an extended analysis of reductionism and biological teleology, Rosenberg (1985, pp. 262–265.) argued on linguistic grounds that intentional statements (for example, "Oedipus desired to marry Jocasta"; and "Oedipus desired to marry his mother") cannot be replaced by descriptions of underlying physical states and processes, and that there are no behavioral regularities or laws about human behavior in this respect. On this argument, a biologist could not define memory and recollection in intentional terms as "true belief about the past which may eventuate in certain actions." However, a Skinnerian behaviorist psychology would provide an acceptable stimulus–response model explaining some features, including differences between short- and long-term memory. Rosenberg concluded from this that a thoroughly nonintentional biological science, which would encompass neuroscience and sociobiology, could never accommodate an irreducibly intentional human biology.

Rosenberg was writing 20 years ago, and with hindsight his argument appears overstated. Elsewhere in this volume (Aunger), it is argued that the neural interpretation of signals from a source initiates the reconstruction of cultural units or *memes*. However, intentionality is a necessary feature of any voluntaristic communicative system if this is to ensure cooperative behavior and resolve uncertainties (Pinker, 1994, pp. 228–229). Communication of the emotions can obscure this where expression is involuntary, as Darwin (1889, pp. 369–379) noted with respect to blushing. There is no epistemological inconsistency between the intentional aspect of communication and functionalist (but not necessarily teleological) accounts of the interplay between social institutions (Leach, 1976, pp. 3–7).

The important philosophical point is that human social worlds depend on communicative activity that, in its creative and imaginative forms, cannot be reduced to simple positivist assertions of truth or falsehood (Austin, 1961, pp. 233–252). Accounts of ceremonial activity in which declarative or "performative" statements achieve well-understood social goals ("I pronounce you man and wife"; "I hereby expel this evil spirit") provide examples of this phenomenon. In some instances, verbal performative acts are extended into long narrative forms accompanied by complex behavioral activities that require substantial investments to learn (see, e.g., Gaenszle 2002). Access to these skills is usually regulated by social structures of authority and power.

In these terms, literal accounts of human motives may suffice to explain behavior that has often been attributed hidden or underlying "symbolical" reasons (Skorupski, 1976). There is an analogy here with some instances of sociobiological reasoning. An illustrative example is provided by the human abuse of chemical substances, and more specifically advertisements for tobacco or alcohol. Diamond (1991, pp. 173–183) has claimed that such advertisements persuade by communicating messages of physical and social superiority that exploit essentially the same instinctive principles as those underlying the function of some physical handicaps in various animal species; that the added cost of the handicap to the animal guarantees the honesty of its signal; but that in humans the costs to health of drinking alcohol or kerosene, or of drug addiction, outweigh the benefits and therefore reliance on such signals is misguided.

However, the epidemiological difficulty of demonstrating adverse costs suggests that the biological argument of handicap signaling may apply only by analogy and not substantively in this case, and that overt intentions suffice to justify such human behavior. Elsewhere in this volume, Berridge and Loughlin note the complexity of factors that, in the second half of the 20th century, influenced change in public health attitudes to smoking and the correlated trends in design of related mass communications. Diamond's argument also pays too little attention to other sufficiently motivating correlates of tobacco smoking, which include wealth, social status, and associated physical attributes such as a slim physique and tall stature (Strickland and Duffield, 1997). For some agents, such as the mildly neuroactive and antihelminthic areca nut in Asia, consumption may have well-understood overt and desirable cosmetic effects linked to sexual behavior apart from any otherwise sought pharmacological properties (Strickland, 2002). Consuming such biological agents, in other words, can transmit social information without a basis in instinctive handicap signaling.

In summary, teleological interpretations of human behavior have a long history in the social and biological sciences. However, these face difficulty where they depend on identifying human purposes of which actors themselves are unaware and that on closer analysis turn out to be unnecessary to explanation. Extending the notion of an ultimate cause to nonreproductive outcomes would help to resolve this issue for both biological and social sciences.

## 16.5 CONCLUSION

Failure to examine the intentionality behind human behavior can result in a misguided interpretation of the data. If they are to integrate in any way, biology and the social sciences need compatible rules of evidence and to have regard to their sometimes dissonant presentational styles. The inference to be drawn is surely that disciplinary boundaries should not obstruct the fruitful interplay of any sciences. The papers in this volume provide a strong basis for further work on the many related issues raised by the social biology of information.

## ACKNOWLEDGMENTS

The author thanks Jonathan Wells and Raquel Ackerman for their comments and helpful advice.

## REFERENCES

Alexander, R.D., *The biology of moral systems,* Aldine de Gruyter, New York, 1989.

Andrews, P., and Martin, L., Hominoid dietary evolution. *Proc. Roy. Soc. Lond. B.*, 334, 199–209, 1991.

Angel, J.L., Health as a crucial factor in the changes from hunting to developed farming in the Eastern Mediterranean, In *Palaeopathology at the origins of agriculture,* Cohen, M., and Armelagos, G., Eds., Academic Press, London, 51–73, 1984.

Aristotle, *On memory,* 451a15–451b5; 452a1–4; 453a6–14; 453b4–6.

Aristotle, *Parts of animals.* 639b11ff.

Aristotle, *Physics.* 199a9ff; 199b32.

Austin, J.L., Performative utterances, In *Philosophical papers,* Urmson, J.O., and Warnock, G.J., Eds., Oxford University Press, Oxford, 1961.

Beattie, J., Objectivity and social anthropology, In *Objectivity and cultural divergence,* Brown, S.C., Ed., Cambridge University Press, Cambridge, 1–20, 1984.

Brosius, P., River, forest and mountain: the Penan Gang landscape, *Sarawak Museum J.,* XXXVI, 173–184, 1986.

Büchel, C., Price, C., and Friston, K., A multimodal language region in the ventral visual pathway, *Nature*, 394, 274–277, 1998.

Call, J., The self and other: a missing link in comparative social cognition, In *The missing link in cognition*, Terrace, H., and Metcalfe, J., Eds., Oxford University Press, Oxford, 321–341, 2005.

Cohen, M., The emergence of health and social inequalities in the archaeological record, In *Human biology and social inequality,* Strickland, S., and Shetty, P., Eds., Cambridge University Press, Cambridge, 249–271, 1998.

Darwin, C., *The expression of the emotions in man and animals,* John Murray, London, 1889.

de Casparis, J.G., The dual nature of Barabudur, In *Barabudur: History and significance of a Buddhist monument*, University of California, Buddhist Studies Series, Berkeley, 47–83, 1981.

Dennett, D., The intentional stance in theory and practice, In *Machiavellian intelligence,* Byrne, R., and Whiten, A., Oxford University Press, Oxford, 180–202, 1988.

Diamond, J., *The rise and fall of the third chimpanzee*, Vintage, London, 1991.

Durham, W.H., *Coevolution,* Stanford University Press, Stanford, 1991.

Feld, S., *Sound and sentiment,* University of Pennsylvania Press, Philadelphia, 1982.

Finnegan, R. *Oral poetry,* Cambridge University Press, Cambridge, 1977.

Foley, R., *Another unique species,* Longmans, London, 1987.

Fragaszy, D., Making space for traditions, *Evol. Anthropol.*, 12, 61–70, 2003.

Gaenszle, M. *Ancestral voices,* Lit Verlag, Münster, Germany, 2002.

Goody, J.R., *The domestication of the savage mind,* Cambridge University Press, Cambridge, 1977.

Hauser, M., Invention and social transmission: new data from wild vervet monkeys, In *Machiavellian intelligence,* Byrne, R., and Whiten, A., Eds., Oxford University Press, Oxford, 327–343, 1988.

Henrich, J., and Boyd, R., The evolution of conformist transmission and the emergence of between-group differences, *Evol. Hum. Behav.*, 19, 215–241, 1998.

Jolly, A., The evolution of purpose, In *Machiavellian intelligence,* Byrne, R., and Whiten, A., Eds., Oxford University Press, Oxford, 1988.

Laland, K., and Brown, G., *Sense and nonsense,* Oxford University Press, Oxford, 2002.

Leach, E.R., *Culture and communication,* Cambridge University Press, Cambridge, 1976.

Lyons, J., *Semantics*, Cambridge University Press, Cambridge, 1977.

Macdonald, A.W., On prajâpati, In *Essays on the ethnology of Nepal and South Asia,* Ratna Pustak Bhandar, Kathmandu, 1–13, 1952/1975.

Macfarlane, A., *Resources and population,* Cambridge University Press, Cambridge, 1976.

Mauss, M., A category of the human mind: the notion of person; the notion of self, *J. Roy. Anthropol. Inst.*, 1938, 68, 263ff, reprinted in *The category of the person — Anthropology, philosophy, history,* Carrithers, M., Collins, S., and Lukes, S., Eds., Cambridge University Press, Cambridge, 1–25, 1985.

Medawar, P., *The art of the soluble,* Methuen, London, 1967.

Menzel, C. Progress in the study of chimpanzee recall and episodic memory, In *The missing link in cognition*, Terrace, H., and Metcalfe, J., Eds., Oxford University Press, Oxford, 188–224, 2005.

Miksic, J., *Borobudur: golden tales of the Buddhas,* Periplus, Hong Kong, 1990.

Milton, K., Foraging behaviour and the evolution of primate intelligence, In *Machiavellian intelligence,* Byrne, R., and Whiten, A., Eds., Oxford University Press, Oxford, 285–305, 1988.

Monod, J., *Chance and necessity,* Collins, London, 1972.

Mus, P., *Barabudur. Sketch of a history of Buddhism based on archaeological criticism of the texts,* transl. Macdonald, A.W., Sterling, New Delhi, 1935/1998.

Nettle, D., *Strong imagination: madness, creativity and human nature,* Oxford University Press, Oxford, 2001.

Phillips, N., *Sijobang: sung narrative poetry of West Sumatra,* Cambridge University Press, Cambridge, 1981.

Pinker, S., *The language instinct,* Penguin books, Harmondsworth, U.K., 1994.

Prentice, A.M., Rayco-Solon, P., Moore, S.E., Insights from the developing world: thrifty genotypes and thrifty phenotypes, *Proc. Nutr. Soc.*, 64, 153–161, 2005.

Richards, P., *Coping with hunger: hazard and experiment in an African rice-farming system,* Allen and Unwin, London, 1986.

Rosenberg, A., *The structure of biological science,* Cambridge University Press, Cambridge, 1985.

Siok, W.T., et al., Biological abnormality of impaired reading is constrained by culture, *Nature,* 431,71–76, 2004.

Skorupski, J., *Symbol and theory,* Cambridge University Press, Cambridge, 1976.

Sontag, S., *On photography,* Farrar, Strauss and Giroux, New York, 1973.

Strickland, S.S., Traditional economies and patterns of nutritional disease, In *Diet and disease,* Harrison, G., and Waterlow, J., Eds., Cambridge University Press, Cambridge, 209–239, 1990.

Strickland, S.S., The Gurung priest as bard, *Kailash,* X, 227–265, 1983.

Strickland, S.S., *Materials for the study of Kejaman-Sekapan oral tradition, Sarawak Museum J., Special Monograph No. 8, Volume I,* 1995.

Strickland, S.S., Anthropological perspectives on use of the areca nut, *Addict. Biol.*, 7, 85–97, 2002.

Strickland, S.S., and Duffield, A.E., Anthropometric status and resting metabolic rate in users of the areca nut and smokers of tobacco in rural Sarawak, *Ann. Hum. Biol.*, 25, 453–474, 1997.

Strickland, S.S., and Shetty, P.S., Eds., *Human biology and social inequality,* Cambridge University Press, Cambridge, 1998.

Thiele, G., Informal potato seed systems in the Andes: Why are they important and what should we do with them? *World Dev.*, 27, 83–99, 1999.

Thorpe, W.H., *Purpose in a world of chance,* Oxford University Press, Oxford, 1978.

Tulving, E., Episodic memory and autonoesis: uniquely human? In *The missing link in cognition,* Terrace, H., and Metcalfe, J., Eds., Oxford University Press, Oxford, 3–56, 2005.

Vitebsky, P., *Dialogues with the dead. The discussion of mortality among the Sora of eastern India,* Cambridge University Press, Cambridge, 1993.

Wolf, E.R., Perilous ideas: race, culture, people, *Curr. Anthropol.*, 35, 1–11, 1994.

Yates, F., *The art of memory,* Routledge, London, 1966.

# Index

## D



## E



## F

## O



## P



## R



## S

## T



## U